COFFEE AGROECOLOGY

Based on principles of the conservation of biodiversity and of equity and sustainability, this book focuses on the ecology of the coffee agroecosystem as a model for a sustainable agricultural ecosystem. It draws on the authors' own research conducted over the last twenty years as well as incorporating the vast literature that has been generated on coffee agroecosystems from around the world.

The book uses an integrated approach that weaves together various lines of research to understand the ecology of a very diverse tropical agroforestry system. Key concepts explored include biodiversity patterns, metapopulation dynamics and ecological networks. These are all set in a socioeconomic and political framework that relates them to the realities of farmers' livelihoods.

The authors provide a novel synthesis that will generate new understanding and can be applied to other examples of sustainable agriculture and food production. This synthesis also explains the ecosystem services provided by the approach, including the economic, fair trade and political aspects surrounding this all-important global commodity.

Ivette Perfecto is George W. Pack Professor of Ecology, Natural Resources and Environment at the University of Michigan, USA.

John Vandermeer is Asa Grey Distinguished University Professor of Ecology and Evolutionary Biology, and Alfred T. Thurnau Professor at the University of Michigan, USA.

COFFEE AGROECOLOGY

A new approach to understanding agricultural biodiversity, ecosystem services and sustainable development

Ivette Perfecto and John Vandermeer

Routledge
Taylor & Francis Group

LONDON AND NEW YORK

earthscan
from Routledge

First published 2015
by Routledge
2 Park Square, Milton Park, Abingdon, Oxon OX14 4RN

and by Routledge
711 Third Avenue, New York, NY 10017

Routledge is an imprint of the Taylor & Francis Group, an informa business

British Library Cataloguing-in-Publication Data
A catalogue record for this book is available from the British Library

Library of Congress Cataloging-in-Publication Data
Perfecto, Ivette.
    Coffee agroecology: a new approach to understanding agricultural
    biodiversity, ecosystem services, and sustainable development/
    Ivette Perfecto and John H. Vandermeer.
        pages cm
    Includes bibliographical references and index.
    1. Coffee. 2. Sustainable agriculture. 3. Ecosystem management.
    4. Tropical crops. I. Vandermeer, John H. II. Title.
    SB269.P44 2014
    338.1'7373—dc23   2014012824

ISBN: 978-0-415-82680-8 (hbk)
ISBN: 978-0-415-82681-5 (pbk)
ISBN: 978-0-203-52671-2 (ebk)

Typeset in Bembo
by Keystroke, Station Road, Codsall, Wolverhampton
Printed and bound by CPI Group (UK) Ltd, Croydon, CR0 4YY

CONTENTS

FIGURES

TABLES

PREFACE

We are inspired by the famous quote from Bertolt Brecht: "The world of knowledge takes a crazy turn when teachers themselves are taught to learn." We are ecologists. We teach ecology. However, in our experiences teaching ecology in tropical America, we were taught by farmers, more specifically coffee farmers. During a 17-year study of the dynamics of tropical rainforests in eastern Nicaragua and teaching ecology in Costa Rica, we began talking to and, unintentionally at first, learning from coffee farmers. They had not heard of the Lotka–Volterra equations, but they understood predator–prey dynamics because they saw it first hand. They knew nothing of the formalities of disease ecology, but they understood how the diseases that attacked their coffee plants spread like epidemics. They had never heard of the neutral theory of community assembly, but they had an intuitive feel for how the vast assemblages of beetles and birds and flies and frogs maintained their amazing biodiversity. And we were taught to learn.

We started that education working in the coffee agroecosystem in the late 1980s while teaching a graduate-level field course on managed ecosystems for Latin American students in Costa Rica. The contrast between the Latin American students and the North American students participating in similar courses was striking. For Latin Americans, having grown up in tropical landscapes, diverse farming systems such as shade coffee plantations were places that maintained biodiversity. For North American students, anything associated with agriculture was already tainted, trashy, not worth preserving. These contrasting points of view were also playing out in the international academic arena in which some of us, as conservationists, were distinguishing between different kinds of agriculture while others continued to insist that conservation must be in areas of pristine (unadulterated) nature. These "fortress conservationists" in the extreme were purchasing large tracts of land, surrounding them with fences, either metaphorical or real, to protect biodiversity by keeping "people" out. For them, people were the problem and could never be a part of conservation areas. We were trying to make the point that 40,000 hectares of banana monoculture is not the same as a 40,000-hectare landscape mosaic consisting of small-scale diverse farms. Yet the fortress conservationists seemed to always try and lump all agriculture as the same, a sort of biological desert that had nothing to do with biodiversity. Fortunately, the discourse about biodiversity conservation has shifted significantly over the last two or three decades and most conservationists now recognize the importance of integrating people and their agricultural systems into the conservation conversation.

At the same time, our education at the feet of the farmers led us to consider the dialectical way in which our insights from basic ecology combined with the pedagogy from the farmers into a synthesis of farmer-based specific and deep knowledge with scientific-based general yet perhaps superficial knowledge. Reflecting on years of collaboration with ecologist Richard Levins, we began a journey of ecological study informed by an explicit sociopolitical framework that included concepts like food sovereignty and environmental justice as much as pest control and biodiversity conservation. This book is a compilation of much of that journey.

Explicitly rejecting artificial disciplinary borders, our research has been quite eclectic, although admittedly our training in ecology shows through in almost all of it. However, we have tried, over all these years, to locate our research in larger sociopolitical realities. Thus, although our studies of the spatial distribution of ants enthrall us intellectually, we also understand and promote the fact that the problems faced with pest control are intimately related to that spatial distribution. Or, our framing of biodiversity as a function of the intensity gradient of coffee production is of immense interest theoretically, yet it is the need of small-scale farmers for just compensation combined with the worldwide concern with biodiversity conservation that gives that theoretical interest its dialectical wholeness. Throughout this book, we have tried to emphasize this interdisciplinary and dialectical perspective.

The structure of the book is straightforward. After an initial chapter sets up a foundational framework, Chapters 2 and 3 concentrate on the politically charged topic of biodiversity conservation. In many ways, these chapters explore some of the questions we formulated in our previous book *Nature's Matrix*, but focusing specifically on the coffee agroecosystem. Chapters 4, 5, and 6 form what might be thought of as the ecological core of the book, elaborating on topics normally encountered in standard ecology courses and books, but specifically as frameworks for the results of the research. Chapters 7 and 8 are more directly aimed at the practical aspects of the coffee agroecosystem, from the more sublime (ecosystem services – Chapter 7) to the more political (farmers' livelihoods – Chapter 8). Finally, in Chapter 9, we present our unique way of visualizing change in agroecosystems, combining both an ecological and an economic perspective on the process.

Many people have contributed to the research that we claim as our own. Most significantly we thank our long-time collaborator Stacy Philpott, whose work is featured repeatedly throughout the book. Stacy remains a major source of ideas and general collaborations. We also thank our recently deceased colleague Russell Greenberg of the Smithsonian Migratory Bird Center, whose collaborations were especially important in the research reported in Chapter 5. Don Walter Peters, formerly the owner of the amazing *Finca* Irlanda in Chiapas, Mexico, while not directly contributing to the research, was a major player in all of this with his remarkable insights about both coffee production and ecological relationships within the coffee agroecosystem. He is one of the major farmers from whom we learned. For the last 15 years, we also had the great fortune of having an excellent team of field technicians – Gustavo López Bautista, Braulio Chilel and Gabriel Domínguez – who assiduously collected some of the core data presented in this book. Without their help, our research would have been impossible. Most of the figures in the book were elaborated by Dave Brenner, and we thank him for his assiduous and wonderful design work. Last, but certainly not least, we thank our students and post-docs, who are the sources of most of the material presented in this book. The list is too long to be included here but all of their work is featured throughout the book. Finally, several NSF grants, plus support from the University of Michigan, contributed to the research reported herein. The writing of the book was supported by an NSF-OPUS grant to Ivette Perfecto.

1

WAKE UP AND SMELL THE COFFEE
(OR A TALE OF TWO FARMS)

Introduction

"Great coffee is this amazing miracle . . . its warm deliciousness in the morning transforms even the most rough-edged of us into intelligent, sparkling, upstanding men and women." This quote, from Peter Giuliano of Counter Culture Coffee, illustrates the love affair people all over the world have with coffee. As we enjoy our morning cup of coffee, few of us think beyond the wonderful aroma of fresh coffee and the transformation that is about to take place in our mind and body as the caffeine kicks in. Few of us reflect on the fact that embedded in that delicious cup of coffee is a fascinating tale of global trade that starts on a farm somewhere in the tropics, travels across the ocean and ends up in our kitchen or neighborhood coffee shop. What happens in between is an amazing story of ingenuity, exploitation and resistance of both humans and nature and the interactions among them. In their book *The Coffee Paradox*, Benoit Daviron and Stefano Ponte examine the trajectory from the bean to the cup, tracing the coffee value chain once the bean leaves the farm.[1] They ponder how we ended up in the paradoxical situation in which there is on the one hand usually a low stock of coffee and on the other hand usually low prices paid to farmers who produce it. They cast their analysis as a particularly interesting example of a more general world pattern.

Here, elaborating on a completely different part of the grand paradox that is coffee, we focus on the first part of the trajectory – how the bean comes into being, or, more precisely, on the many ways the bean comes into being. It is in this part of the tale that the main protagonist is nature itself in all of its complexities. It is also this part of the tale that contains perhaps the most intimate relationship our species experiences – between people and nature.

Coffee is produced by millions of farmers in the Global South. The great majority of them are family farmers engaged in small-scale farming, and only a few of them produce on large – sometimes very large – farms. However, these two extremes reflect far more than just size. They represent two styles, two complex classes of activities and structures, two categories that we refer to as "syndromes of production," characterized by a set of co-occurring ecological, cultural, political and socioeconomic factors that work in consort, sometimes in contradictory consort.[2] This notion of syndromes of production is best introduced by example. To that end, let us take a tour of two farms that represent two very different syndromes of production. During the tour, we will weave ecological and human interest aspects of coffee farming by describing the livelihood of people on these two contrasting farms and their relationship to nature and the commodity they produce. Our intent here, and in the rest of this book, is to present a picture of the very different ways in which a coffee bean becomes that deliciously aromatic beverage that turns us into "intelligent, sparkling, upstanding men and women."

Example 1: the farm as a component of industrial enterprise

A visit to *Finca* Hamburgo in southern Mexico is, in many respects, a visit to Shangri-La; at least that is what the tourism promotion package of the farm suggests. And the truth is that the highly manicured rows of coffee accentuate the rolling hills and mountains on which this farm was built, suggesting the same sort of aesthetics associated with the mythical place of James Hilton's novel *Lost Horizon*, or, when in flower, the peach forest described by the Chinese poet Tao Yuanming in his classical *The Tale of the Peach Blossom Spring* in which a fisherman, after going through the flowering peach forest, encounters a community of people living a happy and simple life detached from the troubles of the outside world. One's first impression of *Finca* Hamburgo might be similar (Figure 1.1). However, with a bit more focus we see the same verdant hills exposed to the inclement action of the climate, unprotected workers spraying pesticides that eventually wash into the creeks, and a massive number of migrant indigenous workers who are paid the equivalent of US$5 for a full day of work. Viewed from afar, it may be Shangri-La, but up close the bucolic scenery dissipates, exposing the realities of a large and intensively managed coffee plantation.

FIGURE 1.1 Panoramic view of *Finca* Hamburgo. Note the rows of coffee unobstructed by shade trees in much of the landscape.

Source: Hsun-Yi Hsieh

Finca Hamburgo is not a true "sun" system, but it has become as close to a sun system as possible, given the constraints of its location on steep mountainous terrain. Several years ago, plantation owners from the region visited the sites in Costa Rica where the international agency USAID had been promoting the elimination of shade trees. They became enchanted with the new, supposedly more modern, technology. They eliminated as much as they could of the shade, but in the end had to revert to a more shaded system, largely because of the soil erosion that rapidly took over when shade reduction was severe. In many ways, it is a picture of good management. By intent, plant biodiversity is extremely low since continual elimination of "weeds" contributes to its impeccable order, and regular spraying of pesticides and application of chemical fertilizers contributes to the unblemished leaves with their deep green coloration.

The coffee from this farm is of very high quality and is produced for export. The size of the "farm" is 900 hectares, in a country in which legally any one property owner can own no more that 300 hectares, but creative accounting disperses 300-hectare segments among various family members. It does not really matter that much since the farm is part of a larger suite of business enterprises, ranging from a hotel and restaurant inserted within the plantation, to city coffee shops, to travel agencies, and more.

A great deal of independent research has been done relative to the biodiversity on this farm. It goes without saying that the overstory tree layer is of very low diversity – that is how it was designed. However, sampling of birds, butterflies and other organisms leaves little doubt that it is an ecosystem with relatively low diversity, much as many conservationists think of agriculture more generally.[3] And indications are that organisms from frogs to mountain lions very rarely venture into the farm, which is to say it seems to form a bit of a barrier to organisms that might need to migrate long distances.

The actual fieldwork on the farm is done by hired labor. Many of the people who work on the farm are permanent residents, some of whom were born there and some of whom are grandchildren of people born there. Its social structure is something of a mixture of the old Spanish hacienda system, which itself resembles European feudalism, grafted onto an industrial social model. *Finca* Hamburgo is, from the point of view of the people who live there, more a community than a farm, and the "monarchy" is the family that owns it. During harvest time, the labor force is at least doubled with migrants, mainly from Guatemala, arriving in droves to pick the 900 hectares of coffee. Pay for the migrants is lower than for the regular workers, but largely respects Mexican laws on minimum wages. Of course, there are other social benefits that go along with being a permanent resident of the farm, including housing and a degree of social security, much as the hacienda overlords of the past (still referred to locally with the Spanish "*patrón*") were held to be responsible for the well-being of their subjects. Without exaggerating too much, a visit to the farm, with an eye focused on labor relations, suggests traveling back to the nineteenth century.

Reflecting not only on this farm, but others in the general region of the southwest of the state of Chiapas, Mexico, we see a general syndrome. It is a whole agroecosystem syndrome, which includes not only the way in which the farm itself is managed with respect to shade, pesticides and fertilizer and the like, but also the way in which the actual human beings involved in the undertaking relate to the land and to each other – relations that are ecological as well as social, economic and cultural. There is an enormous variability, to be sure, but there is also an identifiable set of characteristics: minimal shade coverage, reliance on agrochemicals, a corporatist mentality of the owners and a proletarian attitude of the farm workers. Calling this an "intensive" sun system, although certainly an oversimplification of what actually exists, is a useful shortcut. The category is easily identified by anyone familiar with the on-the-ground operation of coffee farms.

Example 2: the farm as part of nature

Visiting the farm of Edgardo, a traditional Puerto Rican farmer, engenders the same sort of feeling we gain from visiting a nature reserve (Figure 1.2). Ample shade tree cover above the coffee plants gives the farm the "feel" of a tropical forest. The coffee bushes themselves, even as they show obvious symptoms of classical coffee diseases and pest problems, contain leaves of the deep green coloration that comes from rich fertilization, which must come directly from organic fertilization, since no synthetic chemicals are used. The pathways of the farm are dotted with small plaques with appeals to the observers to respect nature and protect biodiversity. Edgardo is perhaps in his late 50s or early 60s and lives on the farm with his wife in a modest dwelling (they have a place in the nearby town also). He talks as much about the "balance of nature" as he talks about coffee, and has a sense of his own identity as a "*Boricua*"[4] and a farmer. He knows every inch of his ten-hectare farm. His intimate knowledge of every corner of the farm enables the strategic planting of a banana tree here and there, a citrus tree between two boulders, or a strategic mixture of shade trees that he knows produce nitrogen for the soil and control the wind that brings fungal spores of the most common coffee diseases in the region.

His farm is affected by some of the same pests and diseases as other farms in the area – the coffee berry borer, the coffee rust (yellow spot disease) – but he claims that these problems are kept in check by good management and the diversity of the farm. He prunes his shade trees, but not excessively, and plants new ones when either old ones fall or new nursery stock becomes available. Some of his trees are there just to shade the coffee, but there is a smattering of the usual fruit trees such as mango, orange and banana as well as the less known *tamarindo*, *quenepa*, *caimito*, *moca* and *cidra*. In addition, there are the leguminous trees that fix nitrogen and fertilize the soils, as well as some timber species such as mahogany and pine.

Like most Puerto Rican coffee farmers, Edgardo sells to the local market; Puerto Ricans demand coffee from the island, although they are loath to pay the price that it actually costs to produce it, which means constant battles with the island's political system to maintain the subsidies that enable farmers like Edgardo to produce coffee at a price that the average Puerto Rican can afford. Unlike other coffee-producing countries, Puerto Rican coffee is not a major export crop today, even though a large

FIGURE 1.2 View from inside the farm of Edgardo Alvarado in Puerto Rico. The insert is one of many small signs placed around the farm. This one says, "Nature is a loving mother, ready to calm our nerves and sweeten our sourness," a quote from Puerto Rican patriot Eugenio María de Hostos.

fraction of the island's agriculture is devoted to coffee production, a reflection of the common pattern of coffee as a major cultural artifact. Puerto Ricans are very proud of the quality of their coffee and they often brag about how in the eighteenth and nineteenth century the Vatican drank exclusively Puerto Rican coffee.

Edgardo and his family do most of the fieldwork on the farm. At harvest time, he always has to hire some extra help, which is always difficult since in order to make coffee profitable at all, the picking labor must be relatively cheap. That is why he prefers to have as much as possible done by himself and his immediate family. Other than the occasional hired hands for the picking season, all other work on the farm is done by family labor.

Edgardo's sense of natural cycles is the basis for his commitment to managing his farm organically, although the farm is not certified due to the cost of certification. He belongs to a small association of organic farmers in Puerto Rico that do organic agriculture more for ideological and philosophical reasons than for economic advantage. His work style may seem intensive in the modern sense, but the actual management of the farm is more in line with what Richard Levins and Richard Lewontin have referred to as "thought-intensive" – that is, intensive in knowledge rather than capital.[5] It, and farms like it the world over, constitute a different syndrome of production from the industrial plantation we described earlier. We call this syndrome the shade coffee syndrome.

The philosophical/methodological approach of this book

The two vignettes above represent extremes, not necessarily along a single contin-uum, but along several continua: large versus small, heavily shaded versus lightly shaded, use of agrochemicals versus organic methods, a corporate business model versus a family business model, major labor management issues versus local minor labor management issues, intent to dominate natural processes versus intent to fit in with natural processes. However, within all this complexity, it is possible to see the idea of "syndromes of production." It seems that at the present time, in coffee production, there are really only a few syndromes, the above two examples sugges-tive of two of them. However, we emphasize that these are not in any way stable configurations. Indeed, the farm on which we have done much of our research is an example of a farm that combines the two previously described syndromes. *Finca Irlanda* is a large farm of approximately 300 hectares that has many of the charac-teristics of its neighboring farm, *Finca* Hamburgo: a large permanent labor force that lives on the farm plus large numbers of hired migrant workers at harvest time, and a business model that incorporates other commercial activities such as tourism and city-based coffee shops. However, for most of its history it was a very diverse organic farm. These kinds of hybrid syndromes can also be seen on small-scale farms that are chemically managed and have no shade trees. Furthermore, there are many reasons to believe that, from an ecological perspective, syndromes of production are rather temporary formulations, perhaps consistently operative over what might seem to us to be a long period of time, but what is likely for many of the other biological organisms involved to seem a short flick of time.

We consider these syndromes as operating in a time frame similar to Braudel's notion of the *longue durée* in history.[6] Certain aspects of the production system are clearly identifiable as operative over a relatively long period of time, say a decade, but are subject to changes, some of which seem to be induced from within, others imposed from outside the system. Our intent is to understand the dynamics that operate in this *longue durée*, but at the same time understand the dynamic processes that inevitably lead to a rupture in the syndrome. A shade coffee plantation suddenly is reduced to a sun coffee plantation. A sun coffee plantation is abandoned. Coffee production from a large area is discontinued completely and suddenly. A pasture is "reforested" by creating a new shade coffee system. These are all possible major transformations of syndromes, and we seek to understand how and why they occur.

Yet in another sense coffee is not an exceptional commodity. Syndromes of production, much like the shade versus sun coffee system, can be seen in almost any rural setting. Indeed, the original elaboration of the idea of syndromes of production by Andow and Hidaka referred to rice production.[7] What we face today in agriculture in general is a *longue durée* that effectively began at the end of World War II. Facing an almost complete collapse of their market, which was heavily dependent on the war, chemical manufacturing companies sought to transfer the demand for products from the military to the agricultural sector. Thus, through a highly successful propaganda (advertising) campaign, the agroecosystem was transformed into a chemically based system. This is the syndrome that dominates agricultural production in many regions of the world today. However, it is good to remember that this particular *longue durée* has only been in action for about 70 years and seems to be coming to a close with more than 300 ocean dead zones worldwide, pesticide poisonings including massive endocrine disruption, increased rates of soil erosion, and so forth.

Even before this industrial agricultural syndrome began its evolution, an alternative was emerging. Its birthplace, at least in the European imagination, was at the beginning of the twentieth century in India, where Sir Albert Howard discovered that traditional Indian farmers had complex systems of production that made the "modern" system he was supposed to be bringing to them seem, at best, irrelevant.[8] Since that time, a drumbeat has kept the cadence continuous and as alive today as it was in the prescient observations of this Victorian agronomist. Today, it goes by many names but perhaps the single most all-inclusive appellation would be "agroecological systems" or, for short, "agroecology." Thinking, as did Howard, of the agroecosystem as just another kind of ecosystem, resisting the domination of any one species (e. g., *Homo sapiens*), the activity of agriculture takes place within the confines of the ecological laws that apply everywhere and the activity of humans within that system is part of the system itself. Agroecology is an alternative syndrome, itself eclectic and diverse, but focusing on some simple ideas: farming with nature rather than trying to dominate it, conserving biodiversity and ecosystem functions, promoting such human values as justice and equality, respecting not only the laws of nature but also the traditions of the human societies and cultures involved in that nature.

It is arguably the case today that there are two (although in the end this is certainly an oversimplification) competing syndromes of production in world agriculture: the industrial model and the agroecological alternative. In conventional agricultural universities, the industrial model forms the backbone of most education and research programs. By contrast, the science behind the agroecological approach has been largely marginalized in these institutions of higher education. However, because the underlying ideas of agroecology are mainly ecological, the general field of ecology provides a legitimate scientific basis for research and development within this syndrome.

The science of ecology is to agroecology as the science of chemistry is to chemical engineering. Furthermore, casting agroecology in modern ecological terms makes far more sense now than it did when Howard correctly identified the Eurocentric "scientific" approach of the times as antithetical to the clearly superior methodology used by traditional Indian farmers. At that time, the field called ecology did not even exist as a science. Now we have ideas like indirect trait-mediated interactions, for example (see Chapter 5), a centerpiece of much of theoretical ecology, having clear implications for practical matters such as pest control (see Chapter 6). A host of other ecological conceptual tools – such as metapopulation theory, food web dynamics, self-organized spatial patterns, trophic cascades and many others – find clear application to the agroecosystem. Studying agroecosystems from the perspective of the science of ecology is effectively trying to understand them in much the same way as we try to understand other ecosystems.

Gaining this level of understanding of the specific agroecosystem of concern, the coffee agroecosystem, carries with it a further important implication. As a system that is replicated across the globe in a similar set of production syndromes, we have what is essentially a replicated experiment in community ecology, of an ecological community that is deeply penetrated by human affairs. Focusing on its ecology thus not only implies an understanding of the coffee agroecosystem, but also may have implications for all human-dominated ecosystems in the tropics.

Our methodology in seeking this understanding can be described most succinctly as the dialectical method. We take it as evident that the traditional agronomist's approach, the maximization of yields, is hardly relevant, yield being only one in a suite of interesting variables in any human-managed system. However, the complexity emerging from the addition of the human species to the system also makes it dramatically more complex than ecosystems not incorporating human relations. Given such complexity, the simplistic Newtonian philosophical position is anachronistic and needs to be replaced with a more holistic and nuanced analysis. In 1985, Levins and Lewontin elaborated on the dialectical approach:

> It is not that the whole is more than the sum of its parts. But that the parts acquire new properties. But as the parts acquire properties by being together, they impart to the whole new properties, which are reflected in changes in the parts, and so on. Parts and wholes evolve in consequence of their relationship, and the relationship itself evolves. These are the properties of things we call dialectical: that one thing cannot exist without the other, that one acquires its properties from its relation to the other, that the properties of both evolve as a consequence of their interpenetration.[9]

More recently, in 2007, Miller and Page placed the same issues within the paradigm of the "new" science of complex systems. For example:

> If parts are really independent from one another, then even when we aggregate them we should be able to predict and understand such "complicated" systems. As the parts begin to connect with one another and interact more, however, the scientific underpinnings of this approach begin to fail, and we move from the realm of complication to complexity, and reduction no longer gives us insight into construction.[10]

Philosophically positioning this book as the intersection of the classical dialectical approach with the more modern complex systems approach provides a framework for synthesis that embraces complexity. During this synthesis process, we intentionally examine issues of historicity, interconnections, heterogeneity and integration of levels in attempting to generate a new understanding of ecological complexity within agricultural landscapes. Our goal ultimately is to provide the reader with an up-to-date picture of what we know of the ecology, broadly conceived, of the coffee agroecosystem.

The coffee agroecosystem as a model system

While our intention in this book is to synthesize and reflect on the work done over the past 20 years on the ecology of this system, it is worth noting that the system itself represents something of an ecological generalization. When Daviron and Ponte published *The Coffee Paradox*, they made it clear that although their specific topic was coffee, their intent was "to use the case study of coffee to recast the development problem for countries relying on commodity exports."[11]

We have a similar goal, although it is an ecological one rather than a socioeconomic one. Our notion of "ecological" is, however, not restricted to the narrow topics of academic ecology. Rather, we take what some have called a "political ecology" approach, incorporating the classical subjects of ecology, to be sure (biodiversity maintenance, species interactions, etc.), but acknowledging the fact that those subjects exist in a socioeconomic and political context. Thus, the dialectical relationship that exists between natural processes on the farm and decisions made by the farmers is part of our analysis. Yet our perspective is clearly clouded by our background, which is biological/ecological. Hence, our point of view on all subjects herein can be characterized as that of political ecology in its large framework, but more focused on the ecological aspects of political ecology in its detail.

The coffee agroecosystem is an ideal system for this sort of analysis. Indeed, were we to adopt a slogan, it might be that the coffee agroecosystem is to ecology what *Drosophila* was to genetics; perhaps a bit grandiose, but worth noting as our personal bias. From a biogeographic point of view, coffee exists in some of the most important biodiversity hotspots in the world, occupying large areas at mid elevations in the wet tropics, as further discussed in Chapter 2. Its replicate state in so many different sites around the globe makes it almost like an experiment in ecology. Take a particular biogeographic unit (mid-elevation wet tropics) and establish a consistent system (coffee farms) that can be used for comparative ecological analysis, and you have an experiment. In this way, a fantastic worldwide ecological experiment has been underway for the past century or more. Understanding how that system (the coffee agroecosystem) works ecologically and how those workings vary across the globe is, in many ways, attempting to understand generalities about ecology. The worldwide network of 50-hectare forest plots has a similar intent, setting up a consistent framework for study in many regions of the world, such that particularities of location could be acknowledged as generalizations emerged.[12] However, rather than experiment station personnel setting up 50-hectare plots, individual farmers have set up those plots over the past century and more. And just as the 50-hectare plot in a dipterocarp forest in Malaysia has very different properties than the 50-hectare plot in the upper Amazon of Ecuador, the coffee farms of the Ethiopian mountains have different properties to those of Colombia. However, it is indeed such variability that allows us to determine where generalizations can be made. From a strict "science" narrative, we could say that an experiment has been set up for us, as the coffee agroecosystem is a replicate ecosystem throughout the mid-elevation tropics.

Our chapters reflect our own interests in ecosystems generally. Chapter 2 treats the topic of biodiversity, focusing on the problem of biodiversity conservation in the context of the agroecosystem, and extending ideas of biodiversity patterns, previously focused on background biogeographic and habitat variables, to the impact of human decision-making on biodiversity patterns. Chapter 3 expands on the subject matter of Chapter 2 by casting dynamic processes of biodiversity maintenance in a larger landscape framework, emphasizing the notion of matrix quality as it affects processes of metapopulation and metacommunity dynamics. Chapter 4 examines the all-important issue of spatial patterning as a result of both habitat variables and biological interactions and as a driver of other ecological processes. Chapter 5 treats the issue of ecological networks, an expansion of earlier notions of food webs, to include indirect and non-linear effects that have been uncovered through the use of the coffee agroecosystem as a model system. Chapter 6 explores the interconnection of two distinct spatial scales, the previously analyzed large scale of Chapter 4 with the related small-scale pattern so evident in the coffee system. Chapter 7 connects the issue of biodiversity with ecosystem services, avoiding the philosophically difficult question of what is an ecosystem service by noting that an agroecosystem is a natural place to search for that human concept of "service." Chapter 8 integrates these mainly academic ecological subjects more into the sociopolitical framework of the realities of farms and farming families. Chapter 9, the final chapter, contains our attempt to look at the overall coffee agroecosystem as a set of dynamic syndromes of production, extant for the past century or so as syndromes in the Braudelian *longue durée*, and, especially, to explore the forces that either promote a rupture with that syndrome or create a source of its apparent stability.

Notes

1 Daviron and Ponte (2005).
2 Andow and Hidaka (1989); Vandermeer (1990); Vandermeer (1997); Vandermeer and Perfecto (2012).
3 Armbrecht and Perfecto (2003); Mas and Dietsch (2003); Perfecto et al. (2003); Philpott et al. (2006a); Dietsch et al. (2007).
4 *Boricua* is the *Taíno* term that refers to people from Puerto Rico. It comes from the *Taíno* name of the island of *Boriquén*, now Puerto Rico.
5 Levins and Lewontin (1985).
6 Braudel (1958).
7 Andow and Hidaka (1989).
8 Howard (1943).
9 Levins and Lewontin (1985: 3).
10 Miller and Page (2007: 27).
11 Daviron and Ponte (2005: xx).
12 Condit (1995).

2

A BIODIVERSE CUP OF COFFEE

Coffee agroforests as repositories of tropical biodiversity

Background to biodiversity

It is difficult to say anything new about biodiversity. In the past 30 years, we have seen article after article, study after study pointing to the same depressing truth: the planet's biodiversity is being lost at an alarming rate. Some have suggested that we are in the midst of a new mass extinction event.[1] However, one need not move to such an extreme position to be concerned about what is evident. Biodiversity is on the one hand important for our species, and on the other hand declining rapidly. To take this point seriously, we need to acknowledge three facts:

1   Research and political action regarding the conservation of biodiversity has emphasized a taxonomically trivial segment of the tree of life.
2   Active research concerning causes and consequences of biodiversity loss has been concentrated in the higher latitudes.
3   Agriculture is considered by many in the conservation community as the anathema to biodiversity conservation.

We briefly consider these three points in the present section and follow the rest of the chapter with a section devoted to the third point using the coffee agroecosystem as a case study.

Taxonomic biases

The human species is the only organism that is capable of being concerned about other organisms. Indeed, it is the only organism that is capable of being concerned, period. The ability to create social constructs that enable us to foretell, with some degree of accuracy, some event in the future has made us into the dominant species we are, with wonderful physical benefits for ourselves, but, so far, with devastating collateral damage to most other species. However, since our ability to foretell is based on our model of how the world works, prejudices about that world shape that model. One – certainly not the only one – of those prejudices is taxonomic. The time and energy we devote to studying and protecting certain kinds of organisms is disproportionate to their relative abundances, and perhaps their potential importance. There are approximately 1.9 million described eukaryotic species on our planet,[2] never mind bacteria and other non-eukaryotic organisms. Estimates of total numbers range anywhere from three million to as many as 100 million![3] Although recent studies place these estimates at the lower range,[4] the thought that, after almost 250 years of collecting and classifying organisms, we still are ignorant of 50 to 90 percent of the species on this planet is mind-boggling. Part of the problem is that we have not been paying a lot of attention to the other species that share the planet with us. However, when we do pay attention, we pay special attention to those species that look more like us, mainly other vertebrate species. This bias is reflected in the fact that even though vertebrate species represent less than one percent of all species, approximately one third of all taxonomists work on vertebrates.[5] Another third of taxonomic attention is on plants, which represent about 10 percent of the world's species. That leaves only about one third of the world's taxonomic work aimed at invertebrates, which are estimated to be almost 90 percent of all eukaryotic organisms! The most recent estimates of the number of species of tropical arthropods, for example, range from 1.1 to 7.4 million, suggesting that 70 percent of the species are still awaiting to be discovered by science.[6] A grimmer estimate puts the percentage of species that still await discovery close to 90 percent.[7]

The vertebrate bias in taxonomy may be excused in conservation circles since it is frequently used to draw attention to broader conservation issues. However, although our aesthetic sense is certainly a legitimate excuse for a call to conserve, there is also a deeper and ultimately more troubling issue. Much of the biodiversity that is generally unnoticed (the beetles, ants, flies, fungi, mosses, bacteria, etc.) is known to have important functional roles in ecosystems.[8] That is, biodiversity provides us with ecosystem services, ranging from the regulation of nutrient cycling to the provisioning of pest control. And much of that ecosystem service function is delivered by the "little things that rule the world," as E. O. Wilson has noted.[9] The fact that the biodiversity of insects and other "little things" seems to be under just as much threat as the elephants and tigers that motivate the general public thus has very clear practical implications.

Geographic bias

The second bias has its roots in political history. As a consequence of Europe's emergence as a political/military power in the last 500 years, political and economic power came to be concentrated in northern latitudes, with the tropics effectively serving as sources of markets for overproduced items or a cheap source of raw materials and luxury items, destined for the Global North. Yet it is precisely in these tropical regions that most of the world's biodiversity actually occurs.[10] For example, there are a reported 1871 species of birds in Colombia,[11] compared to 914 in all of North America (north of the Mexican border),[12] an area almost 20 times as large. In the whole of Great Britain, there are 63 species of ants (11 of which are introduced),[13] while E. O. Wilson reported 43 species in a single tree in a Peruvian forest,[14] and Schulz and Warner reported 161 species in the canopy of 61 trees in a Ugandan forest.[15] In Figure 2.1, we present composite data on species richness as a function of latitude.

It is thus clear that the vast majority of the world's biodiversity is located in the tropical regions of the world. The irony is that the part of the world that is "underdeveloped" is also in this region, as reflected in the proportion of people living in poverty as a function of the latitude (Figure 2.2).

These familiar facts lead us to three important normative conclusions regarding the conservation of the world's biodiversity:

1   There is a need to invest more in the study and conservation of small and noncharismatic organisms that play important roles in providing and maintaining ecosystem processes and functions.
2   Most efforts should be concentrated in the tropical regions of the world, where most of the biodiversity is located.
3   The problem of poverty in the world is intricately linked to biodiversity conservation, and therefore conservation programs must deal with issues of poverty.

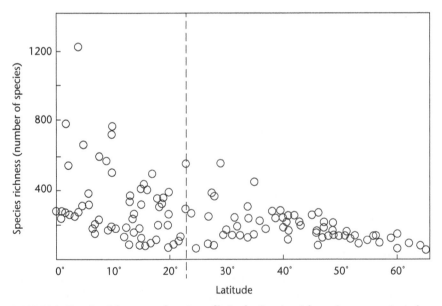

FIGURE 2.1 Species richness as a function of latitude. Species richness is a composite index of plants, birds and mammals. The horizontal dashed line divides tropics (to the left) from temperate zones (to the right).

Source: Modified from Bonds et al. (2012)

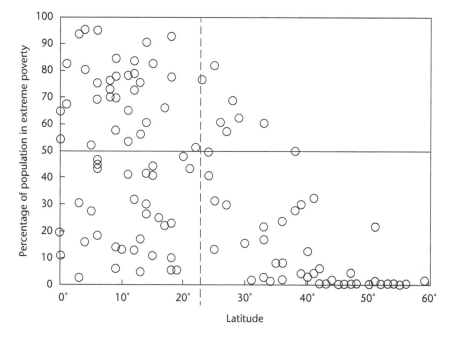

FIGURE 2.2 Poverty and latitude: country data. The horizontal dashed line divides tropics (to the left) from temperate zones (to the right). Note that about half the countries with extreme levels of poverty (above the horizontal dashed line) are located in the tropics.

All three of these points are especially important for the subject of this book, since many of the non-charismatic organisms are essential for the productivity and sustainability of the coffee agroecosystem, coffee is a tropical crop that cannot be cultivated in high latitudes and finally, it is a crop that sustains millions of small-scale tropical farmers and farmworkers, providing one of the essential economic activities so desperately needed in these parts of the world.

The agricultural connection

There has long been a rather negative attitude toward agriculture by folks who are dedicated to the conservation of biodiversity. Driving along any highway in the grain belt of North America, one can view the "amber waves of grain" as the triumph of human ingenuity and provisioning of food security, or as evidence of ecological devastation. We see a parallel here with the general methodology of literary criticism in which a piece of literature must be understood as a dialectical relationship between the text (the actual words in the literary piece) and the reader, which is to say that the literary piece takes on meaning only when viewed in the context of who is reading it. Here the "text", the fields of grain, requires a reader, the human observer. If we force our focus (i.e., if we "read") on the general question of conservation of biodiversity, that sea of majestic grain fields (the "text") looks more like a devastated landscape. Similar trips in the tropics, where most of the world's biodiversity is located, generate the same feelings with banana plantations, soybean fields and overgrazed cattle ranches taking the place of those amber waves of grain. When visitors from the Global North drive through such landscapes on their way to the nearest tropical rainforest reserve, they notice the sharp contrast between the diverse array of odors, sounds, colors, structures and species so evident in the rainforest and the monotony of the pineapple plantation they just drove through (Figure 2.3). From that perspective, it certainly seems that agriculture is the enemy of biodiversity.

Yet such observers certainly realize that people, including those who live in the nations in which those rainforest preserves are located, need to eat, which makes that devastated landscape seem to be a necessary evil. It is at this point that what might be called the devil of linearity seeps into the thought process. If there are ten million people in a tropical nation and each is estimated to eat the equivalent amount of food that can be produced on a hectare of land, then a simple multiplication would suggest that ten million hectares are required to feed that population. If there are 11 million hectares in the nation, that means that one million hectares can be left as natural habitat to protect biodiversity. The activist conservationist then may extend the analysis to note that the population of that nation is growing at a rate of 10 percent per decade, which means that in only ten years' time there will be more people than can be accommodated without eliminating the protected area. In order

FIGURE 2.3 Typical pineapple plantation, La Ceiba, Honduras.

to preserve some of the rainforest habitat, only two solutions occur to the linear thinker: either stop the population from increasing or intensify agriculture to produce more per area. If you are stuck in a linear way of thinking, this conclusion is logical and absolutely correct.

However, we have argued that there are two general categories of error in this way of thinking.[16] The first is related to the assumption that all agricultural systems are deserts of biodiversity; the second is related to a "snap-shot" assumption of biodiversity in the context of the larger landscape. The rest of this chapter deals with the first assumption. The second assumption is the core subject matter of Chapter 3.

Not all agriculture is the same

The depressing vision of industrial agriculture, whether an Iowa corn field or a Costa Rican pineapple plantation, is certainly accurate for most of us. However, when examining the style of agriculture done by many of the world's tropical farmers, we realize that the "farmers" named Chiquita, Dole and Del Monte are actually quite rare in terms of numbers, although they control vast extensions of land. The majority of the farmers in the Global South are small-scale farmers and they do agriculture in very different ways. The vignette of Edgardo's coffee farm presented in the previous chapter is an example. The farm itself is a biodiverse place, with typically three or four main crops grown along with fruit trees, bananas and other diverse species of plants. However, more importantly, those farms also harbor what is referred to as associated biodiversity – that is, species other than what a farmer intentionally introduced, such as birds, epiphytes, ants, bees, spiders, fungi, amphibians and so forth. This kind of small-scale diverse farm can harbor large numbers of such species. Although you are unlikely to find many different kinds of beetles on a pineapple plantation drenched in pesticides, the numbers on a more traditional farm, especially if the farmer eschews the application of pesticides, can be enormous.[17] These kinds of casual observations on tropical farms drive home the point, now pretty much understood, that it is not agriculture *per se* but rather the kind of agriculture practiced that has an effect on biodiversity.[18]

Historical roots of agricultural transformation and biodiversity loss

Harvard biologist Richard Lewontin crystalized the defining feature of the industrial model of agriculture: "Farming is growing peanuts on the land; agriculture is making peanut butter from petroleum."[19] This simple statement summarizes much of the state of modern agriculture as it has evolved since World War II. Because of this structure, farmers, those who actually have contact with the land, are caught in a great squeeze. Their metaphorical petroleum suppliers are large oligopolies whose selling prices are set artificially high, and their metaphoric peanut butter makers, who buy the peanuts they produce, are large oligopsonies whose purchasing prices are set artificially low. The farmer is frequently caught in the middle. Powerless, both economically and politically, he or she can only hope for government

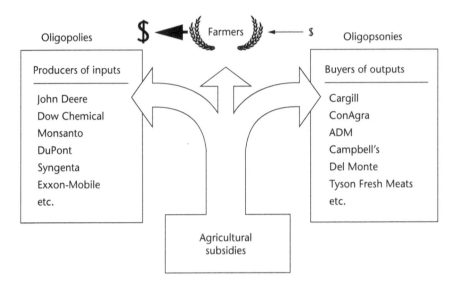

FIGURE 2.4 Structure of the industrial agricultural system after the transformations that occurred after World War II. Farmers are squeezed between powerful oligopolies (a few firms that produce agricultural inputs such as fertilizers, pesticides, fuel for tractors, tractors, etc.) and powerful of oligopsonies (a few firms that buy, process and distribute the produce from the farmers). Agricultural subsidies, therefore, are indirect payments to these oligopolies and oligopsonies.

hand-outs to make up the difference between capital outlay and gross income. In the US and in Europe, those hand-outs come in the form of subsidies, but ironically, the direct subsidies to farmers are effectively indirect subsidies to the large corporations that produce the inputs and the one that buys the output from the farmer (Figure 2.4). The coffee you drink in the coffee shop is filtered through Starbucks or Maxwell House and the coffee farmers in Costa Rica feel that squeeze as much as family farmers in Minnesota or England.

Part and parcel of the transformation of the world system – from one dominated by the production of things because they are needed to the production of things because they can be exchanged – is the commodification of as much human activity as is possible. This was especially important for agriculture, and remains important to the present day. If we assume that, in the capitalist system, society's prime purpose is to make sure that production happens for the sake of profit in all spheres of human activity, it is obvious that we must make all products tradable. If some activities are unmarketable, or produce an unmarketable product, they represent a barrier to this central goal. For example, a farmer who controls potential pests through a specific planting design rather than a purchased pesticide is not contributing to the profits of pesticide manufacturers. There is no commodity produced and thus nothing to buy and sell. Of course, the farmer, as an entrepreneur, may be able to gain more in his or her own transaction, but this assumes that he or she is a key participant in the general system of commodity exchange and control, which in the contemporary developed world is not generally true. The dominant participants, the industrial companies that manufacture the pesticides in this example, would get nothing.

This generalized path of commodifying everything, so clearly followed in industry, was significantly constrained in agriculture at least at first. Agriculture was limited by its articulation with natural processes. While it was within human abilities to substitute all the textile weavers with workers concentrated in large textile mills, wheat producers could not all be brought into a wheat factory. They remained scattered across the landscape, a condition necessitated by the nature of agriculture. Thus, the penetration of capitalism into agriculture was distinct from its penetration, or rather complete take-over, of industrial production. Indeed, the history of agriculture in the developed world during the past two centuries is a history of the penetration of capital in this sense, such that agriculture today bears only a vague resemblance to agriculture of the eighteenth century, in both technological and social terms.

As the industrialization process in agriculture proceeded, the need to homogenize production was key. Much as the philosophy of Taylorism took over the factory system in the developed world, an attempt to make agriculture more "efficient" relied on a particular notion of efficiency.[20] A typical farmer in the eighteenth century would have regarded his or her operation as efficient if a large number of many different kinds of plants and animals were produced. However, the growing ideology of capitalism that was penetrating the factory system was also making inroads into the agricultural system. Systematizing all aspects of production, as Taylor suggested in the factories, became a seemingly obvious goal for the agricultural sector. For our purposes in this book, the aspect of that systematization that is most important is the notion of specialization. A farmer who produces corn, beans, squash, mangoes, citrus, coffee, milk, cheese, bananas and cassava could never really be efficient, according to this analysis. As in the factory sector, in which production was specialized as much as possible, the agricultural sector needed to become equally specialized. Agricultural research and education was put to the service of this agricultural transformation. Technological development focused on narrowly framed questions that were primarily concerned with yield maximization. Those questions shaped the agricultural research agenda, which fed back to the further specialization of agriculture. As with research everywhere, a narrow focus of the questions asked, usually referred to as "reductionism," is sometimes a necessary tactic. Unfortunately, as Richard Levins and Richard Lewontin have pointed out, there is frequently a confusion between the tactic of reductionism (which is necessary for the research process to proceed in the first place) and reductionism as a philosophy.[21] As a reductionist philosophy permeated the agricultural research and education establishment, a research proposal that focused on the interaction of the 10 or 15 plants and animals on the farm and asked how to improve on their synergisms would have been thought foolish. A research proposal that assumed a monoculture of corn and asked what kind of a machine could be invented to harvest it would have been thought the picture of rationality.

The ultimate consequence is that farming in the industrial sector today consists almost entirely of monocultures. Other plants are weeds, insects are potential pests, fungi cause disease. Much as a factory specializes in a single commodity, under this philosophy the farm is supposed to specialize. When translated into the material world of ecology, this means, in part, the massive reduction of biodiversity. Given this historical trajectory, it is not surprising that as agriculture becomes intensified, we see a general loss in biodiversity. That was the goal in the first place, and it has been accomplished remarkably well, although – thankfully – not completely.

Biodiversity on the farm

As noted, one of the main characteristics of conventional intensification is the elimination of crop diversity – in other words, the tendency toward monocultures. This has been justified through a particular interpretation of the concept of efficiency. When only one crop and one variety of that crop is cultivated, one can focus the entire management of the system toward maximizing the yield of that crop and that variety. Unfortunately, that ignores potential facilitative (positive) effects among different species,[22] as well as other important properties of the system, such as stability, resilience and resistance to disturbance, that are associated with diversity and are as important as productivity to most farmers.[23]

When thinking about biodiversity in agroecosystems, it is important to distinguish between "planned" biodiversity and "associated" biodiversity.[24] Planned biodiversity is the diversity of organisms that are purposefully introduced into the system by the farmer: the crops and livestock and any other elements that the farmer decides to use in constructing the farm. This type of biodiversity is sometimes called agrobiodiversity. As briefly mentioned before, associated biodiversity is everything else: all the organisms that spontaneously arrive because the farm or the agricultural landscape provides a habitat for them. Associated biodiversity certainly includes the insect pests, diseases and weeds that afflict the farm, but it also includes the natural enemies of those pests and other beneficial organisms like pollinators, mycorrhizal fungi and free-living nitrogen-fixing bacteria. Furthermore, associated biodiversity also includes wildlife species that may not have any impact on the farm, either positive or negative, but that use the farm as a habitat, temporary or permanent.

Planned biodiversity is the result of farm design and management, while associated biodiversity is the result of the biodiversity that exists in the surrounding environment and that existed in the land before it was converted to agriculture. Since the planned biodiversity includes the crops that form the vegetation structure of the agroecosystem, and represents the main or at least some of the primary productivity of the agroecosystem, it creates the conditions that promote associated biodiversity. In that sense, diversity begets diversity since elements of the planned biodiversity serve as ecosystem engineers that create habitats for other species. Importantly, both the planned and the associated biodiversity are responsible for the ecosystem services generated in the agroecosystem, including – but not limited to – provisional services such as food, fiber, meat and firewood (Figure 2.5).

Although there is minimal debate concerning the important role of planned biodiversity in the productivity, resilience and stability of the farm,[25] there is

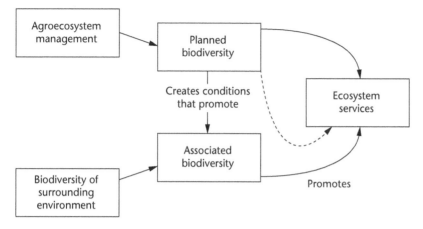

FIGURE 2.5 Diagram of the two components of biodiversity in agroecosystems and how they contribute to ecosystem services. The dashed line represents an indirect effect of planned biodiversity through its effect on associated biodiversity.

considerable controversy over what might be the role of associated biodiversity. As mentioned before, some of that biodiversity is clearly negative, such as pests and diseases. However, a great deal of it is potentially beneficial to the goals of the farm, although the nature of that benefit may be obscure and may require significant research effort to be discovered, a subject we explore in detail in subsequent chapters.

As so much of the world's terrestrial land surface is dominated by managed ecosystems, the chief one of which is agriculture and pastures for livestock, these managed systems are of critical importance for the conservation of biodiversity, including the diversity that may have nothing to do with actual farm activities. If one's vision of agriculture is limited to Iowa corn fields, Central American pineapple plantations or south-east Asian oil palm plantations, the notion that agriculture is the enemy of biodiversity is quite understandable. However, not all agriculture is like that, as anyone who has ever walked through a traditional coffee farm can attest. Nevertheless, the trend since the 1950s has been to move from diverse farming systems to intensive monocultures and homogeneous agricultural landscapes with a concomitant loss of both planned and associated biodiversity. So the question about biodiversity and agriculture, from the point of view of biodiversity loss, is not whether agriculture does or does not have an impact on biodiversity but rather what effect different styles of agriculture have. Specifically, how is extant biodiversity modified as a function of agricultural intensification?

We propose that there are five general patterns of biodiversity loss with agricultural intensification that can be observed, depending on the particular agroecosystem and on the taxonomic definition of the biodiversity that is under consideration, as illustrated in Figures 2.6 and 2.7. The general assumption of most people concerned with the conservation of biodiversity has been that most biodiversity is lost with the initial transformation of a natural ecosystem to an agricultural system (Type I pattern

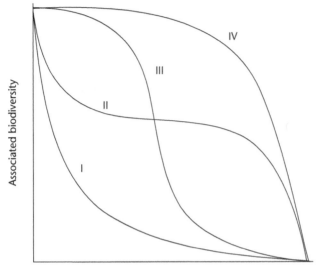

FIGURE 2.6 Hypothesized relationship between intensification and associated biodiversity. Type I and IV patterns are the extremes that might be assumed. Type II and III patterns represent intermediate states.

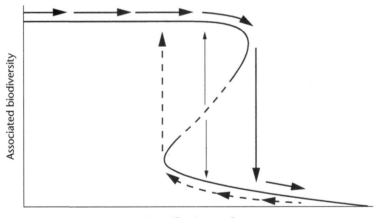

FIGURE 2.7 Pattern of biodiversity loss resulting from relaxing the assumption of monotonicity in the biodiversity–intensification relationship, including a hysteretic relationship between forms of intensification and associated biodiversity. As intensification proceeds, the system is expected to follow the trajectory indicated by the solid arrows. When it reaches the "tipping point", the associated biodiversity suddenly collapses. The reversal of intensification, however, results in a slow increase in biodiversity that passes the state of intensification that led to the collapse (this is the nature of hysteresis) and continues before reaching the reverse tipping point that leads to a sudden increase in biodiversity, as illustrated by the dashed arrows.

in Figure 2.6). This assumption gives rise to the idea that since any kind of agriculture results in dramatic loss of biodiversity, we should intensify agriculture as much as possible in order to produce as much food as possible in as little area as possible and spare land for conservation. This point of view has been called the land-sparing argument.[26] Later on, we discuss this idea in depth and explain why it does not work most of the time.[27] However, there are examples of systems or taxonomic groups that do not show significant reductions in biodiversity until high levels of intensification are reached (Type IV pattern in Figure 2.6).

Furthermore, there is no particular reason for the patterns to be monotonic. Indeed, there are examples of agroforestry systems that show higher levels of biodiversity than adjacent forest fragments, mainly because they maintain most of the biodiversity of the forest and at the same time contain species associated with agricultural systems.[28] Also, there is the possibility that, for a certain range of intensification values, there may be two alternative states of biodiversity in the system (Type II and III patterns in Figure 2.6 and the pattern illustrated in Figure 2.7). The most important consequence of this observation is the potential existence of hysteresis, which is to say that when increasing the intensification of a system, there may be a point at which there is a dramatic drop, or critical transition, in biodiversity. However, reversing the intensification process may not result in the restoration of that biodiversity at the same point (Figure 2.7).

Intensification and biodiversity: coffee as a model system

The intensification gradient in coffee

The coffee agroecosystem is an ideal model system to study how agricultural intensification affects biodiversity. *Coffea arabica*, one of the main species of coffee that is used for producing the beverage we all love, is an understory shrub originating from the highland forests of south-west Ethiopia and south-east Sudan.[29] As an understory plant, it does quite well under a forest canopy (Figure 2.8) and for that reason, when introduced to the Americas, it was frequently cultivated with shade trees.

There is not just one single shade coffee system or coffee agroforest. Indeed, throughout the world you can find a range of systems with greater or lesser shade and with many different tree species, managed in a variety of systems (Figures 2.9, 2.10 and 2.11). The most diverse of the shade coffee agroforests is what in Mexico people call "rustic coffee" (Figure 2.9(b)). In this system, the understory vegetation of a natural forest is replaced with coffee plants, thus maintaining much of the natural forest structure and tree diversity. However, not all shade systems are that diverse and it is common to see shade coffee plantations with a canopy of a single or a few tree species, like *Erythrina poeppigeana* in Costa Rica, or a few species of *Inga* in Mexico (Figure 2.10), or even the non-native *Gravillea robusta* in parts of Guatemala and India. Starting in the 1970s in Colombia[30] and in the 1980s in Central America,[31] there was a strong effort to intensify coffee production mainly by drastically reducing or eliminating shade trees (Figure 2.11), planting higher densities of new coffee

FIGURE 2.8 Coffee plants in an Ethiopian rainforest, original home to the understory species *Coffea arabica*, the plant that changed the world.

Source: Kristoffer Hylander

FIGURE 2.9 Comparison of (a) natural forest and (b) rustic coffee production in Chiapas, Mexico.

varieties and adding agrochemical inputs. This intensification effort led to the "deforestation" of many coffee-growing regions in Latin America,[32] but was implemented to varying degrees through the region, with countries like Colombia intensifying almost 70 percent of their coffee production, while countries like El Salvador maintained 90 percent of their coffee on shade coffee plantations. As a result, you can now find a gradient of coffee intensification throughout the region, going from the rustic system to coffee monocultures, with no shade trees (Figure 2.12). Because of

FIGURE 2.10 Monocultural shade (*Inga* sp.) coffee in Chiapas, Mexico.

FIGURE 2.11 Monoculture of sun coffee in Colombia.

Source: Juan Munduy/CC 2.0 Attribution-NonCommercial-ShareAlike License

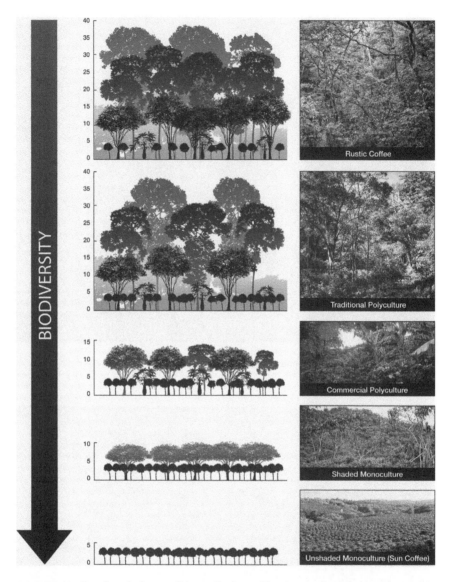

FIGURE 2.12 Gestalt and photos of the coffee intensification gradient and its relationship with biodiversity. Numbers on the y axes represent the height of the canopy in meters.

Source: Modified from Moguel and Toledo (1999)

this variety, the coffee agroecosystem is an excellent model system for studying the impacts of agricultural intensification on biodiversity.[33]

Coffee management systems vary in more factors than just the shade, yet the shade is evident to even casual observers, and extremely important ecologically. The shade trees perform important ecological functions in the agroecosystem, such as providing organic matter and, in the case of nitrogen-fixing trees, increasing the amount of nitrogen in the soil.[34] They are also involved with microclimatic regulation, wind protection, weed suppression, refuge for natural enemies and many other ecological functions and ecosystem services.[35] When the trees are eliminated, the ecological functions they once performed need to be supplemented, frequently with external inputs such as synthetic fertilizers and pesticides. These collateral practices also have an impact on biodiversity and can act as feedback mechanisms that further deteriorate the ecological integrity of the system. As discussed in Chapter 1, we can conceive of a dual categorization, shade versus sun, as two syndromes of production. This distinction is useful to some extent, but should be employed with great caution since the actual pattern in the real world is certainly more complicated. To some extent, as we introduced in Chapter 1 and argue more completely in Chapter 9, it makes sense to think of these two syndromes as having their own socioecological dynamic. In this chapter, however, we focus on the more finely tuned classification of the intensification gradient (Figure 2.12).

Costa Rica, coffee intensification and biodiversity: a case study

We started working in the coffee agroecosystem in the late 1980s while teaching a field course in Costa Rica. Costa Rica is a popular destination for eco-tourists and ecology students from North and South America, at least partially due to strong conservation laws that include the designation of approximately 25 percent of its territory as National Parks or protected areas. Although Costa Rica's reputation as one of the greenest countries in the world is well earned in terms of the protection of natural areas and more recently for its efforts to be carbon neutral,[36] its agricultural policies and focus on monocultural exotic tree plantations in the 1980s left much to be desired and created a negative feedback loop that resulted in the highest deforestation rate in the world.[37]

Costa Rica is no longer the typical agroexporting country that it once was. For once, tourism has surpassed the combined three main export crops together in terms of earning foreign exchange, but also the pharmaceutical and information technology industries have made inroads in the Costa Rican economy. However, agricultural exports are still important generators of foreign exchange, with bananas, coffee and, more recently, pineapples representing the backbone of the agroexport economy.[38] Coffee, in particular, has been iconic in the Costa Rican economy, having contributed to economic development and the establishment of a strong middle class (Figure 2.13).

The 1980s was a decade of economic stagnation and financial crisis for most Latin American countries, including Costa Rica. In previous decades, these countries had borrowed considerable sums of money from the World Bank and other foreign banks and, with economic stagnation, the foreign debt exceeded their earnings. Dubbed the "Lost Decade" of Latin America, this period was characterized by strong pressures to increase foreign currency earnings by increasing exports. Although minor compared to that of countries like Mexico, Argentina and Brazil, Costa Rica's

FIGURE 2.13 Costa Rican five colones currency depicting coffee harvesting and export.

debt nevertheless generated pressure to increase production of agroexport crops to generate foreign currency to service the debt.[39] With help from USAID, coffee intensification (also called "coffee technification") became one of the cornerstones of their recovery plan. Incentives were established to eliminate shade trees and to plant coffee varieties, like *caturra* and *catuai*, that were higher yielding and could better tolerate being grown in the open. This coffee technification program resulted in the deforestation of many traditional coffee plantations in the Central Valley of Costa Rica and other prime coffee-growing regions of the country.[40]

In 1988, when we were teaching a field course to a group of Latin American students, the coffee technification program had reached its zenith in Costa Rica. Having become accustomed to the forest-like traditional plantations that were so common in Costa Rica, we started wondering how much diversity was being lost with the elimination of the shade trees. At that time, most conservationists were focusing on the purchase of land to secure what were thought to be pristine forests and few ecologists were interested in the potential conservation value of agricultural systems. Ironically, ecological theory at the time suggested that intermediate levels of disturbance could actually increase biodiversity,[41] implying that some agricultural systems could somehow benefit biodiversity.

Contrary to this general trend, we saw the coffee agroecosystem as a potential laboratory for the study of biodiversity. In particular, the intensification program that was underway in the major coffee-producing areas of Costa Rica provided us with an opportunity to study biodiversity within an agroecosystem that was changing at a very rapid rate. Inspired by a study conducted by Smithsonian entomologist Terry Erwin,[42] in which he used aerial pesticide fogging to knock down insects occupying the canopies of several rainforest trees in Panama, we set out to conduct a similar study on coffee plantations in the mid-1990s. Erwin and later E. O. Wilson had suggested that the biodiversity of insects in the tropics was much higher than had been previously thought.[43] Although a recent study made more realistic estimates and reduced the uncertainties associated with Erwin's initial estimates,[44] nevertheless the estimates of species richness of arthropods in tropical regions remain astonishingly high. Using the same methodology used by Erwin in his 1983 study, we sampled canopy arthropods on a traditional shade coffee plantation in the Central Valley of Costa Rica (Figure 2.14). Sampling three individual trees – one *Erythrina poeppigeana*, one *Erythrina fusca* located approximately 100 meters away and one *Annona cherimola* located approximately 200 meters removed from the other two – we encountered a surprisingly large species richness of beetles, ants and wasps (Table 2.1). Indeed, our study found beetle species numbers in the same order of magnitude as those reported by Erwin for some of the rainforest trees in the rainforest of Panama.[45] Even more surprising were the low levels of species overlap among the samples. For example, we only recorded 18 percent species overlap of ant species and 14 percent species overlap of beetle species between the samples taken from the two *Erythrina* species.

For comparative purposes, we decided to sample another individual of *Erythrina poeppigiana* on a shaded monocultural plantation and found fewer than half the ant and beetle species than were recorded from the same tree species on a traditional plantation (Table 2.1). Additional sampling in the coffee bushes themselves led to the conclusion that the number of species observed declined as a function of the type of shade the farm contained. It was obvious that the intensification of coffee plantations and the loss of planned biodiversity (the shade trees) led to a concomitant loss of associated biodiversity, at least of the beetles, ants and wasps. The coffee intensification program in Costa Rica led to an agrodeforestation process and was having a similar effect on biodiversity as regular deforestation.

FIGURE 2.14 Funnels located underneath the canopy of trees to collect arthropods after pesticide fogging knock-down.

TABLE 2.1 Number of species (and individuals) of beetles, ants and wasps in the canopy of a single shade tree on two coffee farms in Costa Rica.

Shade tree species	Type of farm	Beetles	Ants	Wasps
Erythrina poeppigiana		126 (401)	30 (333)	103
Erythrina fusca		110 (393)	27 (1105)	61
Annona cherimola		NA	10 (179)	63
Erythrina poeppigiana		48 (107)	5 (64)	46

Source: Data from Perfecto et al. (1997)

Three decades of biodiversity research in coffee agroecosystems

Pioneering biodiversity research in the coffee agroecosystem

In the 1980s, several researchers noted the importance of the coffee agroecosystem in Puerto Rico as a refuge for biodiversity in the face of relentless agricultural expansion.[46] Indeed, as Puerto Rico was becoming more industrialized, it was suggested that the trend of increasing biodiversity following the near-abandonment of the rural economy had its "seeds" within the abundant coffee plantations on the island.[47] However, it was probably the team of ornithologists at the Smithsonian Migratory Bird Center, led by Russell Greenberg, who put the issue of biodiversity in coffee agroecosystems on the map.[48] From a conservation point of view, the research of this team was especially important since their work in both Guatemala and southern Mexico was in coffee forests that were the wintering grounds of migratory birds from North America. There had already been growing concern about the declining populations of these birds, and the Smithsonian team provided an explanation that could hardly be ignored. The decline of North American birds was occurring in Central America. Coffee intensification throughout Mesoamerica and northern South America was resulting in habitat destruction for thousands of migratory birds.[49]

It was at that time that the enthusiasm for technifying coffee production was reaching a fever pitch. With the typical productionist mentality that emphasizes yield above anything else, recommendations flowed from the Global North to cut the shade trees on all the "backward" traditional coffee plantations. One could hardly fail to notice that the deforestation of coffee agroforests in this region coincided with the declining bird populations in North America, and the issue of biodiversity decline as a function of agricultural intensification has ever since been on the minds of thoughtful conservationists. Consequently, since the 1980s, there has been enormous interest in learning about the biodiversity potential of the coffee agroforestry system.[50]

Since many coffee-growing areas overlap with biodiversity hotspots,[51] the importance of coffee for conservation is frequently assumed to be disproportionate to its geographic extension of only 11 million hectares.[52] Historically, coffee has been associated with habitat loss because it is normally established in previously forested areas. For example, in Central and South America coffee farms comprise 54 percent of the perennial crops that have replaced cloud and premontane tropical forests.[53] However, the conservation value of traditional shaded plantations was evident to early naturalists and bird watchers who frequented shade coffee plantations in search for birds and noted that the avifauna within plantations was very similar to that of the intact forest.[54]

Biodiversity loss and coffee intensification: what causes the pattern?

Given this evident pattern, it should come as no surprise that ornithologists were the first to call attention to the potentially detrimental impacts of coffee intensification on biodiversity. In a pioneering study in Colombia, José Ignacio Borrero documented the dramatic reduction in bird species richness associated with the substitution of traditional shaded plantations for what he called "caturrales," sun coffee plantations dominated by the *caturra* coffee variety.[55] Taking note of this and other studies of bird declines in Central America and the Caribbean, the Smithsonian Migratory Bird Center started their research program to study the impacts of coffee intensification on migratory bird species and organized the First Sustainable Coffee Conference in 1996.[56] Since then, many studies and several reviews and meta-analyses have been published confirming the important role of traditional shade coffee plantations as a refuge of biodiversity and its loss with coffee intensification.[57]

Overall, these studies are consistent with the general ecological principle that species richness is related to the structural and floristic diversity of the habitat. Farms with a high planned biodiversity (for example, a high number of species of shade trees) also tend to have a high diversity of associated flora and fauna. These results have been found repeatedly for vertebrates (birds,[58] terrestrial mammals,[59] amphibians and reptiles[60] and bats[61]), invertebrates (ants,[62] beetles[63] and others[64]), plants (trees[65] and epiphytes and ferns[66]) and fungi.[67] Although this is a general pattern, some studies document the same or even higher species richness on traditional shaded plantations compared to adjacent forests.[68] However, most studies also show that community structure and species assemblages differ from forest to shaded plantations. More specifically, coffee plantations tend to have higher dominance of some species and also more generalist species than undisturbed forests.[69]

Although all these studies show that traditional shaded and rustic plantations tend to be more similar to forests than to coffee monocultures in terms of species composition and community structure,[70] it is also the case that coffee systems vary in their suitability as habitats for wildlife. For example, certain shade tree species attract more wildlife than others.[71] Management practices can also affect how suitable a particular plantation is for wildlife. Among the management practices that most negatively affect associated biodiversity are pesticide applications, the regular and heavy pruning of shade trees and the elimination of epiphytes.[72]

Biodiversity within coffee farms varies also with landscape-level factors. In diverse landscapes that have a high proportion of natural habitats, differences in biodiversity among coffee farms of various intensities may be reduced, especially for highly mobile organisms. On the other hand, in a very intensified landscape, where most natural habitats have been destroyed, traditional shaded plantations could be important refuges for biodiversity. These landscape-level factors will be discussed more in depth in Chapter 3.

The question that naturally emerges from the observed pattern of biodiversity loss with coffee intensification is the cause of the pattern. Unfortunately, it is a question too infrequently pursued in agroecology, but fundamental for the sustainable management of agroecosystems. There is an enormous literature on this classical ecological question (see any textbook in ecology). Although debates on the details continue to rage, elementary theory in ecology provides us with a very simple dichotomy of mechanisms. On the one hand, all populations are in some way restricted by limitations placed on them by the physical and chemical environment – which is to say, elements that are not affected by the populations themselves (e.g., if birds nest in trees, the number of trees in a system at least partially determines how many birds can survive). This is the first part of the dichotomy, frequently referred to as the "carrying capacity" of the environment for that particular population. The overall collection of physiochemical factors that limit an organism is sometimes referred to as the "fundamental niche" of that organism, following Hutchinson's original formulation of the niche as a hypervolume (i.e., delimited by ranges in a multidimensional space).[73]

On the other hand, all populations live within the constraints of an ecological community, which is to say that they are forced to interact with the other populations in the ecosystem either directly or indirectly. Mathematically, this idea has been expressed as a subtraction of the effect of other species from the carrying capacity of the environment, or, in Hutchinsonian terms, a subset of the fundamental niche that results from the interactions of the organism with the other biological components of the ecosystem, sometimes referred to as the "realized niche." Thus, when an organism is restricted from a particular environment, it makes sense to ask the question whether that restriction is due to the environment being outside of its fundamental niche or whether other organisms in the ecosystem have excessively reduced its realized niche – in other words, caused it to go locally extinct.

We asked this question in a very simple way with regard to ground-foraging ants on a traditional coffee farm in Costa Rica.[74] Noting that more traditional shaded farms not only had more shade, but that the shade trees also provided another physical aspect to the environment of ground-foraging ants, the leaf litter on the ground, we hypothesized that shaded plantations would harbor a higher diversity of ground-foraging ant species. After establishing that the ground-foraging ant species richness went from approximately 19 species in a more traditional shade system to about six species in a sun system, we sought to determine what fraction of that decline was due to the physical factors of shade and leaf litter. Creating artificial conditions of both shade and leaf litter was a simple exercise, and the experiment demonstrated conclusively that the most important physical factor contributing to the loss of species diversity was the direct effect of sunlight (and most probably the consequent change in temperature at ground level).[75] Many of the species we had previously seen only in shaded areas indeed did respond by moving into the areas we had provided with artificial shade, strongly suggesting that the physical background generated by the intensive coffee management system is at least partially limiting for ground-foraging ants. The actual quantity of leaf litter had little effect on ant diversity, but the strong effect of shade left little doubt that the underlying physical environment was a partial determinant of biodiversity; the fundamental niche certainly matters.

Posing a similar but more complicated question in a study of twig-nesting ants on a Colombian shade coffee farm, Colombian ecologist Inge Armbrecht placed groups of hollow twigs from different species of shade trees on the ground in distinct combinations. In one treatment, eight hollow twigs from a single species of tree (monospecific) were placed in litter bags on the ground, and in the other treatment, eight hollow twigs from eight different species of trees (mixed species) were placed in litter bags on the ground with the experimental set-up repeated eight times. After a period of five months, she discovered a total of nine species of ants nesting in the single-species treatments (regardless of what tree species the twig came from) and 20 species present in the mixed-species treatments (Figure 2.15). It appears that not only does the existence of elements of the fundamental niche matter for species survival, but the diversity of the elements of the fundamental niche can have a significant impact on consequent biodiversity. This result remains rather enigmatic, as the authors themselves emphasize.[76]

Although the fixed resources and environmental gradients of the fundamental niche can be important determinants of biodiversity patterns, the variable resources that give rise to the realized niche are also key. For example, on most sun coffee plantations in Costa Rica there are two ground-foraging ant species that dominate, *Solenopsis geminata*, the tropical fire ant, and *Pheidole radoszkowskii*. The former is an especially aggressive ant, well known among travelers in the Neotropics since it frequently ascends the leg within the trousers and has the irritating habit of stinging *en masse* – i.e., the workers crawl up the pants leg and only sting when a pheromone reaches a critical level, which means close to 20 ants may sting a poor victim simultaneously, making it feel like your leg is on fire (in Nicaragua, this species is

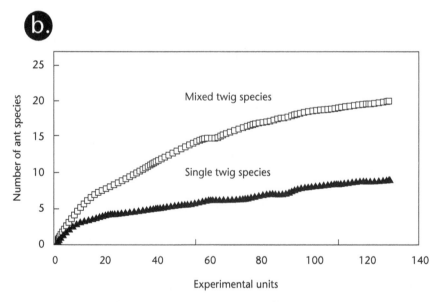

FIGURE 2.15 Study of twig-nesting ant species on coffee plantations in Colombia. (a) Twigs of eight different species used in the study. (b) The results show that more than double the number of species were found nesting in litter bags containing twigs of mixed tree species as compared to the same number of twigs of a single tree species.

Source: Armbrecht et al. (2004)

sometimes referred to locally as "*quita calzón*", "remove your underwear"). Where the fire ant occurs, it tends to dominate any available protein resource, and is thus assumed to be a dominant competitor. However obvious that assumption may seem, it turns out not to be necessarily true; it actually depends on the background environment and the nature of the competing species, as demonstrated with a simple experiment.[77] Having established that a mosaic-like structure of foraging activity occurred on a sun coffee plantation, with fire ants dominating some areas and *P. radoszkowskii* others, we decided to examine the competitive interaction between these two species. To this purpose, we established small plots, 10 × 10 meters, and eliminated one or the other species using poison baits. We then assessed the foraging activities of both species at weekly intervals. What we found surprised us. When *P. radoszkowskii* was eliminated, the fire ant expanded its foraging area, indicating that, when present, *P. radoszkowskii* is able to restrict the foraging areas of the fire ant. In other words, it wins in competition with the fire ant. On the other hand, when the fire ant was eliminated, *P. radoszkowskii* did not respond with increased foraging, meaning that *P. radoszkowskii*'s foraging does not seem to be affected by the presence of the fire ant. In other words, in spite of the aggressive behavior and apparent dominance of the fire ant, *P. radoszkowskii* proved to be the competitively superior species.

These results fit nicely with a framework frequently used by ant ecologists, that of the discovery/dominance trade-off.[78] Some ants are especially good at initially locating a resource that becomes available, while others are especially good at recruiting many individual workers so as to dominate a resource, excluding individuals of other species. It is generally thought that there is a trade-off, with some species being especially good at dominating while others are especially good at discovering, but it is difficult for a species to be good at both. In the context of fire ants and *P. radoszkowskii* at this site (and seemingly other species of *Pheidole* in other systems as well[79]), the former is an excellent dominator and the latter an excellent discoverer. However, this then implies that the ultimate outcome of competition depends on the type of resource that normally becomes available. If resources are normally of a very small size (e.g., small flies that die and drop to the ground), *P. radoszkowskii* will likely discover them first and individual workers simply take them to their nest. However, if resources are normally of a very large size (e.g., a small frog or a large spider that dies and falls to the ground) or can be defended (such as hemipterans tended by ants), individual *P. radoszkowskii* workers will be unable to quickly move them to the nest and the ability of fire ants to recruit workers to the resource and defend it will result in its competitive dominance (Figure 2.16). Thus, competitive dominance depends on the size and distribution of the resources that come into the system. Although here we are talking about only two species, one can easily see how this analysis can be extended to multiple species and how the diversity, abundance and spatial distribution of the resources could influence how many species can co-exist in a particular habitat.

Regardless of the exact causes of the pattern, whether driven by the fixed resources of the fundamental niche or the variable resources of the realized niche – or, for that

FIGURE 2.16 Large tarantula spider dead on the forest floor and being covered with soil particles by a swarm of fire ants. An example of how a dominator can control large resource objects that are input into the system.

matter, something else –, the pattern of biodiversity reduction with agricultural intensification seems to be a general one; rustic or traditional shaded plantations have higher associated biodiversity than shaded monocultures and these have higher diversity than unshaded monocultures (sun coffee). However, it is important to note that the precise pattern of biodiversity loss is not necessarily the same for different types of organisms. For example, in a study that compared the number of species of butterflies and ants at the same sites along a classical intensification gradient, the Type IV pattern (as in Figure 2.6) was suggested for ground-foraging ants, while butterflies appeared to be following the Type I pattern (Figure 2.17).[80] A sensible explanation for this pattern might be that many butterfly species are highly specialized on understory plants in their larval stage. Reducing the density of those understory species, which must be done even in rustic production, reduces a major part of the butterfly's ecological niche, thus the decline in species diversity begins with the introduction of coffee. Ground-foraging ants, on the other hand, require a degree of diversity of habitats to be sure,[81] but mainly require ground on which to forage and in which to nest. So the key resources are removed at a different rate for butterflies compared to ants.

This situation, which now seems quite general, has led to some speculation as to how biodiversity and agricultural yields are related, and what could be done to make biodiversity itself attractive to farmers. That is, how could we give biodiversity a value that could be compared to the value of the agricultural produce on the farm? One way of doing this would be through a concentrated educational effort, aimed at convincing consumers that there is value in biodiversity and that they can contribute to the conservation of biodiversity by consuming coffee that is produced under conditions that maintain that biodiversity. Taking a cue from the highly successful certification programs for organic agriculture worldwide, the Smithsonian Migratory Bird Center embarked on a program of education and certification of shade coffee. Through this program, farmers could be compensated for protecting biodiversity within their farms. Being certified as "bird friendly", "shade friendly" or "biodiversity friendly" could, theoretically, bring an additional income stream to the farmer or provide new markets for their coffee. Thus was born the Smithsonian Bird Friendly certification system (Figure 2.18). The ultimate effect of this and other certification systems in coffee remains contested,[82] but the idea is clear: educate consumers about biodiversity and the way in which the type of production system affects it, and they will demand biodiversity-friendly coffee and be willing to pay more for it, such that the farmers will be encouraged to move to the less intense production system or keep the shade trees on their farms.

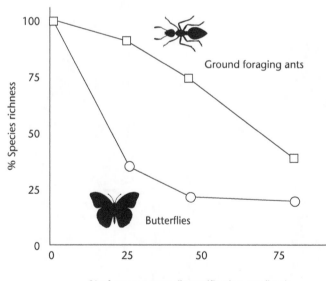

FIGURE 2.17 Pattern of biodiversity reduction for ground-foraging ants and butterflies in a coffee intensification gradient. The percent species richness is calculated based on the number of species found in forest plots.

Source: Perfecto et al. (2005)

FIGURE 2.18 The logo of the Smithsonian Bird Friendly certification program and the artwork used in their iconic poster.

Source: Smithsonian Migratory Bird Center and Julie Zickefoose

Balancing ecological and economic variables: optimality under constant conditions

Whatever the successes of certification systems in the past, present or future, the combination of economic and ecological factors need to be fully understood for efficient implementation of certification programs. There is, we fear, great confusion on this point, leading to some unintended obfuscation in the literature and pointless debate about policy. Two causes seem to be involved here:

1 a failure to take the interpenetration of economics and ecology seriously; and
2 a confusion between static and dynamic conceptualizations of biodiversity issues.

In this section, we intend to clarify what can be said about biodiversity and agriculture if one restricts oneself to an analysis of measurable current data on both agricultural production and biodiversity. It turns out that a relatively standard analysis provides clear understanding of what might be expected from the social behavior of farmers coupled with the ecological behavior of farms and biodiversity, if we assume static conditions. However, we caution, this approach is extremely limited because of the realities of biodiversity dynamics. In the static approach, as taken in this section, we presume that what is measured today represents a constant situation. This approach is clearly a gross simplification since we know that nature is dynamic. However, here we choose to start with a static analysis for heuristic purposes only. In the chapter that follows, we treat the much more difficult topic of landscape dynamics.

In the static approach, we begin by making use of the patterns of biodiversity as explained previously. A graph of biodiversity versus agricultural intensification is generally a decreasing function (although this is not always true), with the highest level of biodiversity at the lowest end of the intensification spectrum. At the same time, it is generally assumed that agricultural production itself is increased as intensification proceeds. There are many exceptions to this second assumption.[83] However, for the purpose of this argument, we will assume that it is indeed true. Also, we presume that both biodiversity and agricultural production are monotonic, either always increasing or always decreasing, giving rise to the two formal equations:

$$B = f(I)$$

and

$$A = g(I)$$

where B is biodiversity, A is agricultural production and I is intensity. The two functions, f and g, then simply specify the output (B or A) from the input I, and have a form as illustrated in Figure 2.19(a).

In Figure 2.19, we see what we refer to as the ecological forces, the response of both biodiversity and agricultural yield to the intensification of agriculture. We call these ecological forces because ultimately the organisms that exist within the agroecosystem as well as the yield that is obtained from a particular agroecosystem are determined by ecological factors. Since almost by definition the agroecosystem involves an economic component, it is essential to relate economic realities to these

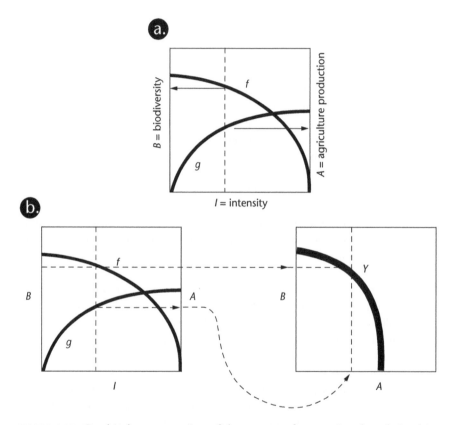

FIGURE 2.19 Graphical representation of the process of converting the relationships among intensity, agricultural production and biodiversity, to the relationship between agricultural production and biodiversity. (a) The relationship between biodiversity and agricultural production as functions of intensity of agriculture (I) [($B = f(I)$, and $A = g(I)$]. The function f relates biodiversity amount to intensification and the function g relates agricultural production to intensification. The dashed vertical line shows the location of the values of B and A at a particular (arbitrary) value of I. Arrows show the values of A and B for that particular value of I. (b) Transferring Figure 2.19(a) to a different representation, where A and B are plotted on the same graph. The two functions f and g then translate into a completely different function, which we label Y, for yield, understanding that this "yield set" refers to the "ecological yield," that is, the amount of both agricultural production *and* biodiversity.

ecological realities. At its most fundamental level, there is some collection of social forces that determines the relative values of both biodiversity and agricultural production. Under a modern system, that "valuation" is usually thought of in monetary terms, which, of course, is an important ideological restriction on the analysis. We skirt that question at this point and simply note that, taking economy at its most general level, there is some process that results in a valuation of both B (biodiversity) and A (agricultural production). Thus, we have:

1   *quantities* of B and A that are ultimately determined by ecological forces; and
2   *values* that are imposed by social rules for both A and B.

To fully understand the system at this level, it is necessary to appreciate that in the real world the quantities and values of A and B are interrelated, which is to say that the ecology and the economy are dialectically related to one another.[84]

We start by making the simplest assumption possible – that is, we presume that the overall economic output of the system is a simple additive combination of the two overall values. In other words, the total value of the system is the value of a unit of biodiversity times the quantity of biodiversity plus the value of a unit of agricultural production times the quantity of the production. This means that B is related to A in a linear fashion, and it is a reasonable assumption that the society in general – perhaps not each and every farmer or consumer, but the overall collective of the society – will attempt to make the total value as large as possible. Given this linear assumption, if we let b stand for the unit value of biodiversity and a for the unit value of agricultural production, the *total* value, T, of a particular quantity of B (biodiversity) and A (agricultural production) will be:

$$T = aA + bB$$

Performing a simple algebraic transformation on this equation, we obtain:

$$B = (T/b) \times (a/b)A$$

which is a linear equation in the variables B and A.

The graphical way of looking at this economic function is shown in Figure 2.20(a). The economic function is linear on a graph of B versus A, and the intercept of that function (the line) is T/b, while its slope is a/b. Since the intercept is T/b, the process of trying to maximize total value is the process of trying to make the economic function intersect the graph as high as possible on the y axis.

The two aspects of this problem, the ecological and the economic, can be conveniently related to one another (i.e., dialectically combined) by simultaneously graphing the ecological function (the ecological yield set, Y, from Figure 2.19(b)) and the economic linear function (Figure 2.20(a)) on the same graph, as we do in Figure 2.20(b), and picking the maximum value of T, which is easily seen to be where the economic function is tangential to the ecological yield set (Y).

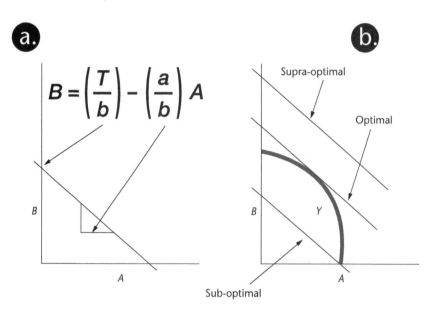

FIGURE 2.20 Optimizing the joint production of biodiversity and agriculture. (a) The linear economic function includes a slope that is the ratio of the value of agricultural product to biodiversity (*a/b*) and an intercept that is proportional to the total value (*T*) divided by the price of biodiversity. Note that seeking the largest value of *T* is the same as trying to place the intercept of the line as high as possible. (b) The ways of placing the economic function on the graph are constrained by the yield set, which stipulates the ecological constraints on the production of agricultural product and the production of biodiversity. If the economic function is "supra-optimal", it would theoretically give a very large total value, but there is no combination of *B* and *A* in the ecological set (the yield set) that could give that value, which is to say the ecological yield set (*Y*) and the economic function do not intersect at any point. The sub-optimal set does indeed intersect the yield set, but at a much lower value of *T* than could be obtained. The point at which the economic function is exactly tangential to the yield set is evidently the optimal solution.

We presume the decisions to be made about intensity of agriculture (*I* in Figure 2.19) will tend toward the maximization of *T*, with the constraints of *Y*, the maximization of total value within the constraints of the ecological system. Note that we are not suggesting a program of active planning, but rather propose that the society in general will tend toward choosing a level of intensity such that total value, which takes into consideration both biodiversity and agricultural production, is maximized. Of course, in much of the world biodiversity is valued at zero or very little, which makes the optimization process one of maximization of agricultural production with no concern for biodiversity. However, here we wish to analyze situations in which society does value biodiversity. When the value of biodiversity increases, the slope of the linear economic function decreases (since the slope is *a/b*, as *b* increases relative to *a*, the slope decreases). The process of economic maximization under ecological constraints is evident with a simple graphical analysis as presented in Figure 2.21. Unsurprisingly, as society places a higher value on biodiversity relative to agricultural production, the point of maximum total value moves toward systems that have lower agricultural productivity and more biodiversity. In general, the analysis says that there is a relationship between planning for agricultural production and/or biodiversity on the one hand, and the valuation given to each of those variables by the society in general on the other. It is a reasonable assumption that the ecological/economic system will move toward that stable point.

Although this analysis is not intended to be a recipe either for planning or action, certain obvious issues emerge. If one desires an increase in so-called wildlife-friendly farming, it would be wise to take sociopolitical action to try and increase the value of b relative to a, for example. And it is actually quite obvious to note that as b is increased, there will be a tendency for the system as a whole to move toward the biodiversity-friendly side of the intensification spectrum; the higher the value of biodiversity, the more the system moves to the lower intensification level (Figure 2.21).

However, this same analysis also suggests other qualitatively distinct possibilities. The construction of the ecological yield function in Figure 2.19(b) is actually only one qualitatively distinct way of constructing this function. Although there are an infinite number of possible configurations for the ecological yield function, it is convenient to consider only two. The first is a convex set, as in Figure 2.19(b). However, a different set of assumptions about the biodiversity and agricultural production curves (in Figure 2.19(a)) can give rise to a concave yield set, as pictured in Figure 2.22.

The qualitative consequences of a concave Y function are dramatically different from those of a convex Y function. As we increase the value of biodiversity relative to agricultural production (economic functions labeled 1, 2, and 3 in Figure 2.22(b)), the same optimal solution results – the entire area should be in the most intensive agricultural production and no biodiversity. However, as we increase the value of biodiversity, the system reaches a point at which there is a sudden jump, a tipping point, and the optimal solution becomes the least intensive system, which, depending on how the analysis is formulated, could be the least intensive agricultural system or a natural forest with no agriculture at all (economic function 4 in Figure 2.22(b)). Such might be the case, for example, if the international price of coffee declines smoothly as the value of biodiversity increases. As the ratio of coffee prices to biodiversity value declines (e.g., in Figure 2.22(b), as we move from economic function 1 to 2 to 3), the landscape will tend to remain in intensive coffee production, but then a critical ratio is reached (economic function 4) when there is a switch and all biodiversity conservation becomes optimal (for example, the transformation of all the farms to the traditional shade coffee system). Thus, it may first appear that the promotion of biodiversity certification is having little effect on the nature of agroecosystem planning. That is, even though the price for biodiversity-friendly coffee increases, there could be little change in agricultural practices. Yet there may be a tipping point at which the whole landscape suddenly becomes transformed

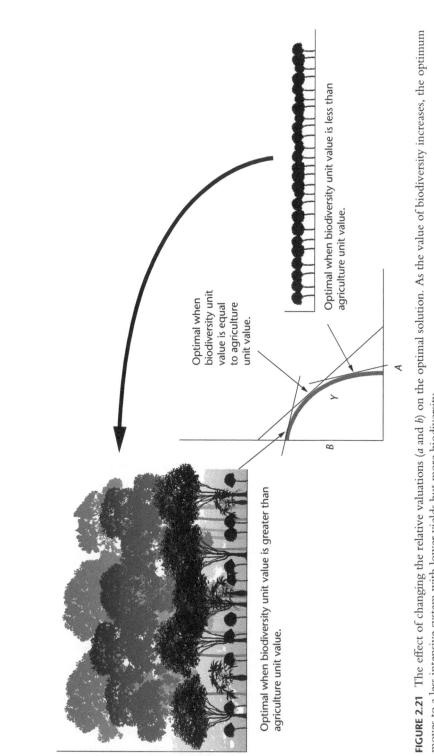

Optimal when biodiversity unit value is greater than agriculture unit value.

Optimal when biodiversity unit value is equal to agriculture unit value.

Optimal when biodiversity unit value is less than agriculture unit value.

B

Y

A

FIGURE 2.21 The effect of changing the relative valuations (*a* and *b*) on the optimal solution. As the value of biodiversity increases, the optimum moves to a less intensive system with lower yields but more biodiversity.

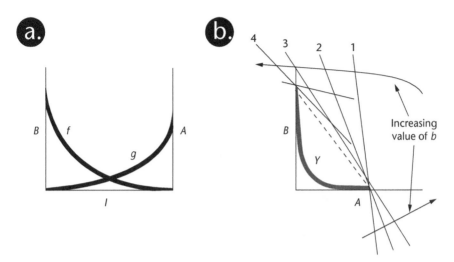

FIGURE 2.22 Construction of the concave Y function and qualitative consequences of different valuation functions. (a) Original curves of biodiversity (f) and agricultural product (g). (b) The resultant Y function with four different economic functions (1 through 4) representing increasing value of biodiversity (b) relative to agricultural production (a).

(Figure 2.22). Unfortunately, we rarely have the foreknowledge of the actual shape of the Y function, so precise prediction will likely remain elusive. Nevertheless, the underlying principles involved tell us clearly what to expect at a qualitative level.

There is another consequence of a concave yield set that is critical. In addition to the actual set itself, it is certainly the case that combinations of patches could occur in a landscape, rather than a landscape that is uniformly good or bad (or in between). Thus, we might have a landscape composed of a large number of farms, 20 percent of which are rustic and 80 percent of which are in full sun. Such a possibility could be represented on a line connecting the rustic and full sun possibilities, as shown in Figure 2.23, for a single combination.

The existence of combinations of production systems when the yield set is concave suggests other outcomes. To see this in its most interesting manifestation, it is necessary to relax the assumption that the economic function will be linear. At first, it seems obvious that linearity should be the rule. That conclusion is based on effectively assuming that the point of decision is at a very small scale and the entrance of the quantity of agricultural product into a market will have little or no effect on its value, which is thus assumed to be constant. The world price of a commodity such as bananas or coffee is largely set independently of what happens on one farm or in one river valley, and thus a linear economic function is probably appropriate. However, when local markets are involved in the dynamics of the system, it is evident that the assumption of constant agricultural product price is unwarranted. That is, if part of the quantities of either biodiversity or the agricultural product influences the valuation of either, that value will be a function of one or both of those variables.

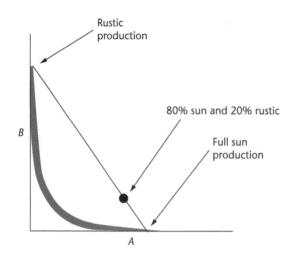

FIGURE 2.23 The construction of the "extended ecological yield set," which includes all possible discrete combinations of possible systems. In this particular example, the points of rustic production and full sun production are indicated, along with the point that represents 80 percent full sun and 20 percent rustic.

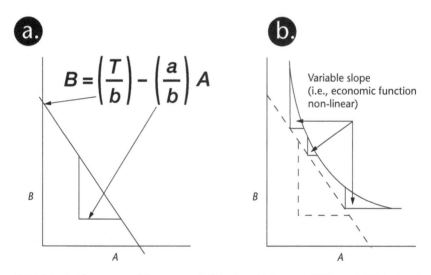

FIGURE 2.24 The nature of the economic function. (a) Repeat of Figure 2.20(a), showing the basics of the linear economic function. The slope of the function is constant (i.e., the function is linear). (b). The expected shape of the economic function with a relaxed linear assumption – that is, with the valuations partially dependent on the quantities produced (i.e., the function is non-linear). Note the changing slope of the function, depending on where it lies in the B–A plane.

To see the general form of relaxing the linear assumption, consider the example of a case in which valuations are themselves functions of the amounts produced (e.g., as the agricultural product becomes locally abundant, society tends to value it lower). This results in a curvilinear economic function, as illustrated in Figure 2.24.

It remains the case that we expect the system to move to its optimum level which is the largest total value possible, and the constraints are with the yield function as before, as illustrated in Figure 2.25.

It is evident that with the relaxation of the linear assumption, the clear distinction between a hard trade-off between either all of biodiversity or all of agricultural product (Figure 2.25(b)), stipulated by the existence of a concave yield set, is lost and it is at least theoretically possible to have an intermediate intensity be the optimal solution for a concave yield set (Figure 2.25(c)).

There is a rather more interesting outcome that emerges from the combination of a concave yield set and a non-linear economic function. The edge of the yield set has thus far been construed as the highest values of all possible combinations of agricultural production (A) and biodiversity (B) evaluated on the same farming site. In other words, the problem has been conceived of as a decision of how to produce in an entire homogeneous area. Yet in the real world, frequently an agricultural area is a mosaic of different types of production, some of which may be good for agriculture but bad for biodiversity, and some of which might be the reverse. If we consider this, rather than the uniform options suggested by the ecological yield set (i.e., any given situation is a point in the set), we construct what we call the "extended ecological yield set" (directly parallel to what Levins refers to as the extended fitness set[85] in evolutionary ecology), all possible combinations of pairs of options. The basic idea is illustrated in Figure 2.26.

If the yield set is strictly convex, the expected landscape might be composed of:

1   all fields dedicated to the extreme end of the ecological yield set for biodiversity; or
2   all fields dedicated to the extreme end of the yield set for agricultural production; or
3   all fields dedicated to some intensity intermediate between the two extremes.

Using the percent shade as an example of a measure of intensity, we might have the whole landscape covered with 100 percent shade or 0 percent shade (no trees at all), or to 40, 50 or 60 percent shade cover. However, if the yield set is concave and the economic function non-linear, we can also imagine a landscape composed of some fraction of fields with each of the two extremes (e.g., 60 percent of the fields with 100 percent shade and 40 percent of the fields with no trees at all). The set of all of these possibilities we call the extended ecological yield set (Figure 2.26). Clearly, the upper edge of that set of points is located on a line connecting the two intercepts of the original yield set and, if such a mixed solution is allowed (i.e., there may be social or ecological reasons outside of this analysis that will not permit such mixed solutions), the non-linear decision function stipulates that the mixed strategy will

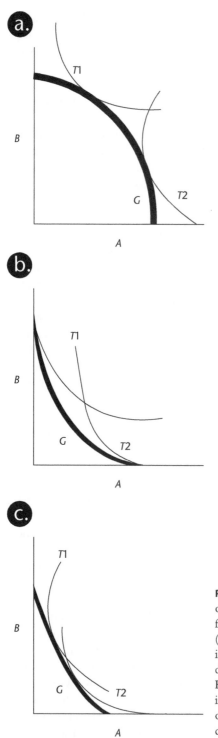

FIGURE 2.25 The three general kinds of optimal solutions with non-linear T functions. (a) Convex G. (b) Concave G. (c) Concave G. (a) and (b) are qualitatively identical to the solutions for a linear decision function (see Figure 2.20). However, (c) shows how an intermediate intensity (intermediate value of I) can be optimum, even when the G function is concave.

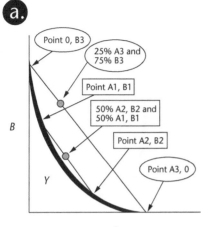

a.

Point 0, B3

25% A3 and
75% B3

Point A1, B1

50% A2, B2 and
50% A1, B1

Point A2, B2

Point A3, 0

B

Y

A

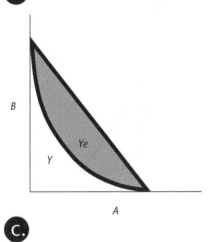

b.

B

Ye

Y

A

c.

B

Ye

Y

A

FIGURE 2.26 Optimization process for the extended ecological yield set. (a) Illustration of the positioning of a 50 percent, 50 percent combination of two points and a 25 percent, 75 percent positioning of two other points. Considering all possible pairs of points positioned in a similar fashion, the entire area between Points 0, *B*3 and *A*3, 0 becomes filled, giving rise to (b) the extended ecological yield set (*Ye*), the set of all possible combinations of all pairs of points in the original set. (c) Optimization on the extended ecological yield set with a non-linear economic function.

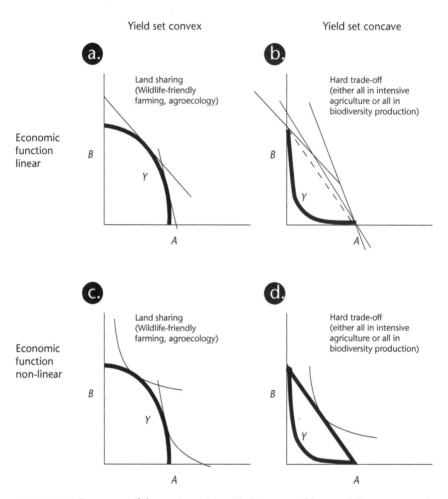

FIGURE 2.27 Summary of the static optima. (a) Convex yield set and linear economic function leads to land sharing as the optimal. (b) Concave yield set and linear economic function leads to a hard trade-off in which either all biodiversity production or all intensive agriculture would be the optimal. (c) Convex yield set and non-linear economic function leads to land sharing as the optimal. (d) Concave yield set and non-linear economic function leads to land sparing as the optimal.

generally be optimal, as illustrated in Figure 2.25(c). In the context of the debate over land sparing versus land sharing for biodiversity conservation,[86] this solution represents the land-sparing option.

In summary, this static approach enables us to say something about the various options that have been suggested and how they relate to both the ecological and economic background. We summarize the four basic qualitatively distinct situations in Figure 2.27, although we caution that the static approach is extremely limiting precisely because it is static.

Both agriculture and biodiversity are very dynamic and complex subjects and this analysis, while useful for initially categorizing possible alternatives, should be viewed with caution. The more dynamic, and we would argue more realistic, approach is to examine landscape dynamics, the subject of the next chapter.

Notes

1 Wake and Vredenburg (2008).
2 Chapman (2009).
3 May (2010).
4 Hamilton et al. (2010); Mora et al. (2011).
5 May (2010).
6 Hamilton et al. (2011).
7 Mora et al. (2011).
8 Hooper et al. (2005).
9 Wilson (1987b).
10 Willig et al. (2003); Hillebrand (2004).
11 Salaman et al. (2009).
12 American Ornithologists Union (1998).
13 Skinner and Allen (1996).
14 Wilson (1987a).
15 Schulz and Wagner (2002).
16 Perfecto and Vandermeer (2010).
17 Perfecto et al. (1997).
18 Perfecto and Vandermeer (2008a); Perfecto et al. (2009).
19 Lewontin (1982).
20 Taylorism is a system of scientific management advocated by Fred W. Taylor to increase the efficiency of each worker in a factory by eliminating unnecessary activities and following a machine-like motion.
21 Levins and Lewontin (1987).
22 Vandermeer (1989).
23 Swift and Anderson (1994); Altieri (1999); Tscharntke et al. (2005).
24 Vandermeer and Perfecto (2005).
25 Vandermeer (1989); Vandermeer et al. (1998); Li et al. (2007); Letourneau et al. (2011); Kremen et al. (2012).
26 Green et al. (2005); Phalan et al. (2011a).
27 Perfecto and Vandermeer (2010); Tscharntke et al. (2012a); Fischer et al. (2011).
28 Perfecto and Vandermeer (2002).
29 Pendergast (1999).
30 Guhl (2008).
31 Rice (1999).
32 Perfecto et al. (1996).
33 Perfecto (1994); Perfecto and Snelling (1995); Perfecto and Vandermeer (1996); Perfecto et al. (1996, 1997); Vandermeer and Perfecto (1997); Greenberg et al. (2008); Perfecto and Vandermeer (2008a).
34 Beer (1988); Beer et al. (1998).
35 Soto-Pinto et al. (2000); Staver et al. (2001); Lin (2007); Lin et al. (2008); Vandermeer et al. (2010a).
36 Marshall (2008).
37 Vandermeer and Perfecto (2005).
38 ICT (2006).
39 Conroy et al. (1996).
40 Perfecto et al. (1996).
41 Grime (1973); Horn (1975); Connell (1978); Denslow (1987); Dial and Roughgarden (1988).

42 Erwin (1983).
43 Wilson (1987a).
44 Hamilton et al. (2010).
45 Perfecto et al. (1997).
46 Brash (1987); Nir (1988).
47 Weaver and Birdsey (1986); Perfecto et al. (1996).
48 Greenberg et al. (1997a, 1997b).
49 Robbins et al. (1992); Perfecto et al. (1996).
50 Perfecto et al. (1996).
51 Moguel and Toledo (1999); Myers et al. (2000).
52 Rice and Ward (1996).
53 Roberts et al. (2000).
54 Griscom (1932).
55 Borrero (1986).
56 Robbins et al. (1989); Rice and Ward (1996).
57 Brash (1987); Perfecto et al. (1996, 2003, 2007); Komar (2006); Moguel and Toledo (1999); Donald (2004); Somarriba et al. (2004); Pineda et al. (2005); Philpott and Dietsch (2003); Manson et al. (2008); Philpott et al. (2008a).
58 Parrish and Petit (1995); Wunderle and Latta (1996, 1998); Greenberg et al. (1997a, 1997b); Estrada et al. (1997); Calvo and Blake (1998); Roberts et al. (2000); Komar and Domínguez (2001); Siebert (2002); Dietsch (2003); Mas and Dietsch (2004); Tejeda-Cruz and Sutherland (2004); Cruz-Agnón and Greenberg (2005); Estrada and Coates-Estrada (2005); Komar (2006); Raman (2006); Dietsch et al. (2007); Gordon et al. (2007); Cruz-Agnón et al. (2008); Tejeda-Cruz and Gordon (2008); Bakermans et al. (2009); Philpott et al. (2009); Bakermans et al. (2012); Philpott and Bichier (2012); Hernández et al. (2013).
59 Gallina et al. (1996); Cruz et al. (2004); Gallina et al. (2008); Rocha et al. (2011).
60 Komar and Domínguez (2002); Pineda and Halffter (2004); Henderson and Powell (2001); Krishna et al. (2005); Pineda et al. (2005); Gonzalez-Romero and Murrieta Galindo (2008); Macip-Ríos and Muñoz Alonso (2008); Santos Barrera et al. (2008); Murrieta Galindo et al. (2013a, 2013b).
61 Cruz et al. (2004); Numa et al. (2005); Pineda et al. (2005); Estrada et al. (2006); Sosas et al. (2008); Saldaña-Vázquez et al. (2010); Williams-Guillén and Perfecto (2010, 2011); Garcia-Estrada et al. (2012); Saldaña-Vázquez et al. (2013).
62 Perfecto and Snelling (1995); Perfecto et al. (1997); Roberts et al. (2000); Perfecto and Vandermeer (2002); Armbrecht and Perfecto (2003); Perfecto et al. (2003); Armbrecht et al. (2004); Armbrecht et al. (2005); Philpott and Foster (2005); Armbrecht et al. (2006); Philpott and Armbrecht (2006); Philpott et al. (2006a); Philpott et al. (2008a, 2008b); Valenzuela-Gonzalez et al. (2008); De la Mora and Philpott (2010); Mera-Velasco et al. (2010); Ottonetti et al. (2010); Philpott (2010); Teodoro et al. (2010); De la Mora et al. (2013); Urrutia-Escobar and Armbrecht (2013).
63 Moron and López-Méndez (1985); Ibarra-Nuñez (1990); Nestel et al. (1993); Perfecto et al. (1997); Estrada et al. (1998); Molina (2000); Arellano et al. (2005); Pineda et al. (2005); Harvey et al. (2006); Halffter et al. (2007); Richter et al. (2007); Horgan (2009); Gordon et al. (2009).
64 Ibarra-Nuñez (1990); Perfecto et al. (1997); Rojas et al. (2001); Botero and Baker (2002); Klein et al. (2002); Franco et al. (2003); Mas and Dietsch (2003, 2004); Tylianakis et al. (2005); Hernández-Ortiz and Dzul-Cauich (2008).
65 Soto-Pinto et al. (2001); Siebert (2002); Cruz et al. (2004); Mas and Dietsch (2004); Bandeira et al. (2005); Mendez et al. (2007); William-Linera and López-Gómez (2008); Ambinakudige and Sathish (2009).
66 Hietz (2005); Solis-Montero et al. (2005); Carreño Rocabado (2006); Garcia-Franco and Toledo Aceves (2008); Mehltreter (2008); Mondragón et al. (2009); Moorhead et al. (2010).
67 Heredia Abarca and Arias Mota (2008).

68 Aguilar-Ortiz (1982); Torres (1984); Ibarra-Nuñez (1990); Stork and Brendell (1990); Robbins et al. (1992); Parrish and Petit (1995); Petit et al. (1999); Greenberg et al. (1997a, 1997b); Estrada et al. (1998); Roberts et al. (2000); Johnson and Sherry (2001); Perfecto and Vandermeer (2002); Gove et al. (2008); Manson et al. (2008).

69 Canaday (1996); Donald (2004); Komar (2006); Manson et al. (2008).

70 Manson et al. (2008).

71 Greenberg et al. (1997a); Calvo and Blake (1998); Wunderle and Latta (1998); Johnson (2000); Johnson and Sherry (2001); Soto-Pinto et al. (2001).

72 Rice and Ward (1996); Philpott (2005a); Cruz-Agnón and Greenberg (2005); Cruz-Agnón et al. (2008, 2009).

73 Hutchinson (1957).

74 Perfecto and Snelling (1995).

75 Perfecto and Vandermeer (1996).

76 Armbrecht et al. (2004).

77 Perfecto (1994).

78 Davidson (1985); Perfecto (1994); Feener et al. (2008); Perfecto and Vandermeer (2011).

79 Perfecto and Vandermeer (2011).

80 Perfecto et al. (2005).

81 Armbrecht and Perfecto (2003); Armbrecht et al. (2004, 2005, 2006).

82 Mutersbaugh (2002); Pagiola and Ruthenberg (2002); O'Brian and Kinnaird (2003); Philpott and Dietsch (2003); Rappole et al. (2003); Giovannucci and Ponte (2005); Martínez-Torres (2006); Philpott et al. (2007); Raynolds et al. (2007); Bacon et al. (2008).

83 Vandermeer (2010); Vandermeer and Perfecto (2012).

84 Vandermeer and Perfecto (2012).

85 Levins (1968); Perfecto (1994); Feener et al. (2008); Perfecto and Vandermeer (2011).

86 Perfecto and Vandermeer (2010); Phalan et al. (2011a, 2011b); Tscharntke et al. (2011).

3

THE COFFEE AGROECOSYSTEM
AS A HIGH-QUALITY MATRIX

The coffee system and biodiversity debates

In the last chapter, we presented a framework for the examination of the basic question of biodiversity in agroecosystems and provided at least a tentative answer to the question, "What is the relationship between agricultural intensification and biodiversity in the coffee agroecosystem?" Although there are exceptions, the shade coffee agroecosystem provides habitat to all sorts of organisms, a habitat worth preserving for its biodiversity value alone. Of course, the caveat is that not all coffee agroecosystems are of this character. The rustic farms of Mexico contain almost as many bird and ant species as the native forests in the regions where they occur,[1] but the sun coffee monocultures of southern Brazil are biodiversity deserts. The message is that it is not the *fact* of agriculture but the *kind* of agriculture that matters for biodiversity, and the many studies that have addressed that question leave little doubt that that is the case.[2]

In general, there has been a great deal of interest in this question during the past three decades, not only in coffee but in all sorts of agricultural systems. The summary put forward in the previous chapter for coffee could be repeated for many other systems with the same general qualitative result – as agricultural intensification (defined as increased dependency on external inputs and tendency toward monocultures) increases, biodiversity goes down, but in a pattern that is not necessarily the same for different kinds of organisms. Knowledge of this phenomenon has led many researchers to the obvious conclusion that one could directly sample the biodiversity in various agricultural systems and associated natural habitats and then make simple calculations to come up with what might appear to be optimum strategies for planning agricultural landscapes. For example, taking the analyses that we presented in the previous chapter about the economic and ecological forces that simultaneously affect agricultural production and biodiversity, one can come up with an optimum strategy for maximizing the total value of the system as a whole, taking into consideration both production and biodiversity. This line of reasoning is what we refer to as the "optimality under constant conditions" argument.

This approach is dominated by an obvious question that is both simple and satisfying: "How much biodiversity is there?" It is a conveniently static approach and is seductive because it allows us to accumulate a large amount of data without much difficulty. We can go into the field and sample organisms from different types of agricultural systems, compare them to the natural systems that are in the vicinity and get a relatively convincing answer to seemingly important questions about biodiversity conservation in agroecosystems. With these sorts of data in hand, it is a simple matter to describe equations that optimize the joint production of biodiversity and agricultural production,[3] an example of which we presented in Chapter 2.

While this is certainly an important exercise, it ignores the dynamic nature of nature. This ignorance can lead to practical recommendations that may sometimes be self-defeating to the goal of biodiversity conservation. A static analysis of biodiversity in agricultural systems could lead us to conclude that a particular system maintains biodiversity simply because we recorded a high number of organisms at one particular point in time from that system. Alternatively, it could lead us to conclude that some agroecosystems are not good for some components of biodiversity because our sampling failed to record them. However, some of the organisms recorded at a particular site may be on their way to extinction (sometimes called the "extinction debt")[4] or alternatively, some organisms that were not recorded from one site may eventually migrate to it and even establish successful populations there. In other words, a focus on the static properties of particular points in space and time can be dramatically misleading. Rather, we need to consider dynamics at whole-landscape levels because we now understand that biodiversity is ultimately determined by dynamic processes (e.g., extinction and migration)[5] that happen over broad regions.

In fragmented habitats, the biodiversity that can persist in the long term is largely determined by the "quality of the matrix." Not only is this because a good matrix is one that can provide habitat for many organisms and sustain high levels of biodiversity, but especially because a good matrix is one that allows movement of organisms among patches of forest or whatever the natural ecosystems are. This is an approach that goes beyond the simple static approach of biodiversity sampling and economic adjustment[6] and takes seriously the fact that both organisms and landscapes in the real world are dynamic and, as with any dynamic system, one can easily be misled by relying on a single snapshot in time. The previous chapter ended with this approach, analyzing the static vision of biodiversity and value, as presented originally in our ecological/economic model.[7] As we noted, some of the contentious issues associated with certain popular debates in the literature, such as the land-sparing versus land-sharing debate, or the forest transition model debate, are laid bare with this approach.

In this chapter, we continue with that analysis, but add to the argument what are generally agreed to be some key dynamic elements, in particular extinction and migration. Recounting earlier work by MacArthur, Wilson and Levins, we modify the basic island biogeography and metapopulation models to take into account the important issue of the quality of the agricultural matrix in a fragmented landscape, expanding on our original approach to this question.[8] Finally, we explore – in a tentative way to be sure – the nature of specific landscape structure, emphasizing the nature of the patchwork that is almost always characteristic of human-dominated landscapes. All of this is discussed in a more sociopolitical framework later in Chapter 8. In the present chapter, we present some of the more analytical details that help anchor the later discussion.

Bringing dynamics into the picture

Foundational arguments

In the late 1960s, the field of ecology was dramatically transformed by two major theoretical breakthroughs, the equilibrium theory of island biogeography by Robert MacArthur and E. O. Wilson[9] and the theory of metapopulations by Richard Levins.[10] As Stephen Hubbell has noted, MacArthur and Wilson's theory represented a radical change in the way in which we look at ecological assemblages.[11] Rather than focusing on the numbers of individuals in a population or the biomass of a population, the new theory looked at the entire assemblage of species and asked us to focus on the rate of arrival of new species to an individual island (the immigration

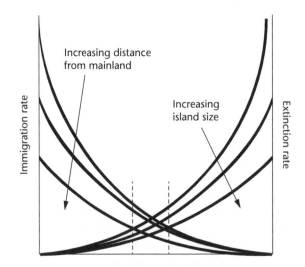

FIGURE 3.1 Classical equilibrium theory of island biogeography. Immigration rate (number of new species arriving to the island within a given period of time) balances with extinction rate (number of species disappearing from the island within the same period of time) to produce an equilibrium number of species. The two equilibria illustrated here are for (1) a small island far from the mainland (left dashed vertical line) and (2) a large island near to the mainland (right dashed vertical line).

rate) and the rate at which species disappeared from an island (the extinction rate). According to this elegant model, the number of species found on a particular island would be a simple balance between extinction and immigration rates. The theory is usually formulated in a graphical form, as we illustrate in Figure 3.1.

Although the theory was developed in the context of real oceanic islands, it is not a difficult conceptual step to think of fragments of natural habitat in a sea of agriculture, a framework that allows a direct application of the theory to conservation. A great controversy emerged from this perspective and, during its day in the sun, occupied numerous pages in numerous journals. Some claimed the theory implied that a conservation strategy should seek a large number of small reserves to cover a broad range of native species ranges, while others argued that the theory suggested that the highest species diversity would occur by creating the largest single reserve area, what became known as the Single Large Or Several Small (SLOSS) debate.[12] It is evident that the theory itself is agnostic to the question, but it definitely produced a great deal of debate on the issue and probably clarified some of the complexities involved.

The second, closely related, theoretical development in ecology emerged from the realization that many, if not most, populations actually exist in the form of

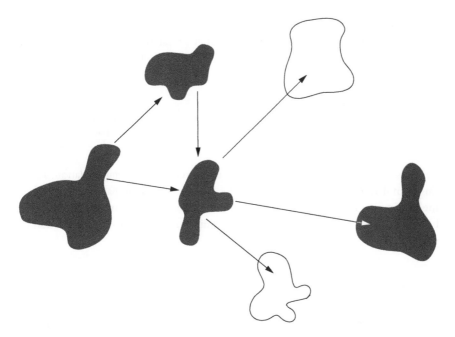

FIGURE 3.2 The classical Levins metapopulation where there are several distinct habitat patches, some of which are occupied by a species (shaded patches) and others that are not (unshaded patches). The proportion of patches occupied (in this case 0.667) will be determined by the process of migration among patches (indicated by arrows) and extinction within patches.

metapopulations. The basic metapopulation theory was developed by Richard Levins in 1969. A metapopulation is a population of populations in which each distinct subpopulation (or local population) occupies spatially distinct patches of habitat (Figure 3.2). The key deviation from classical ecological theory is in the examination of the fraction of habitats (or islands) occupied by a particular species, rather than the population density of that species.

The theory, much like the theory of island biogeography, balances immigration and extinction rates, but in the case of metapopulations, it is a single population that is of concern. The framework is similar to the framework of epidemiology. A disease organism constitutes the population and we do not normally study the number of bacterial cells or virus particles that occupy a human body, but rather look at the number or proportion of individuals who are infected. With this framework we thus ask what fraction of the habitats, or islands, or fragments, are occupied by a particular species. The basic equation is quite simple and stipulates that the rate of change of the proportion of habitats occupied by a particular species is a balance between immigration and extinction rates. The transmission rate of a disease is proportional to the probability that an infected individual comes into contact with an uninfected individual, which, in terms of metapopulations, is the migration rate from one occupied patch to another. This factor is then balanced by the rate of recovery, which effectively means the local extinction of the disease organism (your recovery from a bacterial disease, for example, signifies disaster for that bacterial population). So we can view the rate of change of the number of habitats occupied as a balance between the migration rate and the extinction rate. The basic mathematical form looks, qualitatively, like this:

which is the formal description of the dynamic process of a metapopulation. From this formulation, it is easy to see that if the "migration among patches" is equal to the "extinction from patches," then the "rate of change of occupied patches" will be equal to zero – which is to say, it is at equilibrium. Symbolizing the fraction of patches occupied as p, the formality is,

$$\frac{dp}{dp} = mp\,(1 - p) - ep \qquad\qquad 3.1$$

where dp/dt is the rate of change of occupied patches, $mp\,(1 - p)$ is the migration among patches and ep is the extinction from patches. The constants m and e refer to the *rates* of migration and extinction, respectively.

It is important to note that the extinction rate in this equation is the local extinction rate, not the extinction rate for the species as a whole. Also, the migration rate is similar to the transmission rate in epidemiology, which is why we multiply it by

$p(1-p)$ – in epidemiology, p is the proportion of the population that is sick which makes $1-p$ the proportion that is not sick, and the probability of getting sick is proportional to the "contact rate" of sick (infectious) people with well (susceptible) people, which is thus $p(1-p)$.

A population of any organism with this type of dynamic relationship between extinction and migration is referred to as a "metapopulation," and generally follows the dynamic rules as stipulated in Equation 3.1. Given that this is the case, we can easily calculate what we expect in the long run – that is, what will be the long-term equilibrium condition of the population. Setting the rate of change equal to zero (since by definition, equilibrium means no further change), the equilibrium condition of a metapopulation is,

$$p^* = 1.0 - \frac{e}{m,}$$ 3.2

and remember we are talking about a fraction here, so 1.0 can be thought of as 100 percent. A superficial examination of this relationship reveals the important and unavoidable fact that in order to maintain the equilibrium fraction of occupied patches greater than 0, migration must be relatively large compared to extinction. If migration is very large, the ratio e/m will be almost zero and the fraction of patches occupied will be almost 1.0, but as migration decreases and approaches the value of extinction, the value of p^* approaches zero, which is to say the population goes extinct over the entire region in which the metapopulation is defined.

Metapopulation theory became fairly standard by the 1990s and is now a universally recognized dynamic system that applies to many (perhaps most) organisms. One of the main conclusions of the theory is quite challenging to some of the more traditional ideas in conservation. Early conservationists focused a great deal of effort on possible extinctions of organisms and conservation came to be almost synonymous with the simple avoidance of extinction. Yet metapopulation theory suggests that under perfectly normal and unexceptional conditions extinctions can happen at any given patch and not every habitat patch needs to contain each and every species that it could potentially contain. The basic metapopulation equations (Equations 3.1 and 3.2) make it obvious that under equilibrium conditions some – perhaps many – habitat patches might be void of a species, in the same sense that a disease persists in a population even though not all people are always sick. For example, if the extinction rate is half of the migration rate, we see that only half of the patches that could be occupied will actually be occupied at any point in time.

The ubiquitousness of extinctions

It is critical that the nature of extinction is fully appreciated. It has been estimated that well over 95 percent of all species that have ever lived have already gone extinct.[13]. However, the extinction rates we are talking about in this chapter are not global extinctions of species, but rather local extinctions, which is to say extinctions of populations in habitat fragments. For example, in comparing amphibians occupying isolated ponds and marshes on the E. S. George Reserve near the University of Michigan, David Skelly, Earl Werner and Spencer Cortwright recorded 38 separate extinction events between the late 1960s and the late 1980s.[14] That is almost two extinctions per year! However, those extinction events were balanced by almost the same number of recolonization events, such that the 17 species of amphibians recorded in 1967 all remained in existence in 1987 (Figure 3.3).

In this particular case, the dynamic nature of what is happening is obvious. The amphibians "live" in small temporary ponds, permanent ponds, swamps or marshes. The larval state, the tadpole, is the aquatic form that lives in those ponds and the adults (frogs, toads and salamanders) move out of the ponds and live in the terrestrial

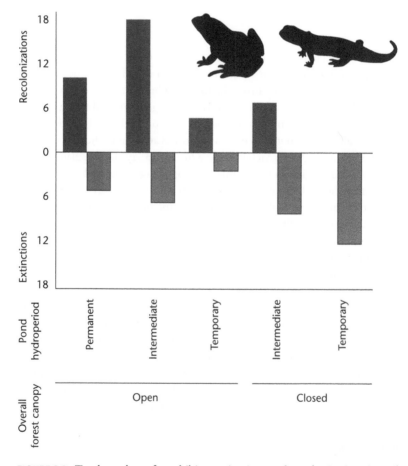

FIGURE 3.3 Total number of amphibian extinctions and recolonizations in each of five pond types categorized according to pond hydroperiod and overhead forest canopy in the E. S. George Reserve, southern Michigan.

Source: Modified from Skelly et al. (1999)

sector of the reserve. With a relatively high probability, a frog emerging from a tadpole in a particular pond will return to that pond for its own breeding. However, occasionally the adult amphibian wanders off and is incapable of finding its "home" pond and winds up in a different pond. From a long-term perspective, this inability to be 100 percent efficient in finding home is extremely important. For a variety of reasons, the population of a particular species in a particular pond may disappear, leading to an extinction event – perhaps due to an explosion of dragonfly predators that eliminates all of them, or perhaps due to a disease that becomes epidemic in that particular pond. Because the fidelity to their home pond is not 100 percent perfect, a pond that becomes devoid of a particular species in a given year will likely receive one of those wandering individuals within the next few years. That is, there is a certain probability of extinction, but also a certain probability of recolonization. The balance between them determines whether or not a population will survive in the long run. If the rate of local extinction is too high, the overall population will go extinct, which is, of course, what the conservation agenda seeks to avoid. However, what we want to emphasize is that local extinctions are natural and impossible to avoid and, indeed, there is usually not much anyone could do to change local extinction rates to anything but a trivial degree.

The importance of local extinctions is universally recognized in contemporary ecology, with much of the relevant evidence coming from the study of islands. As early as Wallace's observations on islands, biologists have appreciated that the impoverished nature of island fauna and flora was at least partly due to a higher extinction rate on islands, especially smaller ones. A particularly instructive study is that of Johannes Foufopoulos and Tony Ives on the diversity of lizards on islands in the Aegean Sea.[15] The standard relationship between island size and number of lizard species is clear in this study. Since there is a near-zero probability that contemporary lizards can swim from island to island, it is very likely that the migration rates are close to zero and any pattern of distribution of the lizards is a function of island size, and thus of extinction rates, only – and this is exactly what the authors find. The whole pattern of biodiversity of the lizards on these islands is a result of the pattern of extinction alone.

It was not much of an extension, especially since MacArthur and Wilson's work, to think of biological preserves and more generally habitat fragments as simply another type of island. This extension of thought then naturally led to the ideas associated with local extinction rates. There is now a large amount of research in all sorts of ecosystems that shows, beyond any doubt, that local extinction rates are normally quite high. For example, William Newmark looked at the extinction records for mammals in national parks in North America and found very dramatic rates of extinction even in the largest reserves (Figure 3.4).[16] Of 14 national parks, only three had not experienced an extinction since their formation.

It is far from controversial to suggest that local extinctions are common.[17] Although the fact of local extinctions is well established, its pattern is clearly not random, and certainly deserves more study.[18] Nevertheless, there is little doubt that amid many complications, populations living in isolated fragments of natural vegetation can expect to experience extinctions, if enough time passes.

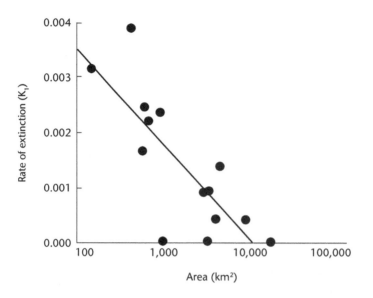

FIGURE 3.4 Extinction records of mammals from North American national parks.

Source: Modified from Newmark (1995)

A further complication may result from spatial self-organization (as we elaborate more fully in Chapter 4). Consider, for example, plant communities in which the constituent species tend to expand in space through seed dispersal, but are attacked by a natural enemy in a density-dependent fashion – in other words, the more dense they are, the more they get attacked by the natural enemy. One of the classical ideas of spatial ecology is the so-called Janzen–Connell effect,[19] in which seeds or seedlings that are concentrated near the parent plant are more likely to be attacked by natural enemies, such as seed predators, herbivores or pathogens, than those that are scattered far away from the parent tree. It can be shown that the combination of seed dispersal and the Janzen–Connell effect will result in the clumping of organisms, even in a uniform environment (see also Chapter 4).[20] Because of the dynamic interplay of dispersal and density-dependent control, any given clump at some point in space is expected to go locally extinct in the long run. In such a situation, fragmenting the continuous habitat does not change much about the local extinction rates, which are a consequence of the density-dependent operation of natural enemy dynamics. However, normal migration through seed dispersal, for example, will be reduced, as we illustrate in Figure 3.5.

Unfortunately, long-term studies that uncover such patterns of extinctions in continuous habitats are not common in the literature. An exception is the work of Thomas Rooney and colleagues from the University of Wisconsin, who found dramatic changes in species composition in plots embedded in natural forest communities in the northern Great Lakes region of the US.[21] Environmental drivers in this case included forces such as deer hunting and invasive species, but one of their key findings was that, even in this unfragmented forest, species loss at a local level was dramatic.

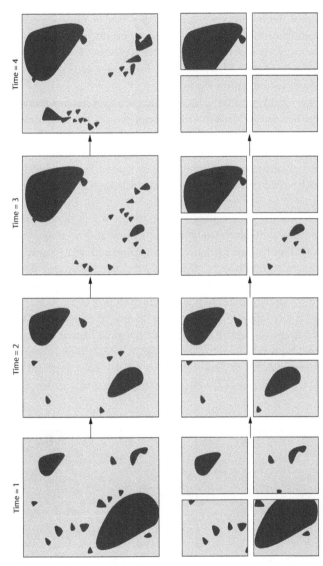

FIGURE 3.5 Diagrammatic illustration of the underlying dynamics of a spatially structured population in a continuous habitat (upper row) and the same population in a fragmented habitat (bottom row). In the top row, over the four time periods illustrated, the population is distributed in a non-random fashion and the patches are dynamically changing all the time, sometimes becoming locally more fragmented, sometimes simply contracting or expanding a particular patch. In the bottom row, we see the same but in a situation in which subpopulations cannot migrate. Clearly the same rules of subpopulations fragmenting, reconstituting, disappearing, expanding and contracting apply in this case. Nevertheless, local extinctions have occurred in three of the four fragments. Clearly the extinction events were common in both the upper and lower panels. However, in the upper panels there were always "spillovers" – which is to say, population migrations – in which small sections that had broken off from larger clusters of individuals migrated into areas where subpopulations had gone locally extinct. This "spillover" effect (migration) is restricted when the habitat becomes fragmented, as is the case in the lower panels.

Interfragment migrations

Conservationists frequently emphasize the problem of extinction and define themselves as part of a movement to avoid extinction. However, as we have argued elsewhere[22] and further argue below, the active movement of organisms in the landscape can be just as important. Indeed, for many years it has been an accepted fact of ecology that the probability that a population will persist in a landscape is a balancing act between extinction and migration. Migration occurs, in principle, in a variety of guises: animals actively move through habitats, seeds are dispersed from habitat to habitat, genes are transferred by pollinators from place to place. All of these phenomena are effectively migration forces.

Recalling the amphibian example from the last section, the 17 species that remain living on the E. S. George Reserve in Michigan live in temporary ponds surrounded by forested vegetation (Figure 3.6(a)). The adults move through those forests and their wandering constitutes the migration potential that eventually counteracts the local extinctions. We can only imagine what would happen if ponds were embedded in a matrix that was more hostile to those migrating individuals, such as a golf course (Figure 3.6(b)).

Albeit of key importance for the sustainability of a population in a metapopulation framework, migration is more difficult to study than extinction and we frequently have to use surrogate variables to gain evidence of migration activity. One such surrogate is species richness (i.e., numbers of species) as a function of distance from a natural habitat fragment, presuming that the community living in the disturbed habitat is a consequence of standard migration and extinction processes, such that biodiversity will decrease as a function of distance from the source of migrants. We used this surrogate to study the difference in migratory potential of two different coffee production systems for ground-foraging ants in Mexico.[23] Sampling ants with tunafish baits on the ground at different distances from a natural forest fragment in two agricultural matrix types (Figure 3.7), we discovered a dramatic difference

FIGURE 3.6 Ponds embedded within two types of matrices. (a) A pond on the E. S. George Reserve near Ann Arbor Michigan. (b) A pond on the Forman Golf Course, in Forman, North Dakota.

Source: (b) Dave Brenner

FIGURE 3.7 Comparison of forest edge abutting shade coffee and sun coffee in the area where we estimated ant species richness.

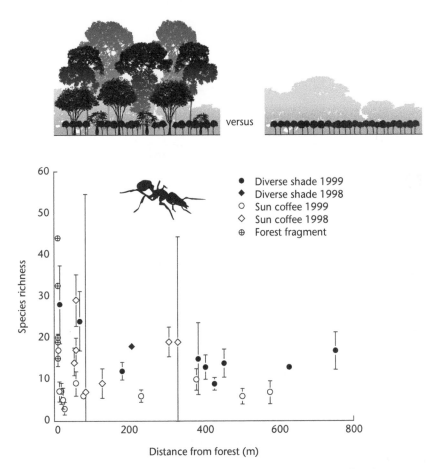

FIGURE 3.8 Ant species richness for 1998 and 1999 on two types of coffee plantations in Mexico as a function of distance from forest.

Source: Perfecto and Vandermeer (2001)

between a commercial polyculture farm and a more intensified sun coffee farm in Mexico. The evidence strongly suggested that migratory potential in the commercial polyculture was significantly different from the more intensified farm, where the species richness of ants dropped as soon as we started sampling within the sun coffee plantation (Figure 3.8).

A similar study with epiphytes yielded similar results (Figure 3.9(a)).[24] In that study, Leigh Moorhead and colleagues were able to demonstrate that forest species

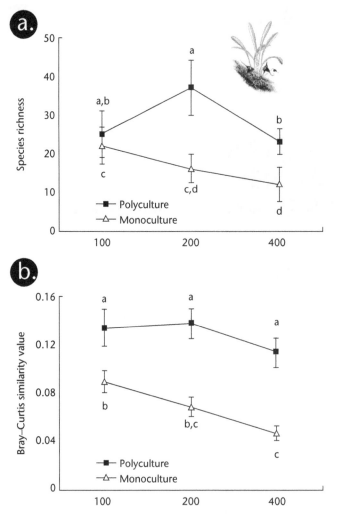

FIGURE 3.9 Species richness (a) and species similarity with forest species (b) as a function of distance from the forest on two types of coffee plantations. Different letters indicate statistically significant differences.

Source: Modified from Moorhead et al. (2010)

were maintained within the diverse shaded plantation while they declined significantly with distance from the forest in the coffee monoculture (Figure 3.9(b)). Both of these studies suggest that the diverse shade coffee plantation is a good-quality matrix that allows migration of forest species and could prevent extinctions even for organisms that cannot establish viable populations in the matrix.

A more direct estimate of interfragment communication is the genetic relationships of a particular species among various fragments. Shalene Jha and Chris Dick conducted a detailed study of the genetic structure of the understory tree *Miconia affinis*, demonstrating the importance of the shade coffee matrix for the genetic diversity of this species.[25] *M. affinis* is widely dispersed by large-bodied birds and small-bodied scrub and forest specialists such as the chestnut-sided warbler (Figure 3.10). It is an obligate outcrosser requiring native bees such as *Xylocopa* and *Scaptotrigona* spp. for cross-fertilization through buzz pollination (Figure 3.10). Using genetic markers to analyze the dispersal history and spatial genetic structure of *M. affinis*, they found evidence of strong dispersal limitation within forest patches and extensive gene flow across the shade coffee matrix.[26] In addition, the genetic diversity of the tree was similar for trees in the forest and trees within the shade coffee matrix. By conducting spatial analysis of pollen dispersal across the coffee matrix, they demonstrated extensive cross-habitat gene flow by native bees with pollination events spanning more than 1,800 kilometers.[27] Pollen was carried twice as far within shade coffee farms as it was within forests, and trees within shade coffee farms received pollen from many more parents than trees within forests (Figure 3.11). The shade coffee farms, by supporting populations of native bees, enhance tree fecundity and genetic diversity. Taken together, these results indicate a high migratory potential for both the bees involved in pollination and the birds involved in seed dispersal, as well as the trees themselves, revealing that the shade coffee plantation represents a high-quality matrix for these organisms.

Seed disperser	*Miconia affinis*	Native pollinators

FIGURE 3.10 The understory tree *Miconia affinis*, its seed disperser the chestnut-sided warbler, and two of its pollinators, the native solitary carpenter bee *Xylocopa tabaniformis* and the native social bee *Trigona fulviventris*.

Source: Modified from Jha and Dick (2008)

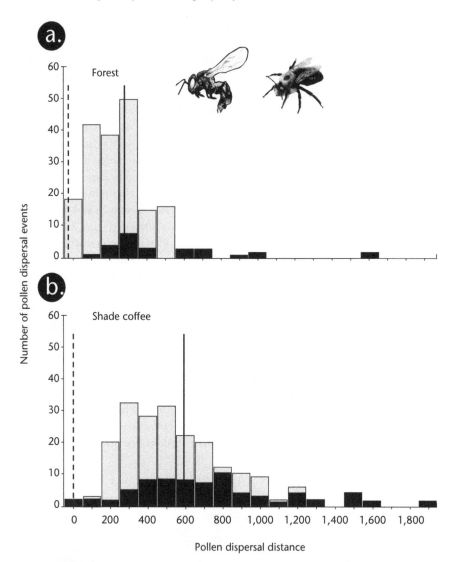

FIGURE 3.11 Pollen dispersal patterns for *Miconia affinis* trees in (a) forest and (b) shade coffee habitats binned in 100 meter distance categories with the proportion of pollen derived from forest (light shading) and shade coffee (dark shading) habitats indicated. Dotted vertical lines represent mean nearest flowering neighbor distances in each habitat, and solid lines represent mean pollen dispersal distances in each habitat.

Source: Modified from Jha and Dick (2010)

The dynamics of extinctions and migrations in fragmented habitats: a theoretical approach

To explore how the dynamics of extinction and migration enrich the analysis considerably, especially if concern is with long-term sustainability and conservation, we begin with a common modification of the basic Levins model (Equation 3.1). Suppose that only a certain fraction of the environment (h) is actually available (in other words, $1 - h$ is unavailable permanently). This means that the assumption that migration is proportional to $p(1 - p)$ is no longer true and must be modified. If we allow h to represent the fraction of habitat that remains occupiable (i.e., $1 - h$ fraction of the habitat has been destroyed), the probability that an occupied habitat comes into contact with an unoccupied (but occupiable) one becomes $p(h - p)$, and the Levins equation becomes,

$$\frac{dp}{dt} = mp\,(h - p) - ep. \qquad\qquad 3.3$$

The basic meaning of Equation 3.3 is illustrated in Figure 3.12.

This form of the model has been quite useful in clarifying the relationship of the probability of regional extinction (the extinction of the species over the entire landscape) to the amount of habitat lost.[28] Its extension to the question of the quality of the matrix is obvious. Suppose that instead of having just undisturbed and destroyed habitat, we also have a modified habitat that can be used by the species in question, such as a shade coffee plantation. We illustrate this idea by modifying the diagram of Figure 3.12, as in Figure 3.13. Here we have three habitat types: the original habitat (forest) in black; a high-quality modified habitat (shade coffee) in gray; and "destroyed" habitat (coffee monoculture) in white.

As can be seen in Figure 3.13, when another type of habitat is added, Equation 3.3 needs to be modified by expanding the migration term (m) to take into consideration the quantity of the different kinds of habitats (h, q_1 and q_2) as well as the different migration rates (m_0, m_1 and m_2) in each of the habitats. It should also be evident that the migration rate in the undisturbed forest (m_0) should be higher than the migration rate in the shade coffee habitat (m_1) and that the migration rate in the shade coffee habitat should be higher than in the coffee monoculture habitat (m_2). The emerging equation takes the following form, as shown in Figure 3.13:

$$\frac{dp}{dt} = (m_0\,h + m_1 q_1 + m_2 q_2)\,p(h - p) - ep \qquad\qquad 3.4$$

To make Equation 3.4 more tractable, we assume that migration in the low-quality habitat (i.e., coffee monoculture) is zero and drop the term $m_2 q_2$ from the equation. The new, more tractable, equation is:

$$\frac{dp}{dt} = (m_0\,h + m_1 q_1)\,p(h - p) - ep \qquad\qquad 3.5$$

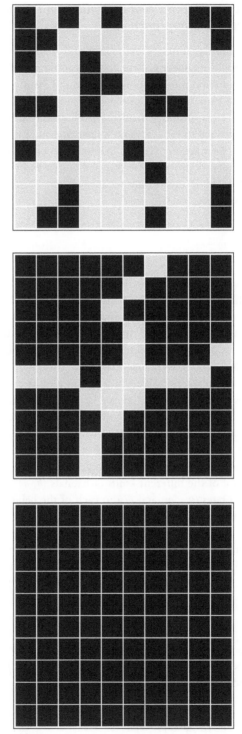

FIGURE 3.12 Idealized landscape of 10×10 cells, with various levels of habitat "destruction," where black cells symbolize original undisturbed habitat and white cells symbolize destroyed habitat, or habitat that is not appropriate for the species in question. In the case of undisturbed habitat $h = 1$, for the 20 percent destruction $h = 0.8$ and for the 70 percent destruction $h = 0.3$.

$$\frac{dp}{dt} = m(h\text{-}p)\, p - ep$$

$$\frac{dp}{dt} = (m_0 h + m_1 q_1 + m_2 q_2)\, p(h\text{-}p) - ep$$

$$m_0 > m_1 > m_2$$

FIGURE 3.13 Idealized landscape of 10×10 cells, with three types of habitat: undisturbed forest (h), shade coffee (q_1) and coffee monoculture (q_2).

With this configuration, there are two conceptually distinct ways of increasing the quality of the matrix. In Figure 3.14(a), the quality of the matrix is increased by increasing the number of high-quality patches. In Figure 3.14(b), the quality of the matrix is increased by increasing the quality of the high-quality patches (as the gray color becomes darker, the quality increases). The overall migration coefficient is proportional to the amount of habitat type (relative number of squares of a particular shading) and the within-habitat migration coefficient (relative size of the arrows).

The equilibrium value of Equation 3.5 is,

$$p^* = h - \frac{e}{m_0 h + m_1 q_1}. \tag{3.6}$$

With this formulation, one can use $m_1 q_1$, the number of habitats of a particular type and the quality of those types, as an index of matrix quality, to examine various management options for the landscape. For example, imagine a landscape dominated by intensive agriculture with a small percentage in forest fragments (e.g., Figure 3.15(a)). Now suppose there is a species of frog that needs forest to survive, but the amount of forest available is not enough to save the frog from extinction. Many conservationists instinctively think that the way to save the frog is by increasing the amount of forest in the landscape. An exemplary manifestation of Equation 3.6 can be used to illustrate the potential consequences of various management options for

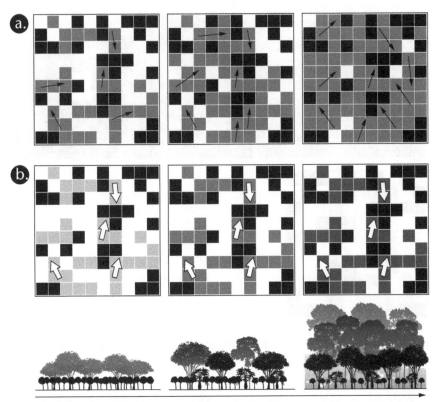

Increasing matrix quality

FIGURE 3.14 Idealized landscape of 10 × 10 cells, with various types of habitat, where black cells symbolize original undisturbed habitat (for example, forest), the white cells represent "destroyed" habitat, or habitat that cannot be occupied by the species in question (for example, coffee monoculture) and the various shades of gray cells represent good-quality modified habitat (for example, various kinds of shade coffee). The arrows indicate migration: larger indicating higher migration, smaller indicating lower migration. (a) Increasing matrix quality by increasing the number of better-quality patches. (b) Increasing matrix quality by increasing the quality of the individual patches (i.e., for example, going from patches of shaded monoculture to patches of rustic coffee agroforest). By convention, we let the proportion of black squares (the original native habitat) = h, the proportion of gray squares = q_1 and the shading of the gray squares (corresponding to the width of the arrows) = m_1.

that kind of landscape (Figure 3.15(b), solid arrow). For this particular manifestation (using particular parameters for m and q), it becomes evident that to bring the frog out of the region of extinction, it would be necessary to increase forest cover from 20 percent to more than 60 percent, something that may be politically or economically unattainable. However, with a small increase in the quality of the agricultural matrix such that it increases the migration rate of the frog and allows it to migrate among fragments, at least occasionally, the frog can be saved from extinction (Figure 3.15(b), dashed arrow). In a situation like this, increasing the quality of the matrix can be achieved in a variety of ways, for example, by eliminating the use of herbicides and other pesticides that have been shown to be toxic to amphibians,[29] or by adding hedgerows or scattered trees within the agricultural landscape.

For another example, consider a landscape with an intermediate matrix quality and little natural vegetation. Suppose that the species of concern is a butterfly that needs more natural forest to avoid extinction. In a situation like this, some conservation biologists would argue that the best way to prevent this species from going extinct is by intensifying agriculture to spare land for conservation, the classical "land-sparing" argument (Figure 3.16 (b)).[30] However, as illustrated in Figure 3.16(a), increasing the amount of forest at the same time as the quality of the matrix declines due to intensification (the two gray arrows in Figure 3.16(a)) will not get the butterfly out of the extinction region; indeed, the attempt to intensify agriculture (the left-facing gray line) so as to "spare" land for conservation (the upward-facing gray arrow) may lead to a Sisyphean dilemma (solid black arrow), in which the perceived line

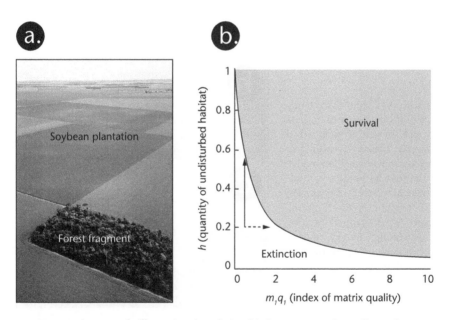

FIGURE 3.15 An example illustrating the relationship between matrix quality and amount of undisturbed (non-matrix) habitat, according to the dynamic metapopulation model, when the starting point is a very low index of matrix quality and 20 percent of forest in the landscape. (a) A soybean plantation with fragments of forests in the state of Mato Grosso, Brazil. (b) A graph illustrating how it would be necessary to increase the amount of forest by a large amount (from 20 to > 60 percent; solid arrow), or alternatively, to increase the quality of the matrix by a small amount (dashed arrow) in order to reach the survival region (gray shading).

Source: (a) John Lee (*Time Magazine*, March 27, 2008)

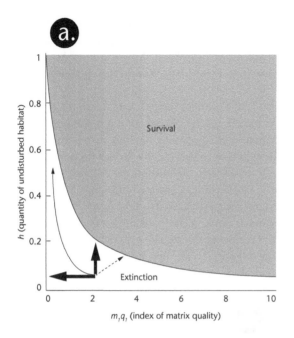

FIGURE 3.16 An example illustrating the relationship between matrix quality and amount of undisturbed (non-matrix) habitat, according to the dynamic metapopulation model, when the starting point is a moderate index of matrix quality and less than 10 percent of forest in the landscape. (a) A graph illustrating how decreasing matrix quality (through agricultural intensification) and increasing amount of forest will fail to take the species out of the extinction region (solid black line), while increasing matrix quality while increasing forest by about the same amount will take the species out of the extinction region (dashed back line). Gray shading indicates survival of the species, and the non-shaded area indicates the parameter range of extinction. (b) A landscape with intensive agriculture and forest fragments (i.e., land-sparing option). (c) A landscape with a diversified agricultural landscape and forest fragments (i.e., land-sharing option).

Sources: (b) Aerial view of landscape near Sentenhart by Hansueli Krapf: http://commons.wikimedia.org/wiki/File:Aerial_View_of_Landscape_near_Sentenhart_15.07.2008_17-12-49.JPG; (c) a Guanacaste portrait from Bella Vista Lodge, Liberia (Costa Rica) by Tizianok: http://commons.wikimedia.org/wiki/File:A_Guanacaste_view.jpg

separating extinction from survival seems to continually move away. On the other hand, increasing the amount of forest by about the same amount while also increasing the quality of the matrix (e.g., by diversifying the agricultural system as in Figure 3.16(c)), thus increasing the migration potential of the butterfly, will take the species out of the extinction region (black dashed arrow).

The qualitative pattern is quite clear from this simple "toy" model. On the one hand, the model says nothing more than could be said with just a qualitative understanding of the processes involved. The more original habitat preserved and the higher the quality of the matrix, the more likely it is that extinction will be avoided. Yet there is something more important here. Rather than simply drawing a straight line on a graph of "original habitat" versus "quality of the matrix", we see an important non-linearity – the border between the extinction zone and the survival zone is bowed inward, and the degree of that bowing depends on the underlying extinction and migration rates that existed in the system before the relevant land use change (i.e., e and m_0) (Figure 3.17). Thus, the degree to which matrix management matters depends, at least to some extent, on the underlying nature of the original habitat in addition to the nature of the organism under study.

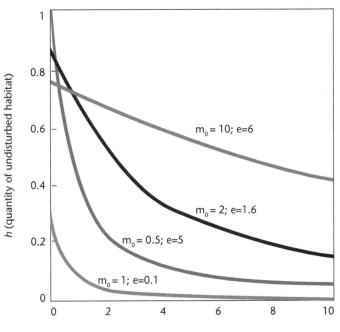

FIGURE 3.17 Various outcomes of the model using Equation 3.6. As the intrinsic migration potential of the species increases, the border between the extinction zone and the survival zone becomes less bowed inward (more linear) and as the extinction rate declines, the region of survival gets bigger.

Landscape structure and interfragment dynamics

In the previous section, we contrasted the static approach of Chapter 2 with the important addition of a dynamic component in terms of extinction and migration processes. From a practical perspective, if our conservation interest is in the short-term issues, as might be warranted in emergency cases when a particularly important habitat or species is on the border of global extinction, then the static approach of Chapter 2 is certainly appropriate, and legitimate questions about land sparing versus land sharing, or forest transitions, may be posed. However, if the interest is in the long-term sustainability of the landscape, at least for the biodiversity conservation component, ignoring the dynamics would be foolhardy. A long-term perspective must look at the "matrix dynamic" aspect of landscape patterns, not just the snapshot one might obtain from satellite imagery or other static data.

When we reach the point of worrying about the ultimate design of the landscape, some things appear quite obvious. For example, if a particular landscape contains two classes of producers – say, slash-and-burn farmers and agroforestry farmers – the important question to be posed is either (1) what fraction of the land is devoted to agroforestry and what fraction to slash and burn, or (2) whether the agroforestry and slash-and-burn systems are organized to allow migrations of organisms of conservation interest. So, for instance, an agroforestry system employing extensive spraying of insecticides could be of less quality than a slash-and-burn system that includes a large amount of time in fallow. Or, contrarily, a slash-and-burn system that burns every year with no fallow could be of lesser quality than a traditional agroforestry system employing organic methods. Both the abundance of particular habitat types and the quality of those types enter into the equation of matrix quality, as explicitly elaborated in the previous section.

The work of Meine van Noordwijk and colleagues is of particular interest in this context.[31] Taking to task some underlying assumptions about forests, forestry and agroforestry, they ask a fundamental question, the answer to which has seemed to be so obvious as to not warrant a query. Yet, as Noam Chomsky noted recently, "[science] begins with the capacity to be puzzled about very simple phenomena."[32] They simply ask: "What is a forest?" The history on this subject is complicated in its detail, but simple in a general sense. The expanding disaster of European imperialism triggered a romanticism that perhaps also grew out of the evident downside to the Industrial Revolution. That romanticism saw nature, and particularly forests, as paradise lost. The consequence was a preservationist attitude reifying pristine forests as a natural category that needed protection from the ravages that had so evidently destroyed them in Europe.[33] In this context, agriculture was seen as the enemy, and the main contributor to paradise destruction. We now know that there are very few places in the world, even in the middle of the Amazon rainforest, that remain untouched by the human species.[34] And, fortunately, there is a growing recognition that not all agriculture has a devastating impact on biodiversity.[35]

Van Noordwijk and colleagues' framework throws away this artificial distinction between forest and agricultural landscapes and proposes the simple idea that in the real world, landscapes range from completely covered with trees to no trees at all. The romantic ideology has tended to dichotomize that fact and, rather than accepting the reality of this gradient, has tended to change landscapes into ones that are either filled with trees or completely absent of trees, "natural" forests versus "agriculture." The development of industrialized agriculture since the Second World War in combination with the emergence of preservationist attitudes has brought us far closer to that dichotomy than ever before, especially in the developed world. However, policies stemming from this vision of the natural world have questionable outcomes, both in terms of alleviating hunger and conserving biodiversity and natural ecosystems.[36] Rather than continuing the historical trend of forcing this dichotomy on nature, we should adopt a more holistic view and embrace the fact that, with few exceptions, landscapes are complex mosaics of trees of varying densities. Sometimes they are so dense that people call them forested landscapes; sometimes they are so sparse that people call them grasslands or even deserts; but variability and continuum is the extant pattern. Van Noordwijk and colleagues' position is that this gradient would be a far more natural and useful way of conceptualizing landscapes and designing management, than the artificial duality of forest versus agriculture. To be sure, a teak plantation is not the same as a forest in anyone's analysis with regard to diversity, ecosystem function and aesthetic content. Yet replacing the highest level categorization from "forest versus no forest" to the quantifiable "density of trees" is a step in the right direction. Indeed, in the end we acknowledge that landscapes are normally dynamic and variable mosaics of complicated organismal associations. Nevertheless, as a first step in classification, density of trees seems better than relying on the romantic residue of "pristine forest."

Expanding on this point of view of landscape as a dynamic and variable mosaic of diverse vegetation associations, is there a way of conceptualizing the dynamic structure of the matrix that would be at once general and realistic?[37] Up to this point, we have been using metapopulation dynamics and island biogeography as convenient platforms for exploring the effects of landscape structure on biodiversity. Another classical conceptualization that might be useful for developing a meaningful theory of landscape structure, especially as it relates to matrix quality, and consequently to the migratory potential of various habitats, is source–sink dynamics.

This concept, sometimes also referred to as the island–mainland model, has long been a subject of investigation in ecology.[38] The fundamental idea is that there is a core population, the source, that sends out migrants from its habitat to smaller habitats within the migratory potential of the organism in question. Those smaller habitats are not capable of maintaining a population in perpetuity, and are thus called sinks. The combination of a source, in which the population could exist in perpetuity, and one or more sinks, in which the population cannot persist for very long, is called a source–sink population.

Although metapopulation structures assume that the subhabitats in which the organisms occur are all more or less equal, most landscapes are composed of sub-habitats that are of variable size and quality, which suggests that many populations will indeed exist in the form of a source–sink population. Yet conceptually it is clear that a collection of sinks, with sufficiently strong migration among them, will be a

Metapopulation   Source–sink population

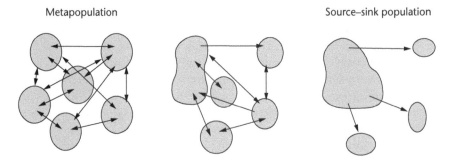

FIGURE 3.18 Conceptualization of the continuum between a metapopulation and a source–sink population.

Source: Based on the original diagrams of Harrison and Taylor (1997)

metapopulation. Thus, we can imagine the idea of a metapopulation and a source–sink population as extremes along a continuum, as suggested in Figure 3.18.

The basic elements of the matrix

Within this framework, we may conceptualize a generalized landscape structure as composed of three key relations:

1   The success of a source–sink population is determined largely by the size of the largest habitat patch (the source). Since the local extinction rate is a decreasing function of habitat size, all else being equal, the average extinction rate will decline with the average size of the habitat patch.
2   The success of a metapopulation is determined by the ratio of the migration rate to the extinction rate (recall Equation 3.2, the equilibrium condition for Levins' metapopulation equation). The migration rate, in turn, is determined principally by the distance between habitat patches, which is determined in part by the number of habitat patches.
3   Regardless of the details, the total number of habitat patches will normally be related to the size of the largest habitat.

In general, we can envision the possible population structures on a simple graph of e (extinction) versus m (migration). From the basic metapopulation equations (Equations 3.1 and 3.2), we see that a collection of habitats will be maintainable as a metapopulation as long as $e/m < 1$, which means that $e = m$ represents the border between the possibility of existing as a metapopulation and inevitable extinction. However, for the population to exist as a source–sink population, the extinction rate of the largest habitat patch must be smaller than a critical extinction rate – call it e_{crit}. Thus, we have two obvious equalities, $e = m$ and $e = e_{crit}$, that divide the space of e and m into well-defined areas that stipulate the potential for a source–sink or metapopulation, as shown in Figure 3.19.

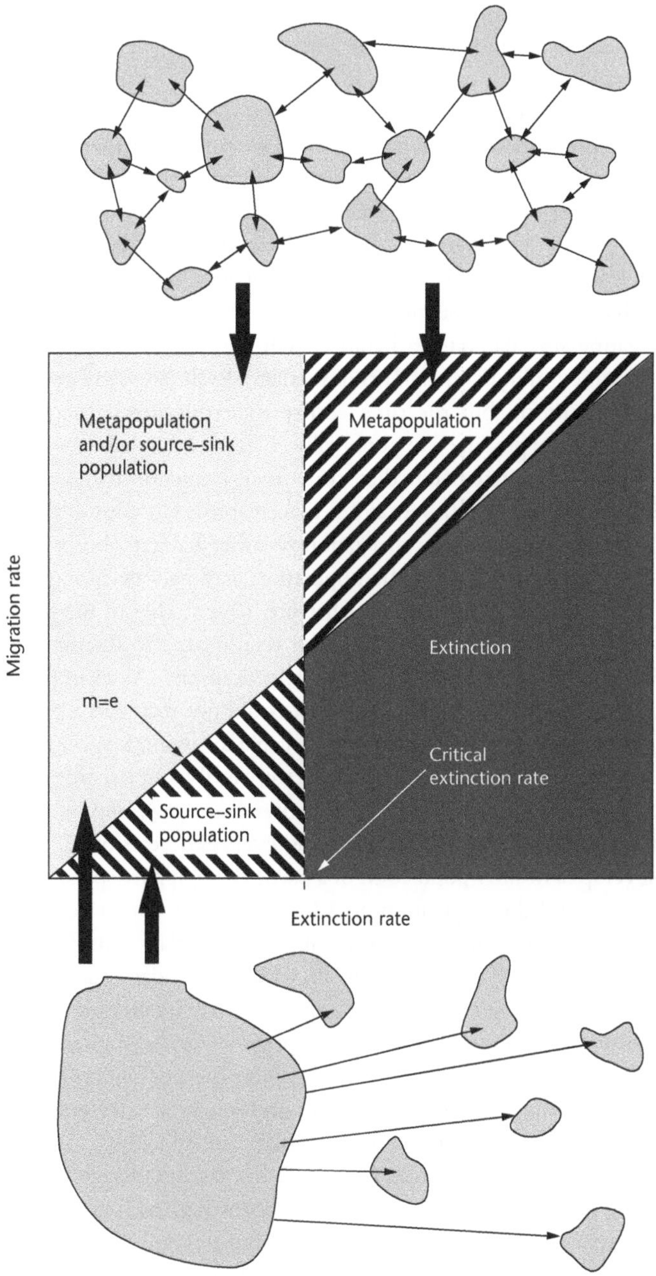

FIGURE 3.19 Population structure resulting from various combinations of extinction and migration rates. The top panel is an illustration of a metapopulation and the bottom panel is an illustration of a source–sink population. The middle panel is a graph of the two critical features ($e = m$ and $e = e_{crit}$) that distinguish source–sink from metapopulation.

Source: Redrawn from Jackson et al. 2014b

We can thus conceptualize spatially structured populations, if they are not composed of individuals all randomly mating with one another (a panmictic population), as either metapopulations or source–sink populations.[39] Within this framework, every subhabitat is either a source or a sink, and if they are all sinks, the population is either a metapopulation or will go extinct. The framework we propose uses both sources and sinks as elements in the construction of a landscape matrix, but with additional characterizations of both, namely with a formality we refer to as either "propagating sinks" or "ephemeral sources," extending previous approaches to the subject.[40] While this use has a formal mathematical definition, it is also a conceptualization that has an intuitive feel under certain conditions of landscape structure.

A propagating sink habitat is a habitat patch that contains a population that is doomed to extinction in that patch (its immediate death rate is greater than its birth rate plus immigration), but the time course to extinction is long enough that propagules from it are produced to potentially colonize other habitats. An ephemeral source habitat is a habitat patch that contains a population that is temporarily "healthy" (its immediate death rate is less than its birth rate plus immigration) and continually produces propagules that colonize other habitats, but whose carrying capacity, or other factor affecting the population, is slowly declining such that the population eventually will expire from that site. Our studies of the biodiversity of ants and epiphytes on coffee farms and their relationship to distance from forests (Figures 3.8 and 3.9) aid in thinking of potential examples. An example of a propagating sink might be a shade coffee farm that could not maintain a population of a particular forest ant species in perpetuity but is able to sustain a colony of the species long enough for it to produce new queens that can fly and become established in other forest patches. An example of an ephemeral source would be a shade coffee plantation that undergoes severe tree pruning every ten years or so. For ten years, the farm develops a healthy population of a particular epiphyte species, but then the trees are pruned and the population goes locally extinct. Similar examples can be envisioned in forested habitats. A propagating sink population might, for example, exist in small secondary forest patches in which secondary succession had not proceeded very far, so nesting sites for a species of cavity-nesting birds would be rare enough that a local population of that species would eventually disappear for lack of sufficient nesting opportunities. However, the population could persist in that habitat long enough to produce some second-generation offspring that migrate to other similar patches of forest. On the other hand, an ephemeral source population might be a bird species that nests in low shrubs in patches of secondary forest, but whose nesting habitat is periodically cleared for grazing. As long as the secondary vegetation patches exist, fledglings can emerge and migrate to other such patches. However, clearing for grazing eliminates the patches of vegetation entirely.

Recalling some basic terms, the "landscape" (sometimes referred to as the "countryside")[41] is composed of two elements, "patches of natural vegetation" embedded in an agricultural "matrix." The matrix can then be of variable "quality" or "permeability" or "connectivity," hence containing "corridors" or "trampolines" or "stepping stones." What is common to all these formulations is that they speak to the importance of the matrix as either a barrier to, or facilitator of, migration among habitat fragments. Here we propose a new framework for the matrix itself, namely as a collection of temporary habitats, some of which are sources, others of which are sinks.

In this landscape context, either subpopulations of a metapopulation (all of which are propagating sinks) or source–sink populations can be thought of as constituting a matrix, through which migration (dispersal) must occur. The sinks are of various qualities such that, even though they are sinks, propagules are able to migrate to another patch before the population goes extinct within that patch, which is to say they are propagating sinks. The sources are also of various qualities such that, even though they are sources and thus contribute migrants to other habitats, populations in them eventually also go extinct since the habitat itself disappears, which is to say they are ephemeral sources. In our original example, the agroforestry system that could not support a population in perpetuity, but provided enough structure for the population to engage in some reproductive behavior before going extinct, represents a propagating sink. By contrast, the slash-and-burn system creates fallow patches that are sources for many populations, but eventually are cleared and burned for the next cycle, thus are ephemeral sources. Clearly, the distinction between propagating sinks and ephemeral sources seems subtle. It is, however, an important distinction that can lead to distinct consequences. A graphic explanation of the distinction is presented in Figure 3.20. Note that the quality of the source or sink is key to the definitions of the two concepts; namely, a high-quality propagating sink is one in which there are generally a large number of existing propagules, whereas a high-quality ephemeral source is one in which the habitat (or population) exists for a longer time than in a low-quality source.

With regard to a collection of sink populations, if all of the sinks are composed of identical populations, with the probability of generating propagules before going locally extinct the same in each, the dynamics of the system are precisely equal to a metapopulation, with each sink a subpopulation of the overall metapopulation. Our purpose in this framework is to analyze the situation in which all the sinks are not equivalent. In the collection of propagating sinks, we pose the existence of various extinction probabilities; indeed, we propose that the extinction probability be the defining characteristic of a sink type. When all of the sinks in the matrix are brought together, the overall probability of maintaining a successful metapopulation structure is an integration of the balance between the overall migration rate and the individual extinction rates of the independent sink types. We argue that many existing landscape mosaics, particularly in areas dominated by agriculture, are collections of propagating sinks for organisms whose original habitat has been largely replaced by agricultural land with only fragments of the original habitat remaining. While it may be of convenience to categorize them as "corridor-like" or "filled with stepping stones" or "of a high quality," their obvious mosaic characteristic suggests that a formulation that specifically recognizes this mosaic structure would be of use, at least for some applications.

In the case of ephemeral source populations, much the same considerations apply, except here we are dealing with subpopulations that become established and regularly

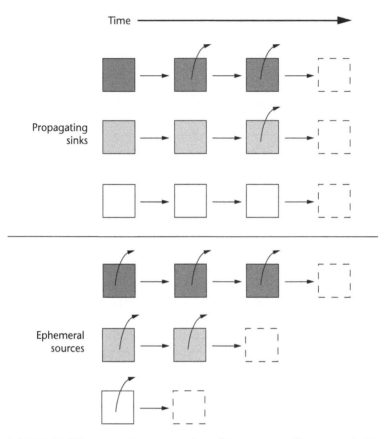

FIGURE 3.20 Diagrammatic representation of the two sorts of temporary habitats, propagating sinks and ephemeral sources. The degree of shading is proportional to quality and a dotted outline indicates an extinct subpopulation. Horizontal arrows indicate change through time, and arrows pointing upward indicate migration out of the subpopulation. The quality of sinks relates to the rate of outmigration, while the quality of sources relates to the rate of disappearance of the habitat (or, equivalently, the rate of extinction of the subpopulations). Note that the lowest-quality sinks represent regular sinks with no propagation potential.

Source: Modified from Vandermeer et al. (2010b)

send off propagules to subpopulations located in other habitats, yet the original subpopulation is doomed to eventual extinction because of the loss of the habitat itself or the deterioration of the conditions within that habitat. The defining feature of quality in the case of the ephemeral source is expected time to extinction of the source subpopulation. Again, a collection of ephemeral source populations will be, formally, a metapopulation if each of the populations is equal. However, if the probability of extinction is variable across the subpopulations, the situation is more complicated.

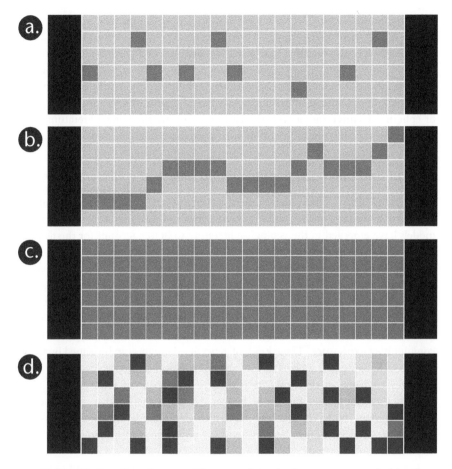

FIGURE 3.21 Various formulations of the type of matrix that separates two habitat fragments. Black coloration indicates "native" source habitat (and, by assumption, extinction rate of zero) and various shades of gray indicate habitats of different quality, defined by an increasing extinction rate or probability (the lighter the color, the higher the extinction probability). (a) Stepping stones or trampolines, in which individuals may jump from habitat island to habitat island, thus traveling from one fragment to the other. (b) Corridors, in which individuals can travel through habitat similar to native habitat to get to the other fragment. (c) High-quality matrix in which the habitats composing the matrix are uniformly of lower quality than the original habitat, but of sufficient quality that although a population could not persist there in perpetuity, it could survive long enough to be able to send propagules to disperse to other fragments. (d) Propagating sink or ephemeral source mosaic, in which habitat types of different extinction probabilities are mixed in a mosaic-like structure.

Source: Modified from Vandermeer et al. (2010b)

While the distinction between a collection of propagating sinks and ephemeral sources is important analytically (see following section), the commonalities in terms of the landscape structure that may result are identical. In Figure 3.21, we illustrate our conceptualization with four general categories of interfragment connections: (a) stepping stones or trampolines, (b) corridor, (c) high-quality matrix and (d) propagating sink or ephemeral source mosaic. It would appear that the propagating sink or ephemeral source mosaic is less well-known in the theoretical literature, even though it is arguably the most common landscape in actual practice.

A mean field approach to propagating sinks and ephemeral sources

Since spatial heterogeneity shapes the dynamics of single populations and affects coexistence of many species within communities and ecosystems, spatially explicit models have emerged in the ecological literature to better account for the fact that a given individual can interact with only a limited number of neighboring individuals and that the local fitness for such individuals can vary across habitat types.[42] However, one of the drawbacks of spatially explicit models is their complexity. While ultimately the details of habitats and the populations contained therein will need to be taken into consideration for any practical applications, frequently we are interested in a more general understanding of the mechanisms that shape populations and communities across space. These kinds of generalized insights can be gained by taking a mean field approach, such as Levins' metapopulation conceptualization. A mean field model describes the population dynamics in terms of some integral characteristics averaged over the entire habitat. In much the same way that Levins' metapopulation model is a spatially implicit model, here we describe a mean field approach to understanding the persistence of populations in heterogeneous landscapes.

There are two conceptual issues summarized in the formulation presented in Figures 3.21(a)–(d). On the one hand, in Figures 3.21(a) and (b), the problem is framed as a binary lattice in which each cell in the lattice is either a good or a bad habitat. Therefore, the problem is really one of the density of good habitats in a sea of bad ones. The likelihood that an individual or its propagules can eventually disperse from one end of the lattice to the other (i.e. migrate from one fragment to the other) depends on whether in a migratory event an individual or propagule can move or jump from one high-quality patch to another. This problem is effectively the same as the forest fire percolation problem; the higher the proportion of high-quality patches, the higher the probability of percolation (this will be further discussed in Chapter 4).[43]

There is an obvious relationship between the stepping-stone formulation (Figure 3.21(a)) and the corridor (Figure 3.21(b)), in that the corridor is simply a "spanning cluster" (percolation) that could easily form spontaneously as habitat patches are added randomly.[44] Thus, it makes sense to see the formulation in Figure 3.21(b) as an extension of the formulation in Figure 3.21(a), explicitly at (or above) the percolation threshold.

A rather different conceptual framing is suggested in Figures 3.21(c) and (d). Figure 3.21(c) is what we have been calling a high-quality matrix, that is, a matrix that resembles the natural habitat in certain key factors that allow migration of the organism in question. Here we have been applying the idea to agroforestry systems that resemble the native forest habitat.[45] This framing seems particularly fitting for the shade coffee system.[46]

However, most landscapes are not like the relatively homogeneous coffeescapes, but rather more like habitat mosaics, with mainly smallholder farms interspersed with fallow fields, small pastures and native vegetation patches. For this sort of landscape, each management type is bound to impose on the population either a distinct extinction probability (ephemeral sources) or a distinct migration probability (propagating sinks), so that the theoretical formulation would be best approximated as a mosaic of different ephemeral sources and propagating sinks, as in Figure 3.21(d). Comparing the matrix quality approach with the ephemeral source/propagating sink mosaic (Figure 3.21(c) with Figure 3.21(d)) is essentially the same as comparing a fine-grained and a course-grained environment.[47] We can even conceive of a gradient of environmental grain that moves from fine grain (Figure 3.21(c)) to coarse grain (Figure 3.21(d)), depending not only on the actual physical structure of the landscape, but also on the perception of that structure by the organism of concern,[48] and, at times, the biological dynamics involved[49].

The modified metapopulation model presented in the previous section can be modified further to deal with both the ephemeral source and propagating sink mosaic formulation. We begin with a collection of ephemeral sources, in which, by definition, each habitat type has a characteristic extinction rate. To make things simple, suppose there are just two habitat types varying only in their extinction rates, and that a population of organisms occupies q_1 proportion of the first habitat type and q_2 of the second habitat type. So we have for each of the two habitat types, qualitatively,

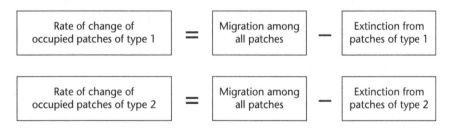

and, following the reasoning we used earlier, these qualitative relations can be translated into a set of simultaneous differential equations, namely,

$$\frac{dq_1}{dt} = m(q_1 + q_2)(1 - q_1) - e_1 q_1 \qquad\qquad 3.7(a)$$

$$\frac{dq_2}{dt} = m(q_2 + q_1)(1 - q_2) - e_2 q_2 \qquad\qquad 3.7(b)$$

The key to this idea is to note that the distinction between the two habitat types is based on the distinct extinction parameters, while the migration parameter is the same for both habitats – the definition of an ephemeral source.

With this simple formulation, the population either exists in perpetuity as a collective metapopulation, or the organism disappears completely from both habitats.[50] We see here an important result concerning variability in quality of habitats – namely that it is not the average of the habitat quality that completely determines the long-term maintenance of the population, but rather a combination of the average and variability. Note that the equivalent single metapopulation equation would be,

$$\frac{dq_1}{dt} = 2mq_1(1 - q_1) - e_1 q_1 \qquad\qquad 3.8$$

with the doubling of the migration coefficient due to the absence of the alternate habitat. From Equation 3.8, we have the standard metapopulation result that the equilibrium condition is $q_1^* = 1 - e_1/2m$, whence the condition for persistence of the metapopulation is $m > 0.5e_1$. Focusing on Equations 3.7(a) and (b), it is easy to construct examples in which $m < 0.5e_1$, which is to say that habitat 1 has two characteristics – first, it is an ephemeral source habitat, and second, a metapopulation existing in that habitat alone would go extinct. Nevertheless, when a second habitat is added, the population is able to persist in both habitats, even though habitat 1 cannot, by itself, form a successful metapopulation.

If the landscape is composed of only the first habitat type, the population will not persist over time. However, adding the second habitat in effect "rescues" the population in the first habitat (Figure 3.22). Thus, neither habitat type alone offers a suitable environment for the population and it would simply die out if no migration were allowed (the basic definition of an ephemeral source habitat). Furthermore, habitat 1 is in a sense even more marginal in that a collection of habitats of this type cannot sustain a metapopulation. Therefore, when the second habitat is added it effectively rescues the population in the first habitat, and the population persists as a fugitive species in both habitat types, which is to say that this is not simply a case of a source population rescuing a sink population.

Turning now to a landscape occupied by a collection of sink populations, we begin as before by assuming there are two habitat types, both sinks, but distinct from one another. Thus, we have the following qualitative arrangement,

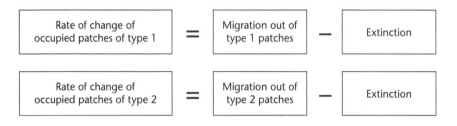

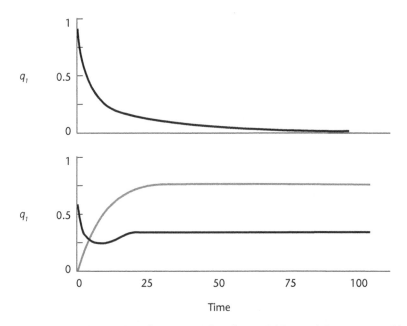

FIGURE 3.22 Time series of Equation 3.8 with $m = 0.15$, $e_1 = 0.31$, $e_2 = 0.05$, black line equivalent to Habitat 1 and gray line Habitat 2. (a) With only the first habitat available, the population goes extinct. (b) With both habitats available, the population persists in both habitats.

The corresponding equations are,

$$\frac{dq_1}{dt} = (m_1 q_1 + m_2 q_2)(1 - q_1) - eq_1 \qquad 3.9(a)$$

$$\frac{dq_2}{dt} = (m_1 q_1 + m_2 q_2)(1 - q_2) - eq_2 \qquad 3.9(b)$$

It is important to note that the distinction between the two habitat types is based on the distinct migration parameters, while the extinction parameter is the same for both habitats – the definition of a propagating sink.

It is not difficult to imagine landscapes with both propagating sinks and ephemeral sources, in which case equations 3.7 and 3.9 become

$$\frac{dq_1}{dt} = (m_1 q_1 + m_2 q_2)(1 - q_1) - e_1 q_1 \qquad 3.10(a)$$

$$\frac{dq_2}{dt} = (m_1 q_1 + m_2 q_2)(1 - q_2) - e_2 q_2 \qquad 3.10(b)$$

and the distinction between unstable and stable can be easily computed as

$$(e_1 - m_1)(e_2 - m_2) > m_1 m_2 \qquad 3.11$$

for the unstable case.

Equation 3.11 can be used to query the stability situation of each option (propagating sinks versus ephemeral sources), by setting the appropriate constants. Thus, for percolating sinks, we have the inequality,

$$m_1 < e - m_2 \qquad 3.12$$

indicating instability, and for ephemeral sources,

$$e_i > \frac{e_j m}{e_j - m} \qquad 3.13$$

for $i, j = 1, 2$ and $i \neq j$, also indicating instability.

In general terms, the ultimate consequence for populations in both types of habitat – propagating sink and ephemeral source – is extinction. However, the elementary source of that extinction is different for each type. While in both cases the habitat itself is "temporary" from the point of view of the population living in it, the reason for the temporary nature for the population is distinct. In a propagating sink, the nature of the habitat itself is such that the population cannot form a successful population within it. In an ephemeral source, the habitat is perfectly suitable for the population, but the habitat itself either disappears or changes such that the population can no longer be sustained therein. Although the distinction seems subtle, the ultimate consequences are substantial.

In the case of ephemeral sources, the conceptualization of quality is based on extinction rates, not migratory potential. Alternatively, in the case of propagating sinks, the conceptualization of quality is based on migration rates, not extinction probability. The implication for conservation management is that policy focus should be on the lowest-quality habitat in a mosaic, rather than on the highest-quality patches. This may seem initially counterintuitive. Much conservation activity is focused on the "natural" patches that remain, which is to say on those subhabitats that are of the highest quality. Considering the landscape as a whole, if the percolation threshold is mainly determined by the extinction rate of the lowest-quality habitat in the mosaic, practical conservation efforts would be best spent focusing on improvement of those low-quality habitats.

For coffee landscapes across the world, this means that a good conservation strategy could be to diversify the coffee monocultural systems by adding shade trees and to reduce or eliminate the amounts of pesticides that are applied. The analysis suggests that by increasing the quality of the lowest-quality farms, extinctions can be avoided. The details of how much shade, what kinds of trees and what kind of management will be best for avoiding extinctions will depend on the organisms in question and the quality of the other habitat patches.

Conclusion

This chapter revisits and expands some of the arguments in our previous book, *Nature's Matrix*, published with environmental historian Angus Wright.[51] At its most elementary level, the argument, we feel, is completely obvious. What seems to have constrained its full appreciation despite its obviousness is initial baggage, both scientific and political – and one piece of that baggage is the idea that biodiversity is static. Thus, as elaborated in the previous chapter, the notion that an ecological function (the yield set) combined with an economic function can generate planning recommendations has been seductive. Given what some see as an emergency regarding the problem of species extinctions, it is attractive to be able to gather data and fit it to optimality models to make recommendations that seem reasonable. We also have been seduced by that approach.[52] Much of that activity can be understood in a very simplified manner by combining a generalized ecological function with a generalized economic function and dialectically relating the ecology and the economy. The various recommendations currently popular in the literature can be easily categorized as one or another of the states emerging from this reasoning, as discussed in Chapter 2.

Yet it is a fundamental mistake to leave the question of landscape arrangements to the static ideas of current agricultural production and biodiversity content, especially if, as it seems, those categories are in reality debris of the lost age of European romanticism. Rather, we argue that the landscape dynamics need to be considered if we want to understand and make recommendations regarding the questions of biodiversity in agricultural landscapes. The ecological notions of metapopulations, source–sink populations and the realities of current patterns of fragmentation, especially in tropical regions of the world, come together to provide a theoretical framework for examining the nature of biodiversity in the matrix within fragmented landscapes. In general terms, what emerges from our analysis is the notion that since local extinctions are inevitable almost always, a focus on the migratory potential of the matrix is essential. With some notable exceptions,[53] the more the agricultural matrix resembles the natural habitat – be it forest, wooded savanna or grassland – the higher that migratory potential will be and the lower the probabilities of regional extinctions.

Recently, there has been an upsurge in the number of empirical studies that examine larger landscape effects on biodiversity.[54] This is a welcome trend. However, given the heterogeneous and idiosyncratic nature of landscapes, a great deal of the literature seeks to understand the factors contributing to biodiversity within each specific landscape, ignoring broader theoretical underpinnings. This is unfortunate since, as in all science, without theoretical generalizations there is a tendency to blindly gather data that may or may not contribute to understanding. A welcome model of detailed study informed by theoretical generalizations is the work done by Gretchen Daily's lab group. Using a ten-year data set of bird distributions in Costa Rica and a beta diversity model, they demonstrated that intensive agriculture tends toward homogenized vegetation at a large scale and reduced beta diversity.[55] In other words, areas that are dominated by intensive agriculture are more similar to each other in terms of the vegetation and therefore increase species overlap and reduce overall species diversity. At a regional scale, they demonstrated that fine-scale forest elements – in other words, trees or groups of trees scattered through the agricultural landscape – substantially support biodiversity outside forests.[56] Similarly, the extensive work done by Teja Tscharntke's group has shown the important role of diverse agroforestry systems in forest margins.[57]

Our theoretical analyses suggest that an initial attempt at generalization might be to ask some simple questions about how patches of landscapes might be conceptually recognized and whether it might be possible to generalize about landscapes using such simplified notions. Thus, we propose that a useful conceptualization is that of a landscape mosaic. Within this mosaic, a dichotomy for landscape patches can simplify the analysis without completely eliminating all levels of complexity. One possible basis for the construction of matrices, we argue, is the distinction between sources that are ephemeral and sinks that are propagating. It is a distinction that at first appears to be subtle, but in the end engenders distinct consequences for matrix structure. We suggest this framework for further development and consideration.

A perhaps yet more radical agenda would be that suggested by van Noordwijk and colleagues, as introduced earlier.[58] Might patches of habitat (farms, field, etc.) be categorized only by the density of trees? Chase Mendenhall's work[59] suggests that there may be much to be gained from such a categorization. In the context of coffee, the paradigm makes a great deal of sense (see Figure 2.12), even if we too frequently talk as if farms easily fall into discrete categories.[60]

Although the topics discussed in this chapter focus on theoretical aspects of bio-diversity in agricultural landscapes more generally, we note that most of this analysis has been inspired by almost 25 years of ecological and biodiversity research on coffee farms in Central America and Mexico. However, most importantly, biodiversity in all its complexity is crucial for the productivity and, especially, the sustainability of any agroecosystem, coffee included. In the following chapters, we examine the complex ecological interactions associated with biodiversity in diverse coffee systems.

Notes

1 Perfecto et al. (1996); Manson et al. (2008).
2 Many examples are presented in Perfecto et al. (2009).
3 Vandermeer (1989); Perfecto et al. (2005).
4 Kuussaari et al. (2009).
5 Perfecto and Vandermeer (2008a); Perfecto et al. (2009); Perfecto and Vandermeer (2010).
6 For example, Green et al. (2005); Perfecto et al. (2005).
7 Perfecto et al. (2005).
8 Perfecto and Vandermeer (2010).
9 MacArthur and Wilson (1967).
10 Levins (1969).
11 Hubbell (2001).
12 Simberloff and Abele (1982); Laurence and Bierregaard (1997).
13 May (1990); Raup (1994); Newman and Palmer (2003).
14 Skelly et al. (1999).

15 Foufopoulos and Ives (1999).
16 Newmark (1995).
17 Newmark (1995); Fischer and Stöcklin (1997); Foufopoulos and Ives (1999); Kéry (2004); Matthies et al. (2004); Williams et al. (2005); Wilsey et al. (2005).
18 Bolger et al. (1991); Brooks et al. (1999); Helm et al. (2006).
19 Janzen (1970); Connell (1971); Hyatt et al. (2003).
20 Vandermeer et al. (2008).
21 Rooney et al. (2004).
22 Perfecto et al. (2009).
23 Perfecto and Vandermeer (2002).
24 Moorhead et al. (2010).
25 Jha and Dick (2008, 2010).
26 Jha and Dick (2008).
27 Jha and Dick (2010).
28 Kareiva and Wennergren (1995); Hanski et al. (1996); Amarasekare (1998).
29 Relyea (2005).
30 Green et al. (2005); Phalan et al. (2011a, 2011b).
31 Van Noordwijk et al. (2011).
32 Chomsky (2011).
33 Oerlemans (2004).
34 Denevan (1992); Whitmore and Turner (1992); Denevan (2001); Willis et al. (2004); Mann (2005).
35 Perfecto et al. (2009).
36 Perfecto et al. (2009).
37 Levins and Lewontin (1980).
38 Holt (1985); Pulliam (1988); Harrison and Taylor (1997).
39 Vandermeer et al. (2010b).
40 For example, Wu and Levin (1994, 1997); Wimberly (2006).
41 Daily et al. (2001); Horner-Devine et al. (2003).
42 Morozov and Poggiale (2012).
43 Turcotte and Malamud (2004).
44 Pascual and Guichard (2005).
45 Perfecto et al. (1996); Perfecto and Vandermeer (2002, 2008a).
46 Perfecto and Vandermeer (2002); Perfecto and Armbrecht (2003); Vandermeer and Perfecto (2007); Perfecto et al. (2009).
47 MacArthur and Levins (1964).
48 Levins (1968); Gillespie (1974).
49 Vandermeer et al. (2008).
50 See Vandermeer et al. (2010b) for details.
51 Perfecto et al. (2009).
52 Perfecto et al. (2005).
53 Lander et al. (2011).
54 Fahrig (2003); Tscharntke et al. (2005); DeClerk et al. (2010).
55 Karp et al. (2012).
56 Mendenhall et al. (2011).
57 Tscharntke et al. (2007).
58 Van Noordwijk et al. (2011).
59 Mendenhall et al. (2011).
60 For a counter-argument, see Milder et al. (2010).

4

SPACE MATTERS

Large-scale spatial ecology within the coffee agroecosystem

What do the spots of the jaguar and the distribution of ants on a coffee plantation have in common?

When you see the spots on a jaguar's coat, the changing shapes of a flock of blackbirds against the evening sky or a visual representation of the links formed through social media like Facebook, a pattern is evident. Yet there is no prime mover causing the spots to be where they are, or the shifting waves of birds, or the knotted tangle of internet connections. Something in the way in which the elements of these systems interact with one another makes a sort of spontaneous pattern pop out. The organization comes from within, and consequently we refer to it as a "self-organized pattern." In the previous chapter, we presented some ideas about spatial patterns, emphasizing the effect that the spatial structure of agricultural matrices has on biodiversity. However, the impact of spatial structure on populations is only one side of the issue. More recently, the reverse process has become a focus of ecological research. Rather than spatial patterns affecting populations, we now understand that populations sometimes make spatial patterns. Rather than organisms responding as innocent pawns to underlying spatial structure, biological forces among them act in such a way as to cause the emergence of spatial patterns.[1] In this chapter, we explore these self-organized patterns and their consequences for the coffee agroecosystem.

Consider, for example, the three illustrations in Figure 4.1. Obviously they are different from one another, yet they also are similar. Levins and Lewontin once made the following important statement about science: "Things are similar; this makes science possible. Things are different; this makes science necessary."[2] In Figure 4.1, we have tried to show how this principle acts in the context of spatial patterns. Obviously, the three pictures are different: a pattern of vegetation, a pattern of the distribution of *Azteca sericeasur*[3] ant nests on a coffee plantation and a big cat's coat pattern. However, our juxtaposition of all three together is meant to imply a central core of similarity, not so much in the similarity of the patterns themselves, but in

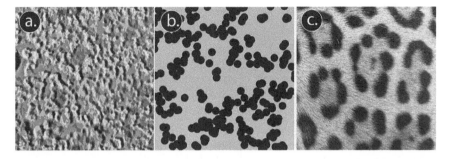

FIGURE 4.1 Three spatial patterns, different yet similar. (a) Patches of vegetation in central Australia. (b) Distribution of ant nests of the species *Azteca* on a coffee plantation in southern Mexico. (c) Spots on the coat of a jaguar.

what seem to be the underlying rules that make those patterns. The patterns in the cat coats and vegetation are popularly believed to emerge from the so-called Turing effect and our claim is that the pattern of the ant nests stems from the same underlying rules.[4] We also argue that those rules are, in a profound sense, generalized rules that apply to many types of spatial patterns, from spots on big cats' coats, to patches of desert vegetation, to distributions of ant nests on coffee plantations.

Much of the literature on spatial patterns is methodological. It is true that the various methods of constructing spatially explicit models or analyzing spatial data have become extremely sophisticated, in part because of the ubiquity of high-speed computers and user-friendly software. It is perhaps the case that we have lost sight of the underlying principles, to the extent that they exist. Have they become a bit lost in the forest of sophisticated methodology? The insights that populations are frequently organized spatially as metapopulations or source–sink populations, or that islands contain biodiversity based on simple processes of migration and extinction, remain important generalizations, to be sure. Yet there is another aspect of spatial patterns that we feel may be equally important, but less well recognized by ecologists. This is the concept of "diffusive instabilities," a remarkable insight of the famous mathematician Alan Turing, thus frequently termed the "Turing effect."[5]

The equations describing this process are perhaps a bit intimidating, so we have not attempted to incorporate them here. The interested reader can consult other sources.[6] However, we can present the idea of the equations here, in brief word form. The qualitative idea of diffusion is probably familiar to everyone. We quote here from one of the early formal descriptions, courtesy of Thomas Graham, published in 1833:

> Fruitful as the miscibility of the gases has been in interesting speculations, the experimental information we possess on the subject amounts to little more than the well established fact, that gases of different nature, when brought into contact, do not arrange themselves according to their density, the heaviest undermost, and the lighter uppermost, *but they spontaneously diffuse, mutually*

and equally, through each other, and so remain in the intimate state of mixture for any length of time.[7]

To gain intuitive insight, imagine that you slowly pour different-colored inks (say, yellow and blue) at opposite ends of an aquarium filled with water. The natural expectation is that the aquarium that begins with a cluster of blue water at one end and yellow water at the other eventually turns into a uniform green coloration throughout due to the diffusion of the pigment molecules. The same idea would apply to a population of animals or plants, except that the population would be reproductive, so we have a situation in which the "ink" is increasing in density at the same time as it is diffusing in the space. The equations, in words, would be something like,

Change in the local population over time = reproduction rate + diffusion rate

Turing's insight was based on the analysis of a set of equations similar to this one. However, rather than just a single population reproducing its numbers, he posited two chemicals, together engaged in a reaction, and both diffusing through some sort of medium. Since both reaction and diffusion are involved, such a system in general is referred to as a "reaction–diffusion" system. Basically, there are two chemicals and they do three things:

1   They change through time.
2   They interact with one another.
3   They migrate, or diffuse, in space.

The basic idea is illustrated in Figure 4.2, replicating the basic form of the classical mathematics of the system.

The adventurous reader can consult other sources[8] to see how this simple arrangement is formulated as a system of partial differential equations, although understanding them is not necessary to appreciate the significance of the insight.

Turing's surprising results were that with two interacting chemicals, even if they were simply diffusing, a pattern with distinct patches of one type in a matrix of the

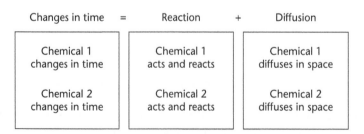

Changes in time	=	Reaction	+	Diffusion
Chemical 1 changes in time		Chemical 1 acts and reacts		Chemical 1 diffuses in space
Chemical 2 changes in time		Chemical 2 acts and reacts		Chemical 2 diffuses in space

FIGURE 4.2 The basic mathematical framework of the reaction–diffusion process.

other could be produced. This is actually quite remarkable since our normal intuition about diffusion is just that – things diffuse around a space such that the entire space becomes a monotonous average (yellow ink at one side plus blue ink at the other produces a green aquarium). Indeed, the "stable" condition of a diffusion process would be that monotonous state, which is why the patterned situation is referred to as "diffusive *instability*."

The process shown in Figure 4.2 is very general and applies to any system in which there are "reactants" and "diffusion". Turing's insight was to see something very special involved when three particular conditions are placed on a reaction–diffusion system:

1 The first agent, or activator, is "autocatalytic," which is to say that it spontaneously forms and increases its density.
2 The second agent, or suppressor, emerges as a consequence of the generation of the first one, but then acts to suppress the further generation of the first.
3 The suppressor diffuses faster than the activator.

With only those three conditions, Turing was able to demonstrate that a spontaneous pattern emerged from a reaction–diffusion system. The basic idea is illustrated in Figure 4.3.

What sort of pattern forms depends to a great extent on the details of both the reaction and diffusion. Since Turing's paper was published, many striking examples have been demonstrated on computers, some examples of which are shown in Figure 4.4.

Shigeru Kondo and Takashi Miura recently argued convincingly that Turing's brilliant but conceptually simple insight could be applied very broadly to many different biological processes.[9] One could apply it, for example, to two interacting species, one of which locally reproduces and disperses, the other of which reproduces positively as a function of the density of the first (it could be, for example, a predator).[10] The first species reproduces independently of the second and disperses locally at some rate. The second species reproduces proportionally to the density of the first and reduces the density of the first. Furthermore, the second species disperses faster than the first. What happens under such a scenario is that both species become distributed in space in a clustered or aggregated fashion, much like the pattern observed in Figure 4.1(b).

In general, much has been written about spatial patterns in ecological systems, and much of that is aimed at understanding the patterns formed spontaneously through biological interactions, the so-called "self-organizing systems". It is our view that many of those systems (perhaps the majority) can be understood as some variant of the Turing process. Sometimes the basic idea needs to be modified considerably, but in the end, the notion of two subjects interacting, one of which is an "activator" and the other of which is a "suppressor", and the latter of which disperses more rapidly than the former, goes a long way toward explaining, at a very general level, the way in which biological patterns are self-organized, from the spots of the jaguar to the distribution of nests of the ant *Azteca* on coffee farms in southern Mexico.

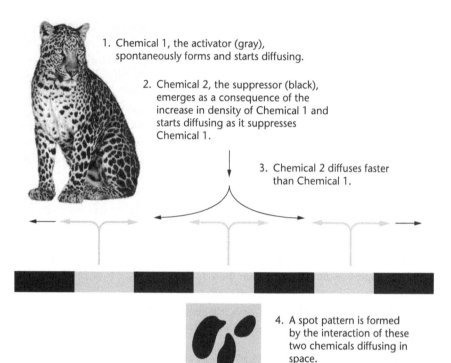

1. Chemical 1, the activator (gray), spontaneously forms and starts diffusing.

2. Chemical 2, the suppressor (black), emerges as a consequence of the increase in density of Chemical 1 and starts diffusing as it suppresses Chemical 1.

3. Chemical 2 diffuses faster than Chemical 1.

4. A spot pattern is formed by the interaction of these two chemicals diffusing in space.

FIGURE 4.3 Description of the process that generates diffusive instability and pattern formation, including the pattern observed in the coat of the jaguar. The spot pattern in the square below the striped line appears in Alan Turing's original paper, published in 1952.

FIGURE 4.4 Classical examples of Turing patterns. Different patterns emerge depending on the values of the parameters in the model.

Source: Modified from Kondo and Miura (2010)

Spatial patterns, power functions and the Turing process in the ant *Azteca*

Spatial patterns: Turing on the farm

To the untrained eye, a forest is simply a collection of trees. A person walking through the forest may notice the beautiful reflection of the light through the canopy or the colorful bird perching on a branch. However, a trained forester or plant biologist walking through the same forest is likely to notice not only the different kinds of tree species that are found in the forest, but many other things associated with them, especially the way in which they occur relative to one another. As we walk through the forest near our home in Michigan, we notice that there seems to be a patchwork of different species of understory trees – some patches are dominated by black cherry trees, while others are dominated by red maple, and yet others are dominated by the small-statured witch hazel. Our casual perceptions are born out by survey data, as revealed in a census we did of a section of the E. S. George Reserve (Figure 4.5). There are clearly distinct patches of those three species and it is perfectly natural to ask: what causes such a pattern?

One of the most obvious answers to this question is that some biophysical factor, most likely having to do with soil characteristics or microclimatic conditions, is responsible. For example, black cherry is thought to do well in dry, sandy soils and red maple is thought to prefer wetter areas; therefore, if the forest environment has some patches of dry, sandy soils and other patches of wetter soils, these two species would be expected to follow a similar patchy distribution according to the type of soil. As these are factors external to the species involved, they are called exogenous factors. However, as noted above, it is also well known that various intrinsic biological dynamics are capable of producing patterns even in a landscape that is homogeneous in terms of the environmental condition. These self-organized patterns are caused by endogenous factors, that is, the intrinsic biological dynamics of the system itself. Thus, the question arises, for any non-random spatial pattern, whether it is caused by exogenous factors (broadly speaking, underlying habitat

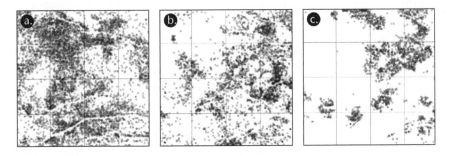

FIGURE 4.5 Forest plot at the E. S. George Reserve, southern Michigan. The spatial distribution of (a) black cherry, (b) red maple and (c) witch hazel in a 16-hectare plot at the E. S. George Reserve.

FIGURE 4.6 Workers of *Azteca*.

Source: David Gonthier

patchiness) or endogenous factors (broadly speaking, biological aspects of the organism independent of the habitat). In the case of the forest of the E. S. George Reserve in Michigan, an extensive analysis of a variety of factors suggested that the observed patchy distribution of the three species was caused by a complex combination of factors, including historical accident, background habitat conditions, interactions among the various species and the Turing mechanism.[11]

Unlike the forests in Michigan, trees in shade coffee plantations may be considered fairly uniformly distributed because they are planted with the intention of having a uniform pattern. We took advantage of this fact to study the spatial distribution of a species of tropical arboreal ant, *Azteca* (Figure 4.6). Since this ant species is also a predator of potential insect pests of coffee[12] (elaborated more fully in Chapter 6), we were interested in knowing how it was distributed in the space of a coffee farm.

Azteca nests in trees and is common in the Mesoamerican tropics, where it is frequently encountered on casual walks in the forest. However, discerning any spatial pattern of its colonies in a tropical forest is inevitably obstructed by the heterogeneity of the habitat. Studying this species on a coffee plantation allows us to investigate basic ecological questions related to the generation of spatial patterns in a more uniform habitat. So in addition to investigating the consequences of the spatial distribution of a predator for pest control, we can also use the coffee agroforests to ask more basic questions about the distribution of organisms in space and what causes those distributions.

To study the spatial distribution of an arboreal species that nests in trees, a relatively large and continuous area is required. Therefore, we asked permission from Don Walter Peters, the owner of *Finca* Irlanda, a large shaded and organic coffee farm in southern Mexico, to establish a 50-hectare plot in which we could map, identify and regularly survey all shade trees for evidence of nesting *Azteca*. With the help of a hard-working group of field assistants, it took us eight months to establish the plot, although in the end it became a 45-hectare plot due to five hectares of very steep topography. After the first survey, we were surprised to find that the *Azteca* colonies were distributed neither randomly nor uniformly within the plantation. Rather, they were arranged in a patchy (clustered) distribution (Figure 4.7). The obvious question that came to mind was: what causes that pattern? And the obvious answer, at least initially, was something in the environment. In other words, we thought that the ants, in choosing their nesting sites, were restricted by some environmental conditions that were themselves patchily distributed. However, examining a variety of

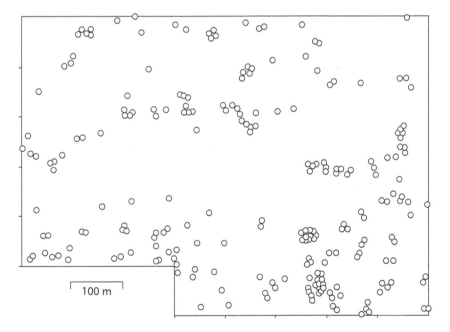

FIGURE 4.7 Distribution of nests of *Azteca* in a 45-hectare permanent plot on *Finca* Irlanda in Chiapas, Mexico in 2004/2005. Note the clear non-random nature of the distribution, with obvious dense clusters of some nests and other relatively isolated nests. Of the 12,227 shade trees recorded in the plot, only 276, or fewer than 3 percent, had an *Azteca* nest.

environmental factors such as slope, species of nesting tree and size of nesting tree, we were unable to identify anything external to the ant population that could explain the particular patchy distribution of the species. Consequently, we looked to the underlying biology of the ant as potentially the source of a self-organized pattern.

The basic biology of the ant is not unusual for a species that has multiple queens. After a queen establishes a colony in a tree, the colony may grow to the point that new nests are established in neighboring trees. What happens is that one of the queens of a multiple-queen colony takes a subset of the brood and workers from the parent colony and establishes another colony in another tree. Evidently, this is one part of the mechanism whereby patchiness is generated. However, unabated new nest formation would obviously result in a continuous expansion of colonies throughout all shade trees on the farm, which means that some force must limit this expansion. This is where Alan Turing's concept of diffusive instability may be important. Using the metaphor of the Turing mechanism, the tendency of the ants to disperse to neighboring trees is equivalent to the "activator" of the system and, therefore, for this to be a self-organized system there must be some "suppressor," otherwise the ants would simply disperse (diffuse) over the whole farm. The ant has several direct and indirect natural enemies, any one of which, or any combination thereof, could form the basis for the control that must occur to prevent the ant from

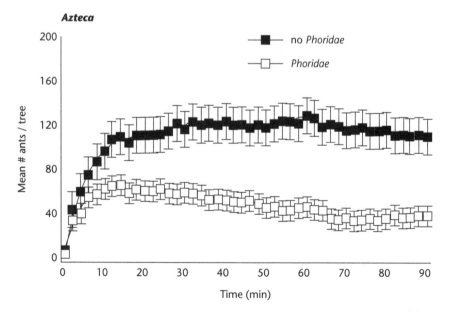

FIGURE 4.8 Effect of phorid flies on the foraging activity (recruitment to tuna baits) of *Azteca*.

Source: Modified from Philpott (2005)

taking over every shade tree on the plantation. One possibility that we have emphasized in the past is a group of parasitoid flies in the family *Phoridae*, known to reduce ant foraging activity. For example, early work by Stacy Philpott examined the foraging activity of these ants in areas with and without the phorid flies.[13] She found that ant activity was reduced by about 50 percent when phorid flies were present (Figure 4.8). Such a dramatic reduction of foraging activity suggests that the phorid flies could indeed cause a colony to disappear, either dying of starvation or moving their nest due to harassment by the flies.

These flies are well-known parasitoids of ants.[14] A female fly lays an egg at the base of the ant's head and the fly larva crawls into the ant's head capsule. As the larva develops it eats away at all the tissue contained in the head capsule. When the larva is ready to transform into an adult fly, the ant's head simply falls off, hence the nickname "decapitating flies." Once a new adult fly emerges (Figure 4.9(a)), it mates and, if a female, starts searching for more ants to attack and oviposit. For reasons that are not completely obvious, the flies are more common and attack more when the ants occur in more dense clusters (Figure 4.9(b)). It may be that the flies are simply more attracted to larger nest concentrations, or it may be that they build up higher population densities locally when there are more nests around.

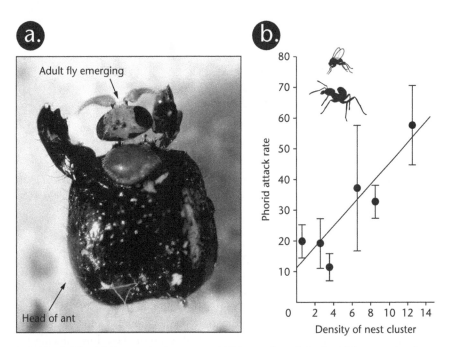

FIGURE 4.9 Phorid flies attack *Azteca* ants. (a) Photo of an adult phorid fly emerging from the head capsule of a decapitated ant. (b) Attack rate of phorid parasitoid flies as a function of local population density of nests of *Azteca*.

Sources: (a) Sanford Porter; (b) modified from Vandermeer et al. (2008)

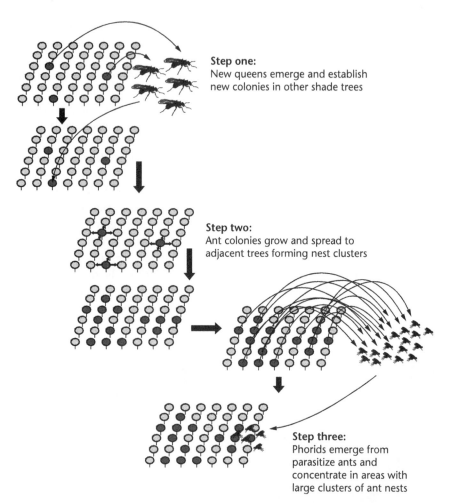

Step one:
New queens emerge and establish new colonies in other shade trees

Step two:
Ant colonies grow and spread to adjacent trees forming nest clusters

Step three:
Phorids emerge from parasitize ants and concentrate in areas with large clusters of ant nests

FIGURE 4.10 Diagrammatic representation of the proposed three-stage process that leads to the formation of ant clusters on coffee farms. Each symbol represents a shade tree on the coffee plantation.

This natural history suggests a three-part dynamic process:

1   Ant colonies are established by fertilized queens.
2   As the colonies grow, they spread to neighboring trees.
3   Phorid parasitoids concentrate on the more dense clusters of ant nests, causing a dramatic behavioral response and possibly direct mortality, thus reducing ant survival in dense clusters of nests (Figure 4.10).

Based on this fundamental natural history, we built a simple cellular automata (CA) model, in which the central cell becomes occupied or dies depending on the eight surrounding cells (called the Moore neighborhood in CA parlance), with the probability of becoming occupied depending on the local ant density (i.e., the number of cells in the Moore neighborhood that are already occupied by ants) and the probability of mortality also increasing with the local density (given the knowledge that the phorids act in a density-dependent fashion, as shown in Figure 4.9(b)). The rules of the CA model are illustrated in Figure 4.11. The values of the parameters of the model (i.e., the slopes and intercepts of the two linear functions in Figure 4.11) were estimated from field data. Output from the model along with observed data from the field are shown in Figure 4.12.

The overall number of *Azteca* nests in the simulations are concentrated between 200 and 500, close to the range of our observations in nature (represented as a horizontal line in Figure 4.12). The model output reflects the erratic nature of CA models, with the same parameters generating a dramatic variability of population densities. However, since the possible range of values could be from 0 to 11,000 ant

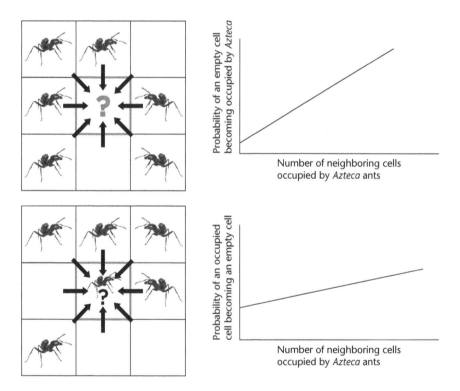

FIGURE 4.11 Basic structure of the cellular automata (CA) model used to study the spatial distribution of *Azteca* nests on coffee plantations. The top half of the figure illustrates the rule to determine whether an empty cell will become occupied by an *Azteca* nest, and the lower half illustrates the rule to determine whether an occupied cell will become empty.

Source: Based on Vandermeer et al. (2008)

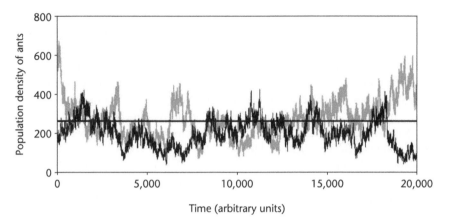

FIGURE 4.12 Simulated population densities of *Azteca* nests from the CA model (two runs, one in black and the other in gray), compared to actual population density in 2004/2005 (horizontal black line). Note that the theoretical potential of the model ranges from 0 to 11,000 on the population density axis, so what is pictured is only a small part of the lower part of that range. The predicted values thus range really quite close to the actual values observed in the field.

nests, the ability of the model, with parameter values within our empirical envelope, to generate population densities and patterns so close to those we observed in the field suggested to us that the basic interpretation of the spatial dynamics was probably correct. In other words, it seems that the non-random spatial pattern observed for this arboreal ant on the coffee plantation is not caused by exogenous factors, but rather it emerges from a self-organization process that includes the expansion of the ant into neighboring trees and the controlling effect of a natural enemy that concentrates in areas of high nest density.

Pattern and power functions

Given that the spatial pattern of the ant nests is non-random, the question arises as to what are the characteristics of that spatial pattern itself. Our interpretation of the underlying biological interactions that give rise to this spatial pattern is similar to those studied by Mercedes Pascual and colleagues at the University of Michigan,[15] broadly interpretable as a Turing structure. The Pascual group, using several different dynamic modeling framings, came to the conclusion that there was a broad range of parameter values for which spatial patterns, similar to the one we observed with *Azteca*, could be described by a power function, an idea that merits some further elaboration.

In his book *How Nature Works* (1996), so delightful in presentation if a bit pretentious in title, physicist Per Bak noted that as a culture we seem to have developed an expectation that the data we examine in nature are "normal." The word "normal" here is meant to be taken in a technical sense, describing the well-known "bell curve," a statistical distribution that follows a particular pattern, with a mean value and tails that extend to either side of the mean value. A great deal of research in all science relies on statistical reasoning to decide whether a research result is as good as it may seem or whether it is simply an example of something that could just happen by chance. The standard procedures usually assume that the data at hand are normally distributed, and the key idea of the "normal data" is that there is a "central tendency" that the data generally follow. For example, the number series 1, 2, 2, 3, 3, 3, 3, 3, 4, 4, 5 has a mean value of 3 and an equal number of smaller and larger values (1, 2, 2 to the left and 4, 4, 5 to the right). The weight of newborn babies is a good example of the normal distribution. The weight of most babies at birth falls near a certain value (the central tendency) while a few weigh less and a few weigh more (Figure 4.13(a)).

However, this assumption of a bunch of numbers with a central tendency is frequently violated in a dramatic way. Rather than the 1, 2, 2, 3, 3, 3, 3, 3, 4, 4, 5 of the above example, we frequently find 1, 1, 1, 1, 1, 1, 1, 2, 2, 3, 7, 20, which is to say a large number of small events and a very small number of very large events (Figure 4.13(b)). Bak noted that the world was filled with such examples, and frequently you could represent them on a graph of the logarithm of the number of observations versus the logarithm of the size of the event and the data would appear to be a straight line (Figure 4.13(c)). A straight line on a log–log plot is, mathematically, a power function ($\log(y) = a + b \log(x)$ is the equation for the line), so if you exponentiate both sides of the equation, you get $y = Ax^b$, where A is e^a and e is Euler's constant – note that the independent variable is raised to a "power," which makes the whole equation a power function. So, for example, if you plot the frequency of earthquakes versus the size of the earthquake, you find that there are a very large number of very small earthquakes and a very small number of very large earthquakes, and that relationship follows, quite precisely, a power function (Figures 4.13(b) and (c)).

What Bak noted in his book is that it appears to be a quite reasonable generalization that self-organizing processes tend to generate clusters that are organized into power functions, sometimes called power laws. This is also the basic finding of Pascual and her colleagues. Given our proposition that the ant colonies are self-organized, as described above, it makes sense to ask whether the distribution of clusters of nests is organized according to a power function, as suggested by the technical literature.[16] Indeed, as expected, the distribution of cluster sizes in our plot does follow a power function (Figure 4.14(a)). Furthermore, calculating the frequency of cluster sizes as generated by the CA model produces a similar power relationship (Figure 4.14(b)). It is also worth noting that the slope of the power function is approximately the same for both the field data and the model (−1.4 and −1.5 respectively).

The importance of these results lies in the fact that a strong spatial pattern is formed in the face of habitat homogeneity, and the power function distribution of cluster sizes in space suggests that the pattern is self-organized. The distribution of the shade trees on the coffee plantation is fairly uniform because they were planted

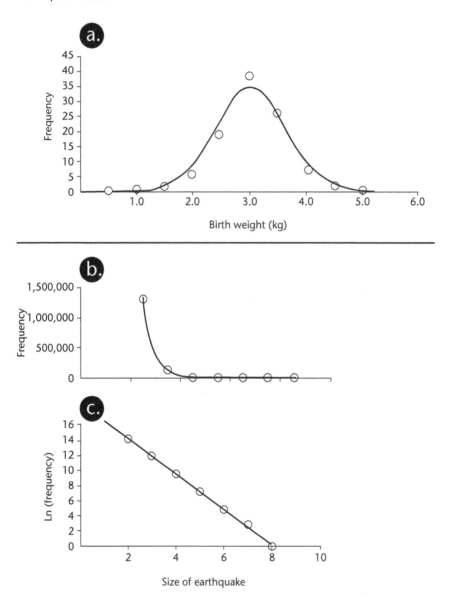

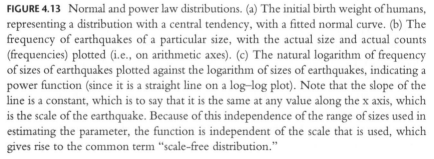

FIGURE 4.13 Normal and power law distributions. (a) The initial birth weight of humans, representing a distribution with a central tendency, with a fitted normal curve. (b) The frequency of earthquakes of a particular size, with the actual size and actual counts (frequencies) plotted (i.e., on arithmetic axes). (c) The natural logarithm of frequency of sizes of earthquakes plotted against the logarithm of sizes of earthquakes, indicating a power function (since it is a straight line on a log–log plot). Note that the slope of the line is a constant, which is to say that it is the same at any value along the x axis, which is the scale of the earthquake. Because of this independence of the range of sizes used in estimating the parameter, the function is independent of the scale that is used, which gives rise to the common term "scale-free distribution."

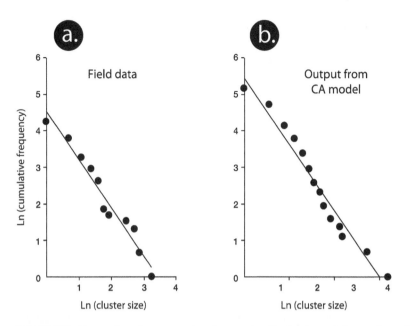

FIGURE 4.14 Power function approximation to the distribution of cluster sizes of nests of *Azteca* in a 45-hectare plot on a coffee plantation for (a) the field data and (b) the CA model instantiated with the field data. Field data are from the 2004/2005 survey based on a 20-meter cluster scale (any nest within 20 meters of another nest is in the same cluster as the latter). (Note that data presented in Vandermeer et al. (2008) are slightly different – those data pooled two different computer runs and two different field surveys, rather than a single run and a single survey as presented here.) The parameter of the power function is −1.4 in the case of the field data and −1.5 in the case of the theory output.

intentionally in a uniform fashion, as far as topography permitted. Nevertheless, the distribution of the ants that nest in those trees shows a patchy distribution and that patchiness emerges from fairly simple ecological dynamics: colonies of ants growing and spreading to neighboring trees, and natural enemies that control the spread in the largest patches. In that sense, we can say that the spatial pattern of the *Azteca* ants on coffee plantations is formed by the same general rules that govern the formation of the spots of the jaguar. As we will see in later chapters, the consequences of that pattern are extremely important for a variety of other organisms and for the stability and productivity of the coffee plantation.[17]

Implications of spatial patterns for system dynamics

Source–sink populations and metapopulations

No organism is an island. All are embedded in complex webs of interactions, and *Azteca* is no exception. Through a decade-long study of its natural history, we have established that a variety of other organisms are closely associated with *Azteca*. For these organisms, their "habitat" on the coffee plantation consists of *Azteca* clusters – that is, patches with nests of *Azteca*. In Figure 4.15, we show eight groups of species that are known to be associated with *Azteca* clusters. Given that several of these "groups" of species are actually known to contain multiple species themselves, there are actually more than 16 species of arthropods and fungi known to be associated with this species.

As was discussed in Chapter 3, key to the study of the spatial dynamics of populations is the nature of the underlying habitat structure in which the population is embedded. Although there are many ways of categorizing spatial population

FIGURE 4.15 Eight of the species groups known to be closely associated with nests of *Azteca*. (a) Parasitoid wasp that attacks the green coffee scale. (b) *Doimus* sp., a beetle that preys on scale insects. (c) *Azya orbigera*, another beetle that preys on scale insects. (d) A typical two-layered web of a lyniphiid spider that concentrates its webs in coffee trees associated with *Azteca* nests. (e) The green coffee scale, *Coccus viridis*, which is a mutualist with *Azteca*. (f) A rove beetle associated with *Azteca* nests. (g) One of the four species of phorid flies that parasitize the ants themselves. (h) The fungus *Lecanicillium lecanii* that attacks the coffee rust disease (orange spots) but also the scale insects. Many of these organisms are involved in the intricate system of autonomous biological control that we describe in Chapter 7.

Sources: (a) PestNet; (b) Aaron Iverson; (d) Linda Marín; (f) K. Taro Eldredge; (g) Sanford Porter

structure, two extreme cases discussed in Chapter 3 emerge as particularly common, the metapopulation and the source–sink population. It is clearly possible to view these two canonical forms as extremes on a continuum (Figure 3.18). For many practical reasons, it is useful – indeed, sometimes absolutely necessary – to know whether a population is a metapopulation or a source–sink population. For example, in the conservation context, a source–sink population commands attention to the location of the source population as the most important target for management activities.[18] By contrast, a metapopulation structure suggests that the overall landscape would be the proper focus of management so as to maintain sufficiently high inter-habitat migration.[19] Here we consider the case of a population that is potentially either a metapopulation or a source–sink population and ask how its nature is fundamentally determined by the way in which the underlying habitat is structured through self-organization, thinking of the nest clusters of *Azteca* as the running example, but insisting that the ideas might apply much more broadly.

Thus, we have a situation in which first, self-organization results in patches of *Azteca* and second, these patches become the habitat for other organisms. But what really is a patch of *Azteca*? Although the question seems fairly straightforward, the answer is: it depends! It depends on the scale relevant to the particular organism. For example, think of a herbivore that is associated with *Azteca* and is able to move distances of the order of 35 meters. Now think of a natural enemy of that herbivore that can move a maximum of 10 meters. For the herbivore, a "habitat patch" consists of clusters of *Azteca* nests that are at most 35 meters apart, while the "habitat patch" for the natural enemy consists of a group of nests that are at most 10 meters apart. We illustrate this difference in Figure 4.16.

We see in Figure 4.16 that the distribution of the cluster sizes seems to be scale-independent in both cases, but that the cluster scale itself dictates the nature of that scale-free distribution. For the organism that moves a shorter distance (Figure 4.16(a)), there are many very small patches and the largest patches are not very large, while for the one that moves a longer distance (Figure 4.16(b)), there are fewer smaller patches and the largest patches are very large. This implies that there are actually two parameters that inevitably determine the final clustering of habitat particles (i.e., nests of *Azteca*), and, therefore, the distribution of patch sizes: the total number of particles and the cluster scale.

Given a particular number of nests, the relationship between the number of isolated nests, also called "singletons," and the largest patch of nests is obviously a negative relationship; as the clustering scale increases, the number of singletons declines, and the size of the largest patch increases. An important fact to recall is that if the population causing the habitat clusters is self-organized, as we argue is the case of the *Azteca* ants, we expect the cluster sizes to be distributed approximately according to a power function. This means that a graph of the log of the number of clusters of a given size plotted against the log of the size of the clusters is a straight line. In Figure 4.16, we have plotted the distribution of the clusters for both situations. As expected, they both are approximated well by a power law, supporting

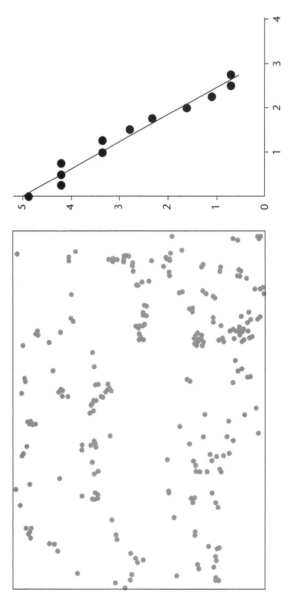

a. *Azteca* patches as perceived by an organism that has a range of movement of about 10 meters.

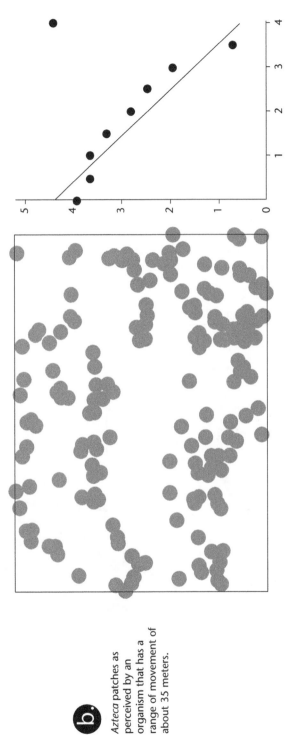

Azteca patches as perceived by an organism that has a range of movement of about 35 meters.

b.

FIGURE 4.16 Distribution of *Azteca* nests in 2004/2005. (a) As might be perceived by an organism that has a short (10 meters) range of movement. (b) As might be perceived by an organism that has a long (35 meters) range of movement. Graphs to the right are cumulative frequency versus cluster sizes, both represented as logarithms. The horizontal axis is the log of the cluster sizes, and the vertical axis is the log of the cumulative frequency of the cluster sizes.

the theory that a self-organizing process generated the patches. However, more importantly for this discussion, they have very different intercepts and slopes.

What does this mean in theory and in practice? In theory, the intercept on the y axis is the total number of the smallest clusters, in this case of singletons, and the intercept on the x axis is the size of the largest cluster. We can see that as the cluster scale decreases, going from 10 meters to 35 meters, the slope of the power function gets less steep (there are fewer small clusters and the largest clusters become larger). This can have very important implications from the point of view of the organisms whose habitat is clusters of *Azteca* nests (Figure 4.15).

As we noted earlier in this section and discussed more fully in Chapter 3, two of the major ways in which populations can be spatially structured are as metapopulations and as source–sink populations. If habitats are relatively discrete entities, as would be the case in any fragmented habitat or in almost any habitat that is self-organized, an important factor to consider is the minimal habitat size that will be able to maintain a population in relative perpetuity. If the habitat in question is self-organized into clusters where each cluster is one of the habitats, and if there is no

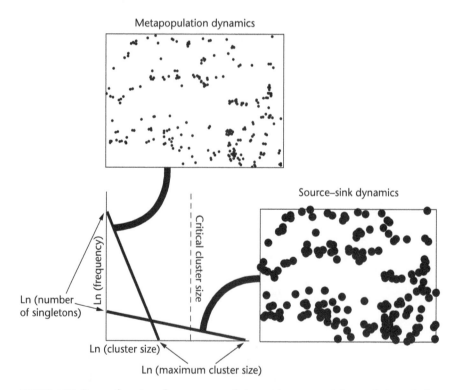

FIGURE 4.17 Power functions for a metapopulation and a source–sink population relating the log of cluster size to the log of the frequency of those sizes. Note that the intercepts of the functions have significant ecological meanings. The critical cluster size is the size for which the organism in question can survive in perpetuity within that habitat size.

cluster size that is at least as large as the critical size for a population to survive (in perpetuity), then in order for the population to persist at all, it must persist as a metapopulation. If, however, there is a single or several habitat patches that are larger than the critical habitat size, then the overall population is likely to persist in the form of a source–sink population. This is illustrated in Figure 4.17 with respect to the *Azteca* example. The important conclusion, then, is that the determination of whether a population is a metapopulation or a source–sink population (assuming that it survives at all) is a result of the self-organizing process of the organism creating the habitat and the scale at which the organism living in the habitat perceives it. This can have important implications for the conservation of biodiversity as well as the success of biological control programs. In the following section, we examine a specific example of a potential pest on coffee plantations.

Coccus viridis: a metapopulation or a source–sink population?

The green coffee scale (*Coccus viridis*) is one of those species closely associated with *Azteca* (Figure 4.15(e)). It is a small sap-sucking insect in the order *Hemiptera* that forms a mutualistic association with *Azteca*. The insect sucks the sap of coffee plants and is protected by *Azteca* against predators and parasitoids. In return, the ants harvest the energy-rich substance (called "honeydew") that is discharged by the insect. Given this relationship, one can say that the ant creates the background habitat into which the green coffee scale must fit. Every dry season, the scale populations drop to very low levels except in some of the ant nest clusters where residual populations persist. Does the population of this insect, a potential pest of coffee, persist as a metapopulation or as a source–sink population? In 2009 and 2010, before the beginning of the rainy season, we surveyed the green coffee scale population in coffee plants close to each *Azteca* nest. If the population is structured as a metapopulation, we would expect to find a random allocation of green coffee scales on patches of various sizes, even after accounting for the fact that large patches have lower (but not zero) extinction rates. If the population is structured as a source–sink population, then we would expect the green coffee scale to be concentrated in the larger patches only. Using a cluster scale of 17 meters, which is the scale at which the ants seem to respond to one another,[20] we plotted the locations where the green coffee scales were found (Figure 4.18). Note that the green coffee scales are concentrated in the larger clusters of ant nests and that the concentrations generally persist in the same nest clusters from one year to the next. This is precisely what would be expected for a source–sink population. Thus, we tentatively conclude that the self-organizing attributes of the arboreal ants create a patch structure that generates a source–sink dynamic for the green coffee scale.

In summary, if the self-organization of habitats by the *Azteca* ants results in a distribution of habitat patch sizes that lacks a central tendency, and may thus be approximated by a power law, or some similar function, any population that requires those habitat patches may exist as either a source–sink population or a metapopulation, conditioned not only by the migration and persistence qualities of that

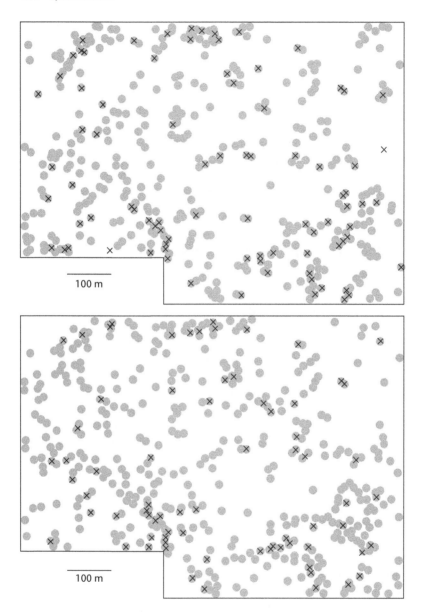

FIGURE 4.18 Distributions of ant nests in the dry season in the 45-hectare plot, represented as 17-meter diameter gray circles, along with Xs marking the locations where at
least one neighboring coffee bush contained green coffee scale insects in (a) 2009 and (b)
2010. Note that the concentrations of green coffee scales tend to occur in the same nest
clusters (defined by gray circles that touch or overlap one another) from year to year. The
distances between the locations of scale insects in 2010 and the nearest scale insects in
2009 are significantly less than would be expected by chance (Jackson et al. (2014a)),
indicating that the clusters of green coffee scales in 2010 generally occur near 2009 clusters,
which is consistent with the persistence of the scale insects as a source–sink population.

population itself, but also by the underlying distribution of the self-organized habitat patches. This framing of population spatial structure redirects the typical focus from one of migration/extinction dynamics alone to one that asks how the spatial structure of the underlying habitats codetermines, along with the migration/ extinction characteristics, the nature of population structure.

However, in addition to the interesting theoretical insights, these results may have important practical implications. It appears that, in the case of the green coffee scale, the habitat structure generated by the self-organization of the *Azteca* ants combines with the dispersal and reproductive characteristics of the green coffee scale to create a dynamic situation of a source–sink population for that herbivorous insect. Knowing that this potential pest has source populations in the largest *Azteca* clusters could help guide pest management programs. We discuss this complication in much greater detail in Chapter 6, where we add to this basic two-population system a much larger system of pests and their controlling agents.

The great transformation

In 2007, about two years after we set up the 45-hectare plot, a new owner took over the farm and decided to change its management to a more intensive system. Intensification in coffee farming usually involves reducing the density of shade trees or eliminating them altogether, as described in Chapter 2. Don Bernardo, the new owner, estimated that the farm had approximately 70 percent shade cover and he decided to reduce that to about 50 percent. As we elaborated in Chapter 2, this decision certainly was not in the best interests of biodiversity conservation and perhaps was not even in the best interests of increasing production, even though that was its intent. However, for the general scientific study of the ecosystem, it afforded us with an opportunity to examine what happens when a well-established agroecosystem undergoes a dramatic forced alteration. In particular, we had been studying the *Azteca* ant for several years and had excellent background information that, we thought, would enable us to make predictions about what would be the system response to the intensification.

Population density

The first and most obvious issue is the population density of the ant. As an obligate arboreal nester, its potential nesting sites would be dramatically reduced, or in terms of classical population dynamics, its carrying capacity would be reduced. Standard population dynamics in its most simplified form is understood to be a balance between the rate of population growth and the set of resources (food, nesting sites, etc.) that are necessary for a population to reach a stable point, but that also set an upper limit to that point, the carrying capacity. At equilibrium, the population may approach that carrying capacity, or it may approach some other value dictated by other ecological interactions such as competition with other species, mortality from predators and parasites, etc. Although the details are complex and specific to the particular system under consideration, the general expectation is that there should be a positive correlation between the attained population density of a given organism and the carrying capacity of its environment. Reducing the carrying capacity by impoverishing the environment should result in a lower population density. In the case of *Azteca* on the coffee farm, the reduction of shade trees, and therefore the reduction of potential nesting sites, was expected to result in a decline in the overall number of nests, a surrogate for population density in ants.

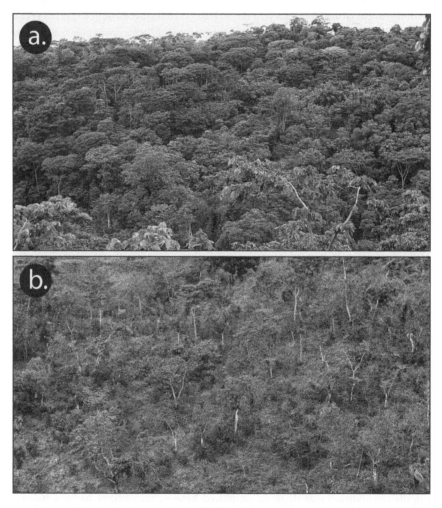

FIGURE 4.19 Drastic shade reduction in 2007 in the permanent plot. What was originally relatively dense shade above the coffee plants (a) has been converted into a landscape with about 30 percent fewer trees and heavily pruned canopies (b). Subsequently many of the trees re-sprouted, but the initial cutting was rather dramatic, as illustrated in this photo.

Source: Stacy Philpott

The transformation of the farm began in 2007 and within two years the shade was drastically reduced (Figure 4.19). In the 45-hectare plot, more than 30 percent of the original 12,227 shade trees were cut down over this two-year period. With this reduction of almost a third of all potential nesting sites, we expected the population of *Azteca* to decline, perhaps dramatically. Indeed, eliminating 30 percent of the trees in our original CA model predicted a dramatic collapse of the population. However, much to our surprise, the population density of ant nests in the field actually *increased*.

The population densities of nests before the tree cutting were 276 and 341 nests (2005 and 2006), during the period of the cutting (2007–2009) population densities went from 435 to 565 to 637, and in the two years after the cutting (2010 and 2011) population densities were 739 and 743 nests. Thus, we observed the rather remarkable fact that after a dramatic reduction of potential nesting sites, this species' population density increased by a factor of 2.7! What could have caused this puzzling result?

To understand what might have happened to the population density of the *Azteca* ants in response to the change in the management of the coffee plantation, we need to take a detour into the realm of regime change.

The idea of regime change

Most people associate the term "regime change" with some kind of change in the political system, such as the changes in France after the French Revolution, the changes in East Germany after the fall of the Berlin Wall or the military coup in Honduras in 2009. However, natural systems can also dramatically change their states, a phenomenon increasingly connected to the metaphor of a human political regime – perhaps not the best metaphor, but one that seems to have caught on. Think, for example, of a terrestrial ecosystem that goes from a grassland to a desert, or a lake that goes from a clear state to a turbid state. We can call such changes regime changes that result as a consequence of some changes in environmental conditions. These kinds of regime changes can happen gradually or they can happen suddenly. The rapid change from one state to another is frequently called a "catastrophic shift," the term emerging from catastrophe theory in mathematics.[21] It describes a situation in which a small change in external conditions generates a very large change in the state of the system, and implies some kind of a threshold and alternative stable states (Figure 4.20).

In more general terms, we can conceive of two alternative states of an ecosystem as both being possible, but dependent on some external driving force. A classical example would be a desert versus a grassland. Depending on the climate, one or the other would be expected. If very dry, we expect a desert; if moderately wet, we expect a grassland. However, what happens when the climate gradually changes from wet to dry? If we plot the probability of a grassland (or desert) as a function of the dryness of the habitat, we can observe three different kinds of transition events. The idea is illustrated in Figure 4.21. We know of many examples in which the transition follows a quasi-linear trajectory (Figure 4.21(a)). As a matter of fact, this is the expected

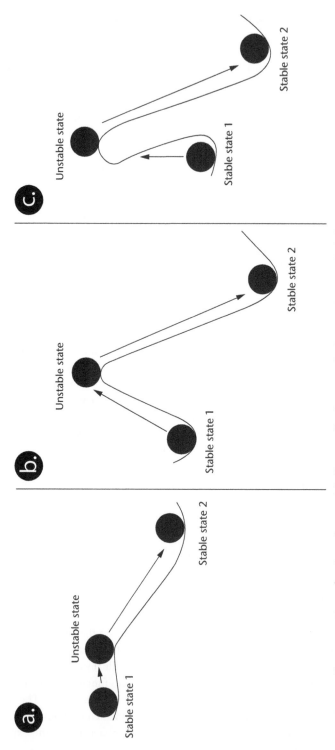

FIGURE 4.20 Representation of regime change. A ball in a valley being pushed up to the top of a hill. Once there, it can roll back down to the same valley (stable state 1) or roll on down to another valley (stable state 2). (a) A case in which it is easy to push the ball into a different stable state. (b) A case in which it is more difficult to push the ball into a different stable state. (c) A case in which it is virtually impossible, without defying the law of gravity, to push the ball into a different stable state. This latter case is frequently called a "critical transition," the transition being the original change from the stable state 1 to the stable state 2.

trajectory for many ecosystem changes and is one that alerts us that the system is changing and that we may have to do something to stop the change or adapt to it.

However, sometimes systems can change in a non-linear fashion (Figure 4.21(b)). In this situation, the system does not respond much over certain ranges of conditions, but responds strongly when conditions approach a critical level or threshold. A pest outbreak is an example of this kind of regime change. The population of the pest starts at a very low level and the damage to the crop is minimal, but because the pest population grows exponentially, it reaches a point at which there is a population explosion and the crop is devastated. In practice, that threshold level is used in integrated pest management to determine when a control measure should be taken to prevent the outbreak.

Finally, a third kind of regime change can occur when the system undergoes what is called a "catastrophic transition" (Figure 4.21(c)).[22] In this case, the response curve is folded backward, implying that for certain environmental conditions two alternative stable states are possible. Among the published examples, one can find the regime change of a grassland to a desert,[23] the change of a forest to a savanna,[24] the degradation of a healthy coral reef to one overgrown with macroalgae,[25] and the change of a clear, shallow lake dominated by submerged vegetation to a turbid lake dominated by phytoplankton.[26] In Chapter 2 (Figure 2.7), we suggested that biodiversity might exhibit such a change in the face of agricultural intensification.

Over the last century, and in particular since World War II, humans have induced more frequent and higher-intensity changes in ecosystems than at any other previous time in the history of humanity.[27] Deforestation, overfishing, pollution, the alteration of geochemical cycles, and climate change are all dramatically altering our environment. One of the salient questions that emerges from this realization is how ecosystems respond to these changes and what kinds of changes we might expect. Within the literature, it has been important to:

1 document that alternative regimes actually exist;
2 characterize the shift from one regime to another as quasi-linear, non-linear or critical (the three possibilities illustrated in Figure 4.21);
3 determine the mechanism causing the regime change; and
4 in the case of non-linear and catastrophic transitions, try to find early warning signals that might alert us to the fact that a major shift is on a near horizon.[28]

In the case of the *Azteca* system, we asked if the observed change in population density is an example of regime change. If so, what kind of trajectory does it follow? And finally, what mechanism generated such an unexpected shift?

Since the *Azteca* system is a self-organized system, a regime change might imply some sort of change in the rules of self-organization. For example, if the system is truly an example of Turing dynamics, the regime shift may be simply due to a quantitative change in dispersal rates or interaction terms. However, it also could be due to a change in the nature of the suppressor (or the activator). In the *Azteca* system, the regime change observed was a change in the population density of ant nests and

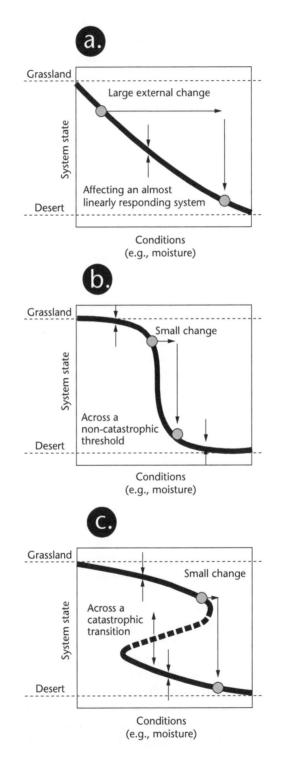

the input-forcing parameter (the environmental condition that caused this change) was the change in number of potential nesting sites, the carrying capacity of the environment. As mentioned above, the original expectation was that the population density of *Azteca*, as measured by the number of nests, would decrease with the reduction in the number of trees on the farm; yet we found the opposite, an increase in the population. In trying to understand this change, we can graph the observed population density and see how it corresponds, qualitatively, to the expected trajectory given elementary ecological theory and to the three possible trajectories if there is regime change (Figure 4.21). We thus obtain the graphs in Figure 4.22.

The logic of basic ecological theory fails us here, since the field data are the exact reverse of what would be expected according to standard theory (compare the data points to the expectation in Figure 4.22(a)). Furthermore, as we see in Figures 4.22(b), (c) and (d), the population data, although roughly similar to the catastrophic transition hypothesis, are not abundant enough to help us determine the nature of the transition. They could fit, at least superficially, all three trajectories. We need to take a different approach.

If indeed the system is organized according to the basic Turing process, it makes sense to ask whether a change in some aspect of that process is responsible for determining the trajectory of the regime change. Based on field data, we know that there are several natural enemies that can negatively impact the ant population, not just the phorids that we discussed earlier,[29] but other possibilities, further discussed below. This means that there are several potential Turing suppressors in the system. Therefore, we hypothesized that the reduction in the number of shade trees may have triggered a switch from one to another less effective suppressor, causing the regime change in the ant's population density.[30] Thus, we can conceive of a second variable, the relative effect of suppressor A versus suppressor B. The qualitatively distinct transition modes can then be represented in a three-dimensional graph visualizing the achieved population density as a function of the relative suppressor effect, as illustrated in Figure 4.23.

We can see how several trajectories can be expected depending on the effect of other species in the system. Those trajectories can lead to either the expected decline in population density with a decline in carrying capacity, or a counterintuitive increase

FIGURE 4.21 (opposite)   Qualitative illustration of ecosystem transitions resulting from external inputs, using moisture as an example of conditions and vegetation as an example of a system (with the two extreme states, desert and grassland). (a) A linear or quasi-linear response in which a large (or small) change in the input variable results in a large (or small) change in the ecosystem state. (b) A non-linear response in which a small change in the input variable results in a large change in the ecosystem state. (c) A catastrophic transition in which a catastrophic shift occurs with only a minute change in the input variable, and in which a "hysteresis" is evident (a range of input values in which the system becomes "stuck" in one or another of the regimes, such that moving left may not have the same result as moving right had had previously).

Source: Redrawn from Schaffer et al. (2001)

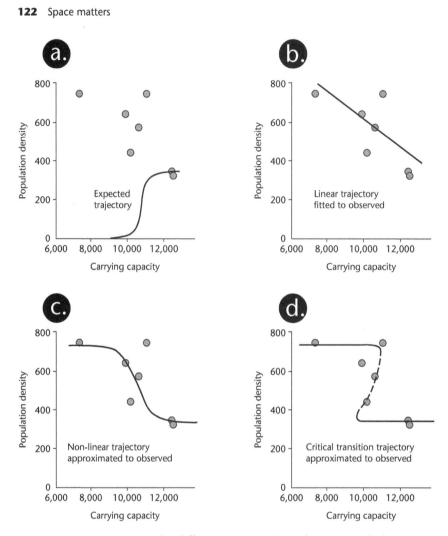

FIGURE 4.22 Expectation under different scenarios (smooth curves) and observations (small circles) of population density response to change in the carrying capacity of *Azteca* in the 45-hectare plot under the different carrying capacity of the environment, over the seven-year period from 2005 to 2011. The carrying capacity is indicated by the number of potential nesting trees. The expected response are the bold lines; the observed data are the shaded circles. (a) The bold line is the expected response under the standard cellular automata model (Vandermeer et al. (2008)), eliminating cells in proportion to the number of shade trees pruned in the experiment. Points are observed from the 45-hectare plot. (b) The linear transformation (least squares regression). (c) Eye-fit curve suggesting a non-linear transition. (d) Eye-fit curve suggesting a catastrophic transition.

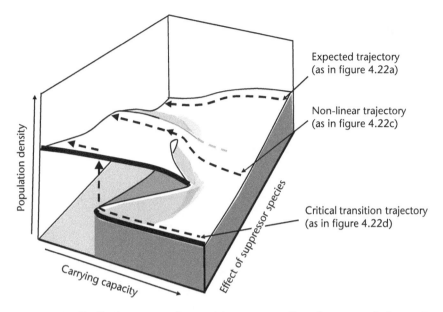

Expected trajectory
(as in figure 4.22a)

Non-linear trajectory
(as in figure 4.22c)

Critical transition trajectory
(as in figure 4.22d)

FIGURE 4.23 Qualitative nature of potential response surface: the expected changes in population density as a function of both its own carrying capacity and the combined effects of the proposed suppressor species in the system. Note that we did not include the linear trajectory (Figure 4.22(b)) to simplify the figure.

in population density with a decline in carrying capacity (non-linear and critical transition trajectories in Figure 4.23).

Changes in spatial patterns of Azteca

Beyond the evaluation of the general population density, it is quite evident that the distribution of the ants (Figure 4.24), both before and after the dramatic population flush, is clustered in space and does not have a clear central tendency to be some specific cluster size. That is to say that the population is not normally distributed in space. As we explained earlier, inspired by the many self-organized processes so commonly observed in nature, it has become common, where distributions lacking a central tendency are either expected or observed, that the parameter of a fitted power function is used to characterize the distribution.[31] Consequently, in studying the changes in spatial patterns that accompany changes in population density, it is most natural to use the paradigm of the power function, which is how we proceed here.

When examining the distribution of clusters of anything, there is a critical idea that frequently takes center stage in spatial analysis, the "percolation threshold." The idea emerges from a variety of metaphors, the most common of which is the propagating forest fire. Imagine an ideal forest of ideal trees placed at random in an area. If you set fire to one of the trees on the edge of the area, what is the probability that

…

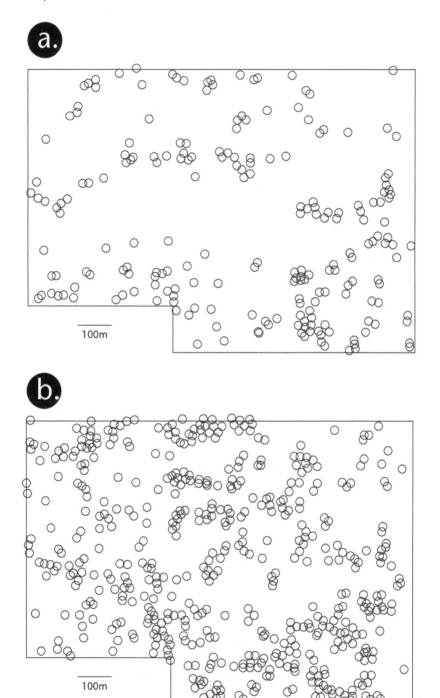

FIGURE 4.24 Distribution of the populations in (a) 2005 (same as Figure 4.7) and (b) 2011. Note the dramatic (almost threefold) increase in the population density.

the fire will spread all the way to the other side? Obviously, that probability will be a function of the total number of trees in the forest. The basic idea is illustrated in Figure 4.25. As illustrated in the figure, the percolation threshold refers to the situation when a cluster of points expands from one edge of the lattice to the other. That situation is also called a "spanning cluster." We discussed this briefly in Chapter 3 in the context of the connection between two forest fragments.

The percolation threshold concept applies in all sorts of contexts, not just forest fires. When dealing with clusters of points, like the ant nests in our plot, the conceptual utility of a percolation threshold may emerge, for example, in considering a natural enemy that might attack the ants. If there is a "spanning cluster," the natural enemy can penetrate over the entire area, easily moving from one nest to another nearby. The percolation threshold thus takes on a special meaning in a spatial analysis of clustering.[32]

When analyzing the distribution of clusters of points in a two-dimensional plane, such as the *Azteca* nest data we presented in Figure 4.24, we encounter a technical problem from the start. Since the nest positions are represented as points in space, precisely what constitutes a cluster of nests depends on the scale of influence surrounding the nest itself. Clearly, two nests located 500 meters from one another are not in the same cluster, and two nests located one meter from one another are indeed in the same cluster. But what is the relevant scale within which nests are thought of as being in the same cluster? This is what we refer to as the cluster scale, as we discussed in a previous section of this chapter (Figure 4.16). For example, using the field data from 2011, cluster scales of 24 meters and 38 meters are illustrated in Figure 4.26. Obviously, the two distributions are dramatically different, yet they result from the same basic data, the positions of each nest in the 45-hectare plot.

Note that at a scale of 24 meters, there is no single cluster that spans the entire plot (Figure 4.26(a)), but at 38 meters, there is one giant cluster that spans top to bottom and left to right (Figure 4.26(b)). The cluster size distribution will depend on the scale that is chosen to decide which nests belong to the same clusters. So, for example, comparing Figure 4.26(a) with Figure 4.26(b), the cluster scale is the diameter of the circle surrounding each nest. To examine the difference between observed spatial patterns and a random expectation, we generated 100 artificial nest distributions based on the nest densities in each year for which we had data (from 2005 to 2011), but positioning each nest randomly. After calculating the critical cluster scale – that is, the smallest scale that gives a spanning cluster – for each year based on random positioning of the nests in space, we computed best-fit power functions for each of the artificial distributions. Those power functions (the collection of gray lines), along with the field data are presented in Figure 4.27. Note that all power function fits are based on the critical cluster scale. It is evident that the two "regimes," nest densities before and after the tree cutting, are distinct from one another and the three-year transition period moves rather smoothly from one to the other, both in the parameter of the power function (the slopes of the lines) and the position of the power function relative to the random expectation. Furthermore, they are quite distinct from what would be expected from a random distribution.

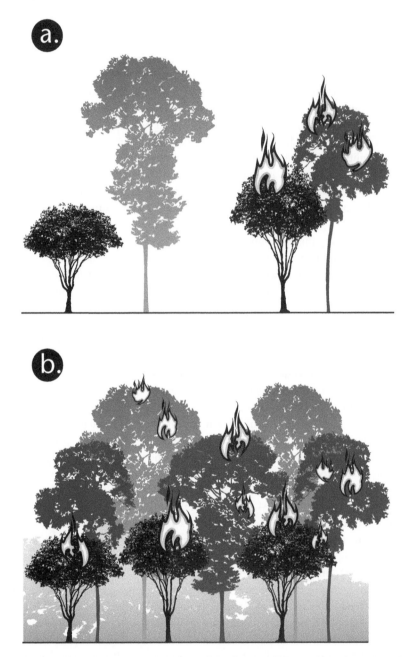

FIGURE 4.25 Illustration of the idea of a percolation threshold. (a) This forest has a low density of trees and thus a fire started at the edge of the plot will be extinguished before it penetrates too far. (b) This forest, by contrast, has a population density of trees that is large enough so that a fire started at one edge of the forest can "percolate" all the way to the other edge. The critical density of trees where a fire is almost certain to penetrate from one edge to the other is referred to as the percolation threshold.

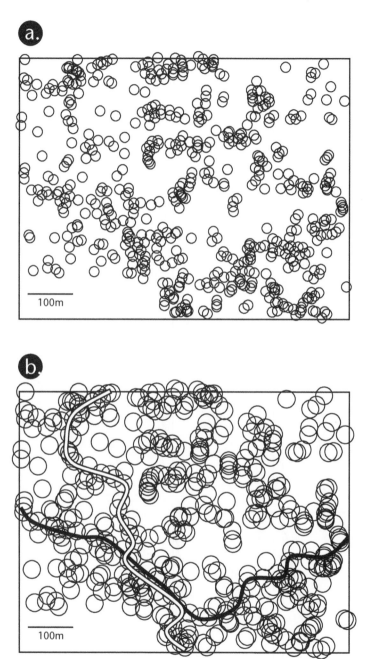

FIGURE 4.26 Distribution of ant nests in the 45-hectare plot in 2011 at two different cluster scales. (a) Cluster scale of 24 meters. (b) Cluster scale of 38 meters. Note that at the 38-meter cluster scale there is clearly a spanning cluster (percolation); indeed, 38 meters is the critical cluster scale at which percolation occurs. Two percolating pathways are illustrated.

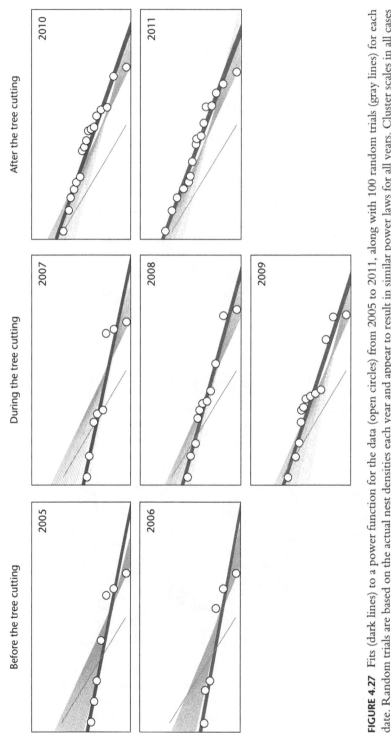

FIGURE 4.27 Fits (dark lines) to a power function for the data (open circles) from 2005 to 2011, along with 100 random trials (gray lines) for each date. Random trials are based on the actual nest densities each year and appear to result in similar power laws for all years. Cluster scales in all cases are based on the spanning cluster scale (the percolation point). Constant lines (power function exponent = −1.0, with intercept fixed at 2011 level) are constant over all panels, and represent a point of comparison from panel to panel.

Source: Reprinted from Jackson et al. (2014b)

These spatial pattern changes also correspond to what would be expected if there had been a change in the Turing suppressor (i.e., a shift along the effect of suppressor species axis in Figure 4.23).

This rather detailed analysis of the spatial population dynamics was unable to clearly distinguish among the various possibilities regarding the nature of the regime change (i.e., the patterns in Figure 4.21). However, the good fits of the data to a power function during all years (Figure 4.27), along with dramatic changes in the parameter of that function (the slope of the line on the log–log plot), suggest that something new has been generated during this regime change. Comparing the power function fit from 2005 to that of 2011, we see a change in the slope of the power function from a less steep to a more steep slope (in 2005/2006 the slope was about −0.3 and in 2010/2011 it was reduced to about −0.8).[33] Recalling Figure 4.17, when the slope of the power function moves from a less steep to a more steep slope, the expectation is that any population using the self-organized habitat will tend to move from a source–sink population to a metapopulation. This tendency will be partly counteracted by any change in the total population density of the self-organizing species (i.e., in this case, the *Azteca* ants), but the point is that the combined changes in the power function parameter and the population density imply possibly dramatic changes in the nature of the engineered habitat. We can see this if we superimpose the graphs in Figure 4.27 on top of one another, as we do in Figure 4.28.

Thus, it appears that the transition has changed the basic pattern from a probable source–sink population to a metapopulation. Clearly, this is odd since the overall size of the *Azteca* population has increased. However, ironically, the largest cluster size (according to the power function approximation) has actually decreased during the transformation, even as the overall number of singletons has increased. If the critical cluster size for any of the organisms living within the *Azteca* patches (see Figure 4.15) is located between the two largest cluster sizes (the dotted line in Figure 4.28(b)), its characteristic as a source–sink population will have changed and it will now be a metapopulation. Note that such a change would have happened even if the organism in question itself did not change any of its vital statistics (birth rate, dispersal rate, etc.). It results only from the spatial structure generated by the

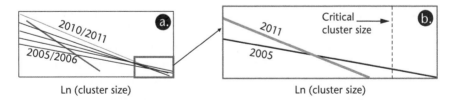

FIGURE 4.28 Superimposition of the power functions of Figure 4.27 on top of one another. (a) All years, where it is evident that the intercept systematically increases from 2005/2006 through 2010/2011. (b) Insert from (a), where it is evident that there has been, theoretically, a decrease in the expected largest cluster size over the years of the transition.

self-organizing process of *Azteca*. Furthermore, this dramatic change might in turn imply that there could be a change in the nature of the suppressor agent that is involved with the creation of the spatial pattern in the first place, a subject to which we now turn.

Regime change and the assumed Turing suppressor

Based on the observed changes in the number of nests of *Azteca* on the coffee farm, and the spatial analyses described above, it is certainly arguable that there was a regime change, although the change itself was contrary to the original expectation. And from our detailed spatial analysis, it seems that the system changed in a more general way than just a decrease in carrying capacity. Could that change have involved a change in the nature of the Turing suppressor in the system?

Since the CA model we had developed for this system was available and since we already had good estimates of the parameters, it was a simple matter to imitate the natural experiment by removing the same number of the cells in the system as the trees had been eliminated. It could have been that the model would reflect the unusual result of an *increase* in the population density when faced with a dramatically *decreased* carrying capacity. However, when we performed this experiment with the model, the overall population collapsed rather quickly (Figure 4.22(a)). The transformation of the farm had effectively provided us with an experiment that we could use to test our original model of the system, and we found that the model predicted a population collapse whereas the field data reported more than a doubling of the population.

The contradiction between the theoretical expectation and the field data forced a rethinking of the model itself. Based on observations that, under stress, *Azteca* could move its colony further away than the closest neighboring trees (as assumed in the original CA model), we modified the model such that individual nests could migrate further than the immediate surrounding cell (the Moore neighborhood), as illustrated in Figure 4.29. With this modification, it was simply a matter of finding parameter values that could produce the observed trajectory: a reduction in carrying capacity resulting in an increased population density, and that could generate the same qualitative behavior of the power function exponent that we observed in nature.[34]

With this simple modification, we were able to replicate the observed increase in population density with a reduction of carrying capacity (Figure 4.30). It appears that the reduction in the density of potential nesting sites causes the ants to move their budding nests, on average, further from their original site. Inadvertently, the ants thus form clusters that are not as dense as previously, thus partially "escaping" the spatially explicit density dependence of the controlling agent.

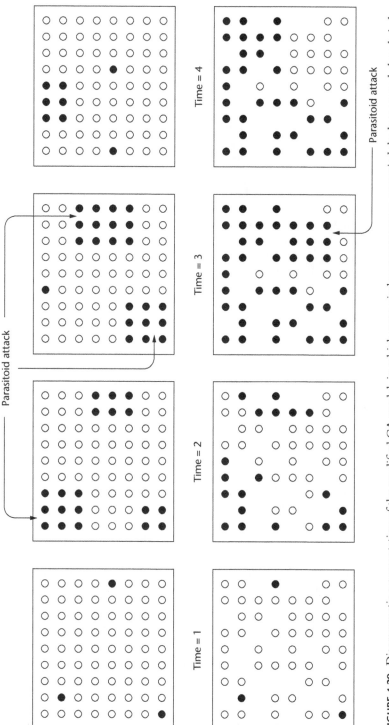

FIGURE 4.29 Diagrammatic representation of the modified CA model, in which open circles represent non-occupied shade trees and closed circles represent shade trees occupied by an ant nest. The top row is before the thinning of the shade trees; the bottom row is after a 30 percent reduction of shade trees. The parasitoid finds the ants only when they occur in relatively dense patches. With fewer trees (lower row), the ants need to move further in order to expand their colony, but in doing so, they create more diffuse patches that the parasitoids are unable to find as efficiently. The consequence is a higher population density even though the background habitat has actually been reduced.

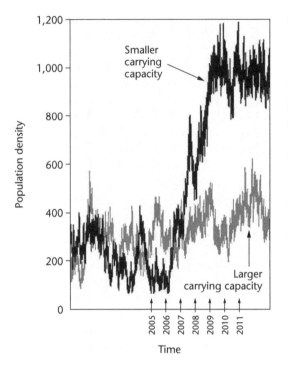

FIGURE 4.30 Results from the modified CA model. The original model was modified to allow for further expansion of nests in space. The gray run is with the original, larger, carrying capacity. The black run begins with the larger carrying capacity with 30 percent of the trees eliminated halfway through the run. The position of the years is an approximate representation of the correspondence of model output with the qualitative nature of the field results.

In the end, this analysis suggests that:

1   There was indeed a regime shift in the *Azteca* system, triggered by the cutting of the trees on the coffee farm.
2   This shift occurred relatively quickly, going from a stable state in 2005/2006, transitioning from 2007 to 2009, and arriving at another stable state in 2010/2011.
3   The regime change followed a relatively smooth non-linear trajectory similar to that in Figure 4.22(c).

There may have been a change in the suppressor involved in the Turing process. In a classical complex systems dialectical fashion, it could be the case that the change in management of the farm pushed the system in the direction of changing spatial structure, which induced a change in the force responsible for the generation of the spatial pattern, which in turn created the final spatial pattern within which it (the organism exerting the "force") must live, perhaps now as a metapopulation.

Alternatives for the suppressive force: food web elements

Our original proposal was that the phorid parasitoid fly was the key natural enemy that acted as the suppression agent for the functioning of the Turing-like spatial pattern formation.[35] Subsequent work (some of which was described in the previous

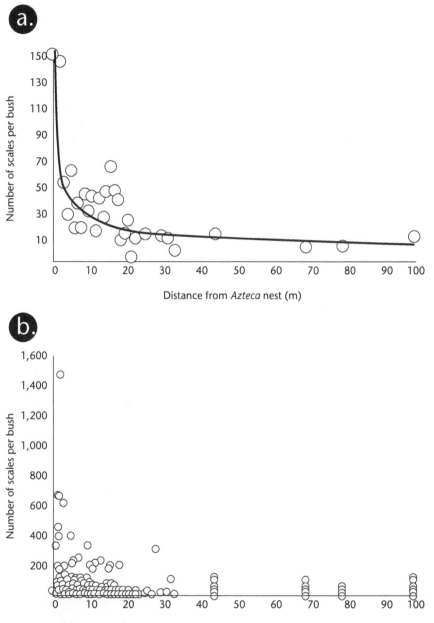

FIGURE 4.31 Scale population densities as a function of distance from an *Azteca* nest. (a) Mean values for 1-meter intervals (the curve is the best fit power function). (b) Population densities for individual trees. Note the pattern of very large numbers on some trees and very small numbers on others. Also note that the low numbers (approximately 50–100 individuals) remained common even at a distance up to 100 meters from the central nest.

section) suggested different possibilities, two of which have been since studied intensively,[36] and both of which differ from the phorid in that the control they exert on the *Azteca* ants is indirect, through an attack on the scale insects, which are effectively the main food source of the ants. The distribution of scale insects is strongly affected by the distribution of the ants.[37] In one study, we counted the number of scale insects on coffee bushes at increasingly large distances from a central ant nest, as shown in Figure 4.31. Note that all bushes with more than 300 scales are located within 5 meters of the ant nest. Indeed, there seems to be a pattern in which the especially dense populations of scales is within 5 meters, very low densities (but never converging on zero) are further than 20 meters, and intermediate densities between 5 and 20 meters, corresponding to a hypothesized pattern of active tending by *Azteca* (within 5 meters), tending by other ants plus dispersion from the *Azteca* zone (between 5 and 20 meters), and only tending by other ants (further than 20 meters). Thus, we see a small zone around the *Azteca* nest (about 10 meters in diameter) in which the scale insects are extremely common (and thus a likely target of predators and diseases), and a zone away from the *Azteca* nest with much lower abundances.

Within this pattern, we propose that there are two natural enemies of the scales that could conceivably act as the Turing suppressor. One is a fungal disease, the other a myrmecophilous beetle.

The effect of a fungal disease on spatial patterns

The white halo fungus, *Lecanicillium lecanii*, is a fungus known to attack a variety of arthropods, many of which are important agricultural pests,[38] including the green coffee scale, *C. viridis*, in coffee.[39] When attacking the green coffee scale, the fungus creates a "halo" of white tissue around the edge of the insect, whereby it gets its common name (Figure 4.32). It often creates local epizootics (epidemics), killing nearly all of the scale insects on a coffee bush or a small group of neighboring bushes. Thus, this fungus may reduce the amount of carbohydrate food available to an ant colony, resulting in an indirect negative effect on colony survival and/or the tendency to move.

Knowledge of the spatial dynamics of the fungus, in terms of incidence and severity, is important for assessing its potential to influence the self-organization process. The hypothesis is that as scale insects become locally very dense (which they only

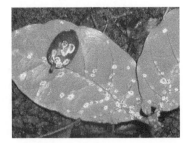

FIGURE 4.32 White halo fungus (*L. lecanii*) attacking the green coffee scale insects (*C. viridis*). The circle of white fungal tissue around each individual scale insect gives the fungus its name.

do when ants are vigorously protecting them), they are subjected to a higher probability of attack from this fungal disease. Yet the source of the fungus must be nearby to initiate a local epizootic. Thus, how the spatial distribution of the fungus changes throughout the course of a local infection is a basic component of *L. lecanii*'s natural history and a clear determinant of its impact on the spatial dynamics of the ant. Observations on the spatial distribution and its changes over time were made at two sites, both of which were located in an area where several nests of the *Azteca* ants had been located previously.[40]

The distribution of both scale insects and the white halo fungus disease for the two sites is shown in Figure 4.33. From previous sampling, we knew that the abundance of scale insects decreased as a function of distance to the tree in which the *Azteca* colony was located (Figure 4.31). At a distance of more than 10 meters, the majority of coffee bushes have only a few scale insects, with only an occasional tree containing a larger cluster (actually, these are always tended by a different ant species,

SITE A   SITE B

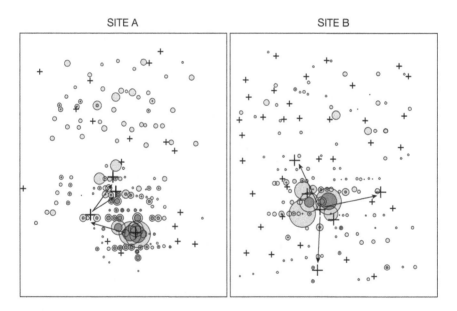

FIGURE 4.33 Representation of two intensively sampled sites in the study area (final sample in June/July 2008). Site A was occupied by *Azteca* at least since 2004, while site B was newly occupied some time within the past three years. Lightly shaded bubbles are proportional to the number of scale insects per branch of a coffee bush located at that particular coordinate. Darkly shaded bubbles are proportional to the intensity of the white halo fungal disease (caused by *L. lecanii*) on that bush. Large crosses indicate the position of shade trees occupied by *Azteca* and small crosses indicate position of unoccupied (and presumably occupiable) shade trees in the system. Arrows indicate the known direction of the spread of the ant colony from our own records. Plots are both 40 × 50 meters.

Source: Modified from Jackson et al. (2009)

although never at the level reached under the protection of *Azteca*). The pattern of white halo fungus distribution, coupled with the pattern of migration of the *Azteca* nests in these two plots, strongly suggests that the ant nest moves partially in response to the fungal infection of its main food source, leaving a trace of scale populations devastated by the disease near the locus of the original ant nest site, and scale populations built up but not yet infected nearer to the more recently occupied shade trees.

From this data, a general picture emerges of *Azteca* establishing a nest in a new shade tree and searching out local coffee bushes for local concentrations of *C. viridis*. Having encountered local concentrations, the mutualistic effect of the ant permits the scales to build up to extremely high population densities in bushes near to the shade tree containing the new nest. However, these extremely large populations of scale insects become targets for the epizootic development of the white halo fungus, which, once established in an area, appears to become endemic, following the ant colony around as new shade trees are occupied, perhaps eventually resulting in the death of an entire cluster of ant nests, or a large-scale abandonment of the area and migration to some more distant site. In this sense, the fungus acts as the suppressor in the Turing-like process.

To test the plausibility of this hypothesis, we modified the ant CA model described earlier in this chapter[41] to include the spatial distribution and dynamics of *L. lecanii*. Each cell in the modified model can be in one of three states: empty, occupied by an ant nest whose scale insect populations are free of *L. lecanii* or occupied by an ant nest whose scale insects are infected by *L. lecanii*. The key difference between this model and the previous one is that the ant nest mortality, which in the original model was a probabilistic function of the number of neighboring ant nests, is now a function of *L. lecanii* infection. If infected, the probability of ant nest mortality is equal to the virulence of the fungus; otherwise, the probability of nest mortality is zero. If a nest at an infected site survives one time step, the site remains infected in subsequent time steps until the nest dies. In other words, there is no recovery.

Employing a genetic algorithm,[42] we searched the parameter space of "transmissibility" and "virulence" of the fungus, searching for outcomes of the CA model that correspond to the population density and spatial pattern observed in the field. Interestingly, a relatively narrow band of combinations of these two parameters provided us with the approximate distribution seen in the field, with most combinations either leading to the extinction of the ant or a combination of population density and spatial distribution that was clearly distinct from what we observed in the field (Figure 4.34).

Although the range of the disease parameters (virulence and transmissibility) that provide for an approximate fit to the field data is relatively narrow, nonetheless the basic natural history (as represented qualitatively in the CA model) is *capable* of producing the self-organized clustering pattern of ant nests that we observed in the field. In Figure 4.35, we show a snapshot of the results of the model with parameters in the high-fitness region (of Figure 4.34). As can be seen, the qualitative nature of the nest clustering we observe in the field (as, for example, in Figure 4.24) is

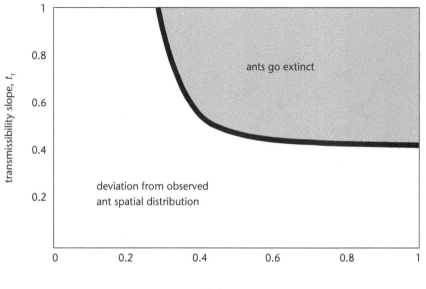

FIGURE 4.34 Graphic illustration of a range of parameter values and their outcome in a CA model. The thick black line represents the high-fitness region (according to the genetic algorithm methodology) in v, t_1 parameter space in which it is possible to generate spatial patterns of ant nests that are qualitatively and quantitatively similar to the pattern observed in the field. Away from this region, it is not possible to generate the observed spatial distribution using the basic parameters of the original model.

Source: Modified from Jackson et al. (2009)

reproduced with this mechanism added to the basic CA model, instead of the direct density-dependent mortality factor we originally used to approximate the action of the phorid fly.

It is also worth noting that the dynamics and propagation of *L. lecanii* may in fact create the conditions for its own survival. Since epizootics only occur when the scale insect population reaches a critical size, and since that critical size only occurs when ants are tending the scales, it is clear that ants are necessary for the production of the epizootics. If the fungal pathogen drives the shifting pattern of the ant–scale mutualism, it could be said that the fungus creates the background conditions that are necessary for its survival because of its potential to influence the spatial distribution of *Azteca* nests.

The effect of a myrmecophilous beetle on spatial patterns

Of particular interest is yet another alternative element of the basic ecological inter-action web, the myrmecophilous beetle *Azya orbigera*. The details of the population biology of this species will be extensively covered in Chapter 5. For the moment,

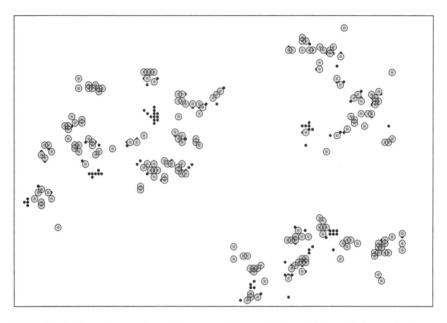

FIGURE 4.35 Example snapshot of the ant and fungus CA model. The black dots indicate the locations of ant nests. The shaded circles indicate nest sites infected by *L. lecanii*. Note that the model only considers the presence of the fungus and not its intensity. Note how the model reproduces the clustered nature of the distribution of ants.

we just note that this beetle is a major predator of the green coffee scale. However, for reasons of its complicated life cycle, it requires the patches of ants for its regional persistence. Adults fly far and feed on relatively isolated scale insects, but the larvae require the concentrated scale insects that occur only when the ants are present. Due to its voracious consumption of scale insects, this beetle could also act as the suppressor of the ants in the postulated Turing dynamic process that generates the observed spatial pattern of *Azteca*. Given that, it is possible that – much like we propose for *L. lecanii* – this predaceous beetle contributes to the formation of the spatial pattern that it needs to survive.

This basic idea brings together two relatively distinct literatures. On the one hand, there is a substantial literature showing that local interactions can sometimes create complex and surprising self-organized spatial patterns.[43] For example, the distribution of species in simple predator–prey and parasitoid–host deterministic models in homogeneous environments can form spiral waves, clustered distributions, crystal lattices and chaotic patterns.[44] On the other hand, and mainly in a separate literature, it has often been noted that spatial heterogeneity can allow coexistence of predator–prey and parasitoid–host associations that are otherwise prone to local extinctions.[45]

Following the basic research strategy described above for *L. lecanii*, we modified the basic CA model, this time in a relatively more complicated fashion, including larvae and adults of the beetle as separate categories. In our model, the beetle acting

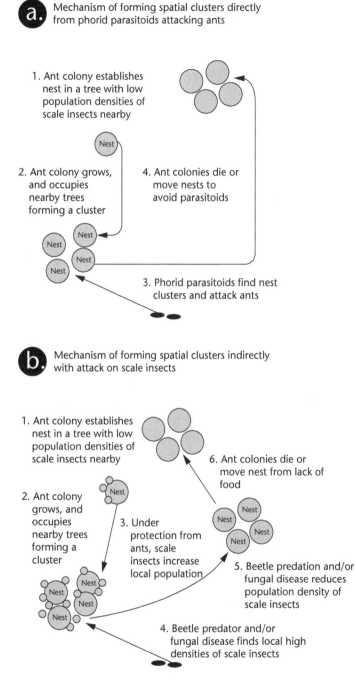

a. Mechanism of forming spatial clusters directly from phorid parasitoids attacking ants

1. Ant colony establishes nest in a tree with low population densities of scale insects nearby

2. Ant colony grows, and occupies nearby trees forming a cluster

4. Ant colonies die or move nests to avoid parasitoids

3. Phorid parasitoids find nest clusters and attack ants

b. Mechanism of forming spatial clusters indirectly with attack on scale insects

1. Ant colony establishes nest in a tree with low population densities of scale insects nearby

2. Ant colony grows, and occupies nearby trees forming a cluster

3. Under protection from ants, scale insects increase local population

4. Beetle predator and/or fungal disease finds local high densities of scale insects

5. Beetle predation and/or fungal disease reduces population density of scale insects

6. Ant colonies die or move nest from lack of food

FIGURE 4.36 Summary of the proposed mechanisms of suppression of colony expansion. (a) Direct mechanisms involving attacks on ants. (b) Indirect mechanism involving attacks on scale insects.

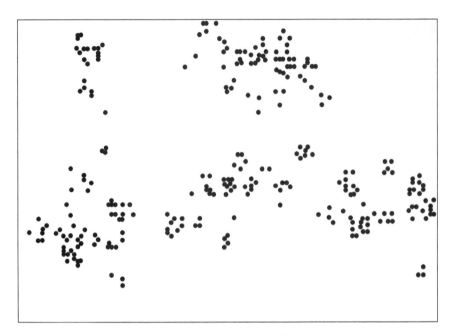

FIGURE 4.37 Model distribution of ant nests under the density-dependent control of the predaceous beetle *A. orbigera* in a modified CA model.

as an indirect cause of ant nest mortality through its predatory activity on the scale insects is the control force that counteracts the expansion of ant nests by satellite nest formation, and contributes to the formation of the clustered spatial distribution of ant nests (Figure 4.36). Thus, our model can qualitatively generate the cluster formation of ant nests in nature, one snapshot of which is shown in Figure 4.37.

In this model, the scale populations in areas with ant nests can increase to relatively high values compared to areas with no ants. In areas with ant nests, where the protection by ants excludes most other natural enemies except the larvae of *A. orbigera* (because it is protected from ant predation), scale populations thrive and grow in size. This is followed by an increase of the beetle population, which eventually imposes sufficient predation pressure to cause scale populations to decrease. These dynamics leave ants without their main carbohydrate source, and in doing so, increase the probability of mortality. Due to the build-up of beetle populations in areas with ants, these areas act as sources for beetle adults that then disperse to the rest of the farm and contribute to the maintenance of low scale populations, as happens in the real world in addition to the model. Due to the tendency of beetles to concentrate in areas with large clusters of ant nests (see Chapter 5) and the diffusive nature of beetle migration, there is a stronger effect of beetle predation in larger clusters of ant nests, resulting in a density-dependent effect. The ability to respond in a density-dependent fashion to ant nests renders this beetle potentially important for the formation of ant nest clusters. Furthermore, the very heterogeneity of the beetle's food source (the green coffee scale, which is concentrated in clusters of

Azteca nests) is extremely important for the population of the beetle to persist.[46] However, in this case, it is the action of the predator itself that causes the heterogeneity of the prey, which is to say that the beetle causes the formation of a spatial pattern in its prey, the very spatial pattern needed for that beetle to survive.

Summary

In this chapter, we presented a case study of the ant *Azteca* to demonstrate why space matters in ecology. First, we showed that *Azteca*, an important predator in the coffee agroecosystem, is distributed in a non-random fashion on the coffee farms. Then, with the help of a cellular automata (CA) model, we argued that the observed aggregated pattern in *Azteca* is the result of a process of self-organization generated by a Turing-like process of activation (the budding of ant nests) and suppression (the density-dependent mortality caused by natural enemies).

However, this species, like all species, does not live in isolation; it interacts with a whole suite of other organisms. In fact, the *Azteca* spatial pattern creates the conditions in which other species (insects, spiders, fungi, etc.) live, and dictates to some extent the nature of the population dynamics of those other species, providing a source–sink background for some and perhaps a metapopulation structure for others.

The decision of the farm manager to intensify the farm and cut 30 percent of the shade trees provided us with the opportunity to study what happens to a self-organized population when its carrying capacity – in this case, the number of potential nesting sites – is reduced dramatically. Based on simple ecological theory, we expected the *Azteca* population to decline dramatically, but surprisingly, the number of *Azteca* nests tripled! We then examined this change in the context of regime change and asked whether the transition from low population density to high population density follows a linear, non-linear or critical transition trajectory. Although the type of data we had was not detailed enough to determine precisely the type of trajectory, by examining the power law relationship between the cluster size and their frequency, and in particular the slope of the power law, we were able to tentatively suggest that the increase in the number of ant nests followed a smooth non-linear trajectory. Furthermore, this counterintuitive result made us reconsider our initial model. We modified the original CA model, incorporating natural history observations about how far the ants move to establish new nests and the effect of other natural enemies that affect the ants indirectly by attacking their scale hemipteran mutualists. With these modifications, we were able to replicate the counterintuitive results that were observed in the field.

Our studies on the spatial ecology of *Azteca* also suggest that some organisms can cause a spatial pattern formation that is essential for their own survival. An example of this is the coccinellid beetle *A. orbiguera*, which preys on the green coffee scale that is tended by the *Azteca* ants. By reducing the population of the hemipterans, this beetle acts as a density-dependent suppressor, *à la* Turing, therefore contributing to the formation of the clustered distribution of the ant. However, it is precisely this spatial distribution that is responsible for the survival of the beetle, since its larvae can live and grow only in the ant patches where its prey is abundant (and, as we will see in Chapter 5, the ant protects the beetle larvae from parasitoids). In addition to being a fascinating story in spatial ecology, it turns out that this dynamic relationship is important for the biological control of the green coffee scale at the level of the entire farm. Space matters! Organisms are not only constantly surrounded, affected and shaped by it, but they in turn affect and shape space for themselves and other organisms in constantly shifting dialectics.

Notes

1 For example, Rohani et al. (1997); Rietkerk et al. (2002); Rietkerk and van de Koppel (2008).
2 Levins and Lewontin (1985: 141).
3 This species was previously identified as *Azteca instabilis* and many of the publications on this species use that name. To avoid confusion, the species will henceforth be referred to simply as *Azteca*.
4 Vandermeer et al. (2008).
5 Turing (1952). As a sad historical footnote, it is worth noting that Turing contributed in a major way toward the defeat of the Fascists in Europe by cracking the Nazi secret code. His heroic work in this regard could not shield him from lethal homophobia. As a gay man, he was hounded by prejudice (Hodges (1992)). The homophobic culture of England in the 1940s was unable to fully recognize this genius due to his sexual orientation, and eventually drove him to suicide, thus depriving us of the many other insights that might have come from his brilliance, reminding us of the danger of zealotry combined with ignorance, a danger we still face today.
6 For example, Tilman and Kareiva (1997); Vandermeer and Goldberg (2013).
7 As quoted by Philibert (2005: 1, emphasis added by Philibert).
8 For example, Tilman and Kareiva (1997); Vandermeer and Goldberg (2013).
9 Kondo and Miura (2010).
10 Alonso et al. (2002).
11 Allen (2012).
12 Vandermeer et al. (2002b); Perfecto and Vandermeer (2006); Philpott et al. (2012).
13 Philpott (2005b).
14 Feener (2000); Philpott (2005b); Mathis et al. (2011); Hsieh and Perfecto (2012).
15 Pascual (1993); Pascual et al. (2002a, 2002b); Pascual and Guichard (2005).
16 For example, Bak (1999); Pascual et al. (2002a, 2002b); Kéfi et al. (2007, 2011).
17 Vandermeer et al. (2010a).
18 Pulliam (1988).
19 Perfecto and Vandermeer (2002); Perfecto et al. (2009).
20 Jackson et al. (2014a).
21 The classical reference on catastrophe theory is Thom (1972).
22 Scheffer et al. (2001, 2009).
23 van de Koppel et al. (1997); Kéfi et al. (2007).
24 Noy-Meir (1975); Walker (1989); Staver et al. (2011).
25 Done (1992); Knowlton (1999); McCook (1999).
26 Scheffer et al. (1997); Carpenter et al. (1999).
27 Vitousek et al. (1997).
28 Scheffer et al. (2009, 2012).
29 Perfecto and Vandermeer (2008b); Vandermeer et al. (2010a).
30 Jackson et al. (2014a, 2014b).
31 Pascual et al. (2002a, 2002b); Newman (2005).
32 Pascual et al. (2002a, 2002b).

33 These slopes are obviously dramatically different from the ones we reported in 2008 (Vandermeer et al. (2008)), the reason being that here we are using the cluster scale that results in a percolation threshold (or a spanning cluster) for each year.
34 Jackson et al. (2014b).
35 Vandermeer et al. (2010a).
36 Liere and Perfecto (2008); Jackson et al. (2009, 2012b); Liere and Larson (2010); Hsieh and Perfecto (2012).
37 Vandermeer and Perfecto (2006).
38 Jackson et al. (2009).
39 Uno (2007); Vandermeer et al. (2009); Jackson et al. (2012b).
40 Jackson et al. (2009).
41 Vandermeer et al. (2008).
42 Jackson et al. (2009).
43 Rohani et al. (1994, 1997); Hassell et al. (1995); Vandermeer and Yitbarek (2012).
44 Hassell et al. (1991); Comins et al. (1992); Rohani et al. (1994); Rohani and Miramontes (1995).
45 Nicholson and Bailey (1935); Noy-Meir (1975); Tilman and Kareiva (1997); Dieckmann et al. (2000); Gurney and Veitch (2000); Murdoch et al. (2013).
46 Liere and Larson (2010); Liere et al. (2012).

5

WHO'S EATING WHOM AND HOW

Trophic and trait-mediated cascades
in the coffee agroecosystem

Birds: from icons of biodiversity to functional components of agroecosystems

One of the more important observations about biodiversity and coffee agroeco-systems was that of Russell Greenberg of the Smithsonian Migratory Bird Center, connecting the decline of eastern songbirds with the transformation of coffee farms in Central America, as we noted in Chapter 2. The conservation community could hardly ignore coffee intensification when one of the most iconic symbols of bio-diversity, birds, was under threat in this evident albeit indirect fashion. However, birds are more than just biodiversity emblems. Early studies on the effects of birds on arthropods in temperate ecosystems concluded that birds exert little control over arthropods.[1] However, more recent studies and meta-analyses have shown that vertebrate insectivores, including birds, can be important in reducing arthropod populations.[2] Furthermore, recent studies in tropical forests and coffee agroforestry systems found that the effects of birds and bats,[3] and of birds and lizards,[4] is additive. These and other studies argue that vertebrate insectivores perform important functions as regulators of arthropods in these systems and may play an important role in the control of herbivores in forests and agricultural systems.[5]

Since most species of vertebrate insectivores eat many species of arthropods (i.e., are polyphagous), they usually can also consume arthropod predators, (i.e., are omnivorous).[6] Note the difference between the categories polyphagous, which simply means eating many different species of food, and omnivorous, which is usually taken to mean eating at distinct trophic levels. Most vertebrate insectivores are omnivores. Indeed, it is difficult to imagine a bird, bat, lizard or frog somehow systematically avoiding spiders, which are almost always predators, while at the same time eating insects, many of which are herbivores. Consequently, it is possible that any direct negative effects these vertebrate predators have on herbivores, and consequent indirect positive effect on plants, could very well be counterbalanced by the reduction of arthropod predators, especially spiders.[7]

Nevertheless, studies conducted on coffee farms in the Neotropics have shown that birds are responsible for the reduction of the main insect pest, the coffee berry borer. In a study of an 18-hectare coffee plantation in Jamaica, Mathew Johnson and his students estimated the economic value of the reduction of the coffee berry borer by birds in a growing season to be US$310 per hectare, or a total of US$5,580 for the whole farm.[8] Furthermore, using experiments in which birds were excluded from certain areas and allowed to forage in others, our team in Mexico demonstrated that birds can prevent pest outbreaks in the coffee agroecosystem.[9] One possible explanation for these results is that insectivorous birds, though omnivores, may have a density-dependent response to prey. That is, they consume whatever prey is at highest densities. Since insect herbivores are bound to have higher densities than insect and spider predators, birds will tend to focus on them or respond to their rapid increase in populations. Thus, predatory omnivores may be "switching" from a focus on herbivorous insects to insect and spider predators depending on relative population densities. Clearly, omnivory in simple food webs can have important implications for biological control. Consequently, its systematic study, as has been the focus of pure ecology in recent years, is warranted for practical reasons. We thus begin by conceptualizing it as a generalization of various classical food web structures.

Omnivory and its place in food web structure

Theoretical framework: omnivory and its relatives

Sometimes omnivory is evident through nothing more than casual observation.[10] For example, it is common knowledge that small vertebrates such as birds and lizards are voracious consumers of both predatory spiders and herbivorous insects. Furthermore, the phenomenon has been documented in a variety of ecosystems,[11] with experimental studies showing it to be generally important in nature.[12] Given its evident existence in many ecosystems, ecologists have pondered its significance with regard to a variety of ecosystem properties. In this context, something of a debate has emerged as to whether omnivory is a stabilizing or destabilizing force in ecological systems.[13] The details of this debate are complex and eclectic, but basic theory suggests that omnivory can be either stabilizing or destabilizing, depending on background conditions.[14] The particular case of insectivorous birds, which invariably also eat spiders, in the coffee agroecosystem occupies an obvious niche in this theoretical edifice.

Omnivory is a concept that has three qualitatively distinct underlying structures, all emerging from classical three-dimensional systems. Consider the case of insectivorous birds mentioned above. They generally eat arthropods, which includes spiders, which are also predators; thus, the implication is that bird predation will always be an example of omnivory. However, much depends on the details, and sometimes those details can be important for understanding the dynamics of communities. For example, if the herbivores in question are very small (e.g., aphids, early instar bugs or small caterpillars) their body size may be too small for birds to bother with, but hunting spiders can catch them easily. Thus, the system would be effectively a trophic chain with the herbivores being eaten by the spiders, which are in turn eaten by the birds. Alternatively, the spiders may be very large and intimidating for birds, such that both spiders and birds eat the herbivores, but there is no consumption of spiders by birds. The birds and spiders would thus be related as classical competitors for the common resource, herbivores. Thus, we can easily imagine a range of food web structures ranging from trophic chain to classical resource competition.

There is a third structure that emerges from simple topological considerations. Suppose the herbivores are very large and the spiders are unable to capture and eat them. In this case, the birds may be predators on both spiders and herbivores, but there is no trophic connection between the spiders and the herbivores. The important distinction is between polyphagy (eating different things) and omnivory (eating at different levels in a food web). Perhaps the distinction is a bit subtle, but the dynamic consequences can be important.[15]

Omnivory is thus seen as the topological connection between three distinct three-dimensional forms: trophic chain, competition and polyphagy. In Figure 5.1, we illustrate these three structures. Furthermore, particular circumstances can result in the transition from one structure to another – that is, from competition to polyphagy, from competition to trophic chain and from polyphagy to trophic chain, and vice versa. In Figure 5.1, we use the size of the organism as a way to illustrate the difference between the various structures. Size is here metaphorical in that it is certainly not the only factor that determines whether one organism can or cannot eat another. Chemical defenses and behavior, among other things, can also influence what a particular animal is able to eat and therefore can also influence the particular structure of the three-dimensional system.

The theoretical literature on omnivory has been concerned with whether omnivory is a stabilizing or destabilizing force in food webs – whether the addition of an omnivorous component will tend to destabilize a food web that has been previously stable, or whether it can stabilize a food web that has been previously unstable. The problem has always been that this simple question is not sufficiently refined for systematic study of the phenomenon. There are two issues that need to be addressed in refining the question. First, what is the meaning of stability and second, how is omnivory manifest? Regarding the meaning of stability, confusion is not likely to be resolved soon. The engineer's "stable point" has been classically the focus of much of theoretical ecology and is undoubtedly a good place to start. However, in most ecosystems, it is not the place to end. It is not what most ecologists have in mind when they speak of stability. Rather, the "tendency to persist" is a more likely metric of what is usually meant by stability of ecosystems. Here we take stability to mean the long-term persistence of a system.[16]

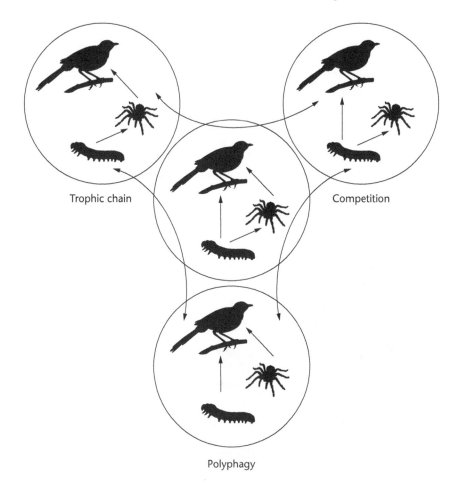

FIGURE 5.1 Graphic representation of the three possible three-dimensional situations united through the concept of omnivory. Omnivory is represented in the center circle, with the bird eating both the spider and the caterpillar (which represents a herbivore) and the spider also eating the caterpillar (see Vamdermeer (2006a) for more analytical details).

The particular way in which omnivory is manifest is often not well defined. For example, with reference to Figure 5.1, in the case of a trophic chain, the predator may become an omnivore simply by increasing its consumption of the herbivore. However, it matters greatly whether that increase is relative to (1) its consumption of the other predator (the spiders) or (2) the consumption of the herbivores by the spider (henceforth, we will refer to birds, spiders and caterpillars, with the understanding that they are simply metaphors for more general categories). In the first case, the trophic chain structure gradually transforms toward a competition structure, while in the second case, it gradually transforms toward a polyphagy structure. In either case, just imagine one of the arrows in the omnivory circle (the center circle in Figure 5.1) slowly disappearing. Casting omnivory as a point on the gradient between two basic three-dimensional cases was pioneered by Kevin McCann and Peter Yodzis, who explicitly studied the part of the general gradient from trophic chain to competition (Figure 5.1).[17] Our approach here is a bit more general.

With the classical ecological equations,[18] the basic questions relating omnivory to stability can be asked generally for all three structures, or to be more accurate, for all three transitions from one structure to another (Figure 5.1). Recall that there has been quite a debate about the role of omnivory in stabilizing ecosystems. Part of the reason that debate remains unresolved is because the question has not been formulated properly. We suggest that the formulation we present in Figure 5.1 represents

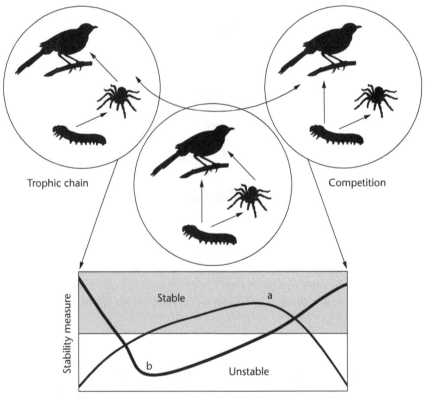

FIGURE 5.2 Illustration of how to interrogate the stabilizing properties of omnivory. A measure of stability is plotted against some parameter that causes the system to move from a trophic chain to competition. Trajectory (a) goes from an unstable trophic chain to an unstable competition, but in between goes through a zone of stability. In this case, it can be said that omnivory is actually a stabilizing force. Trajectory (b) goes from a stable trophic chain to a stable competition, but in between it goes through a zone of instability. In this case, it can be said that omnivory is actually a destabilizing force.

a more proper general framework for posing the question in the first place. So, for example, we see that the simple question "Is omnivory a stabilizing force?" is actually nonsensical without stipulating a reasonable counterfactual – that is, stabilizing "relative to what?" It might be, for example, that an omnivory system that emerges out of competition would be stabilizing, but one emerging out of polyphagy is not. In Figure 5.2, we illustrate what we regard as the proper way of asking the question in the particular case of the transition between trophic chain and competition.

Given the structures proposed in Figure 5.1, and the analytical framework proposed in Figure 5.2, there are two ways of exploring the question of how omnivory relates to ecological stability:

1   If one of the baseline structures (either trophic chain, competition or polyphagy) is inherently unstable, can omnivory act to stabilize it?
2   If the baseline structure is stable, can omnivory act to destabilize it?

A related, but somewhat more complicated, query is whether omnivory is a phenomenon that permits a smooth and continuous transition from one state to another. In other words, is it possible to go from polyphagy to competition without a loss of stability at some intermediate state of omnivory? It turns out that such a smooth transition is possible for the polyphagy–competition and polyphagy–trophic chain transitions, but appears to be impossible for the trophic chain–competition transition, the reasons for which are beyond the scope of this text.[19] That is, in reference to Figure 5.2, moving along the x axis (the "driving parameter"), if you begin and end with a stable system, somewhere in the course of moving from one end to the other, the system will become unstable.

It can also be shown that in each formulation, the two ends of the continuum are united by the concept of omnivory (Figure 5.1), which is to say that the "middle" of the continuum is omnivory. Using this conceptualization, it is possible to enter the long-standing debate about the nature of omnivory as a stabilizing or destabilizing force and conclude that it can be both stabilizing and destabilizing, a finding that applies to all three cases (transitions from trophic chain to competition, from competition to polyphagy and from polyphagy to trophic chain). Setting the two ends of the continuum so as to be stable, omnivory can be destabilizing, and setting the two ends of the continuum so as to be unstable, omnivory can be stabilizing. We conclude that the consequence of omnivory depends on the stability conditions of the parent systems from which it derives.[20]

Theoretical framework: coupled oscillators

There is a basic fact about the general structure of a three-species consumption system, the sort of system pictured in Figure 5.1. At its most elementary level, predators (for example, lions) eat herbivores (for example, zebras) and you can imagine a kind of equilibrium condition for both predators and herbivores, a "just right" state of equilibrium with just the right number of predators to keep the herbivores under control and just the right number of herbivores to keep the predators well fed. However, looking just a bit more closely at the fundamental relationship and going beyond our tendency to look at averages, the truth is that there sometimes may be too many lions, which leads to more zebra deaths than normal, and some other times when there may be too few zebras, leading to more lion deaths (through starvation) than normal. Thinking this idea through, it seems that it should be the case that a predator–prey system will normally bounce back and forth between a state of too many predators versus a state of too few predators. Actually, we normally have a kind of four-part play in which (1) too many predators results in the death of many prey, which leads to (2) too few prey resulting in the death of many predators, which leads to (3) too few predators resulting in many new prey, which leads to (4) many prey resulting in too many new predators, and on and on in a cyclical pattern. In other words, the system oscillates between a state of too many predators with too few prey and a state of too few predators with too many prey. These oscillations may not last very long and the two populations could very well both approach some sort of balance, but the underlying ecological rules result in a situation in which the population numbers of predators and prey oscillate with respect to one another, the classical predator–prey oscillations.

The fact of oscillations potentially resolves a famous and persistent ecological conundrum. Esteemed ecologist G. Evelyn Hutchinson pointed out an evident, yet sometimes ignored, contradiction in basic thinking about ecology. One part of that basic thinking is the so-called Gause's principle (after the Russian ecologist G. F. Gause), which derives from, on the one hand, some very simple equations describing competitive interactions and, on the other, some very simple common sense. It says that two species cannot survive in the same place over the long term if they both do the same thing, which is to say, live in the same niche, which results in the summary statement "No two species can occupy in the same niche." So, for example, the competition situation pictured in Figure 5.1 could not persist in perpetuity since both carnivores (birds and spiders) eat the same thing (occupy the same niche). Hutchinson challenged this basic and elementary ecological rule with a basic and elementary observation – phytoplankton all seem to do pretty much the same thing (photosynthesize and float), yet in a given lake or given part of the ocean there are usually many species apparently coexisting. It was paradoxical, said Hutchinson, that they all seemed to be occupying the same niche yet there were so many of them, giving rise to the title of his famous article "The Paradox of the Plankton."[21]

An elegant resolution to the paradox was provided by a simple extension of the basic equations that describe consumption.[22] If we relax the assumption that consumer (predator) and resource (prey) are linearly related to one another (i.e., that the per capita growth rate of one is a linear function of the other), we first wind up, under some circumstances, with the tell-tale oscillations of consumer and resource. Then, if we put two of these oscillations together in a model of two consumers and a single resource, we might anticipate the possibility that as oscillations of the first oscillator reach their low point (i.e., low value of consumer number one), that may provide an opportunity for the consumer in the other oscillation. Not surprisingly,

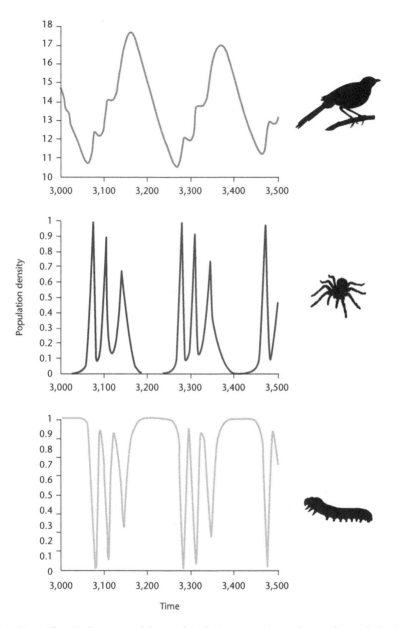

FIGURE 5.3 Chaotic dynamics of the trophic chain system. Complex oscillatory behavior of population density results from coupling two oscillatory systems together. Note that the bird and the spider form relatively long-term oscillations (which, obviously, can also be detected in the caterpillar population), while the spider and the caterpillar form relatively short-term oscillations (which, obviously, can also be detected in the bird population). The system as a whole can be viewed as two oscillators (one: bird–spider; two: spider–caterpillar) that, metaphorically, compete with one another for dominance. The chaotic pattern results from this coupling of the two oscillators (see also Hastings and Powell (1991)).

the equations turn out that way. Indeed, the equations, in this case, preceded the qualitative intuition. What happens is that because of the up-and-down changes in population densities, the two consumers can effectively partition the different population stages of the resource oscillations. Thus, relaxing the linear assumption makes the entire edifice of Gause's principle fade away, and two consumers can indeed coexist on a single resource.[23] This does not mean that the principle itself is incorrect, or that it should fall into oblivion just because it was based on overly simplistic assumptions. It still remains an important point of departure for thinking about interacting consumers. However, the enigma of the original paradox of the plankton is dramatically reduced, and a variety of subsequent theoretical studies add significantly to our understanding of the basic processes.[24] Furthermore, an even more interesting issue emerges if we combine these oscillators in a different way.

A top carnivore eating a carnivore is one oscillator. A carnivore eating an herbivore is another oscillator. Rather than coupling the oscillators horizontally by consumers eating the same resource, we can couple them vertically with a top carnivore eating a carnivore which, in turn, eats an herbivore, a classical trophic chain as pictured in Figure 5.1. This conceptualization generates an especially interesting pattern when released from its linear constraints.[25] Using the pictorial representations of Figure 5.1, we see that the caterpillar is eaten by the spider, which implies one oscillator (spider and caterpillar oscillate with respect to one another), but the spider itself is eaten by the bird, which implies a second oscillator (bird and spider oscillate with respect to one another). And the two oscillators are connected, or coupled. Examining the details of this system, we find that these two oscillators sort of compete with one another for dominance and, under some circumstances, neither one is able to dominate, such that the system looks like oscillations between the bird and the spider dominate for a while but then switches to looking like oscillations between the spider and the caterpillar dominate for a while and then switches back again, as illustrated in Figure 5.3. The disturbing feature of this is that the switches are not predictable. It is formally a case of chaos.

As a general rule, when these oscillators are connected to one another, as they are in each case of Figure 5.1, complications arise, and frequently chaotic dynamics (Figure 5.3) and other even more complicated behaviors result.[26]

Herbivores and their arthropod and vertebrate predators

Conceptually, many terrestrial food webs fall within the framework described in the previous section, a herbivore class and two potential carnivore classes, one of which may or may not be a top predator. For example, vertebrate predators (mainly birds, bats and lizards) may eat significant numbers of invertebrate predators (e.g., spiders) relative to the insect herbivores (e.g., caterpillars), or they may not, depending in part on the specific nature of the invertebrate predators in the system (e.g., too small to worry about, too big to deal with). If the system is characterized by competition (upper right circle in Figure 5.1), increased action of vertebrate predators will result in a reduction of herbivores. On the other hand, if the system is characterized by a trophic chain (upper left circle in Figure 5.1), increased action of vertebrate predators will result in an increase of herbivores, since vertebrate carnivores reduce invertebrate carnivores, releasing herbivores from predation.

Trophic structures have been a fundamental framework in ecology since Elton's[27] and Lindeman's[28] classical formulations of the food chain and trophic dynamics.[29] Yet a proper analytical framework for dealing with them remains contested.[30] In some situations, it seems obvious that there is a trophic chain – for example, in the African savannas where lions eat zebras, which eat grasses, or in a prairie where birds eat grasshoppers, which eat the prairie vegetation. That conceptualization gave rise to the "green world hypothesis" of Nelson Hairston and colleagues,[31] who postulated that terrestrial herbivores consume relatively low plant biomass because they are held in check by their natural enemies (predators, parasites and diseases). However, nature tends to be more complex, and in many situations it is hard to distinguish among these clear-cut trophic levels, giving rise to the idea of a food web (rather than chain) in which carnivores and top carnivores are fluid categories.[32] The "primary producer" and "herbivore" categories are well defined and rarely thought to be anything other than Lindemanian (belonging to a specific and well-defined trophic level), but higher trophic categories can be more enigmatic[33] (e.g., the birds eating both caterpillars and spiders in Figure 5.1).

Part of the confusion derives from the fact that food webs in most terrestrial ecosystems are characterized by a structural constraint. They are dominated by size-structured predator–prey interactions.[34] On the one hand, arthropods tend to be small in size, but are major components of both herbivore and carnivore categories. On the other, vertebrate predators are normally larger and eat both herbivorous and carnivorous arthropods. The consequence is that arthropod predators do not generally eat vertebrate predators (except for those that are parasites) but vertebrate predators generally eat both herbivorous and carnivorous arthropods.[35] This elementary observation leads to that persistent question, reflecting the contrast between the Lindemanian "food chain" versus the more complex "food web" concept about terrestrial trophic interactions: how many "effective" trophic levels are there in a particular system?[36]

The ecological literature seems to have converged on a simple scheme, despite the immense complexity of terrestrial food webs.[37] We formulate this simple scheme as a gradient between two of the extremes described in the previous section. Depending on the extent of feeding bias in the vertebrate sector, the food web can be characterized as carnivore competition – three trophic levels if you include the plants (right circle in Figure 5.2), trophic chain – four trophic levels (left circle in Figure 5.2), or something intermediate between these two – omnivory (center circle in Figure 5.2). Whether terrestrial ecosystems, excluding ecosystems dominated by large grazing herbivores and their predators, are characterized by three or more trophic levels is an interesting question and has been the focus of considerable empirical and theoretical work.[38] The question also has obvious practical significance.[39] For example, it has been argued that birds and bats could be beneficial because of their potential for controlling insect herbivores in forests[40] and agroecosystems.[41]

Such an argument effectively assumes that the terrestrial food web has three levels (plants, herbivores and predators), which is to say that the effect of these vertebrate predators on herbivores is direct and negative, but their effect on the plants is indirect and positive (by reducing the herbivores – an enemy of my enemy is my friend). However, if there are also invertebrate predators in the system (e.g., spiders), the vertebrate predators also may eat the spiders, thus generating an indirect and positive effect on the herbivores by reducing the invertebrate predators, an arrangement frequently termed "intraguild predation," since the invertebrate predators are in the same "guild" or trophic category as the vertebrate predators. In that case, the vertebrate predators may have either a net negative or net positive indirect effect on the plant, depending on the relative strength of the intra- and interguild predation. While these theoretical ramifications are clear conceptually, rarely have they been subjected to careful empirical study.

Teasing out the trophic structure in the coffee agroecosystem

Many years ago, while walking on a shade coffee farm in Mexico, we encountered what seemed to be an outbreak of caterpillars devouring some of the trees used to shade the coffee. We asked the farmer if he was concerned that these caterpillars would kill his shade trees. He smiled and told us not to worry: "In a few weeks the migrant birds will arrive and they'll take care of the pest." Sure enough, in a couple of weeks, the first bands of migratory birds arrived and the caterpillars were gone. This observation triggered our curiosity about the role of birds in pest control in agroforestry systems generally. It also made us question what would happen if the farmer decided to intensify his coffee farm and eliminate the shade trees. Do birds provide significant pest control services on coffee farms, and if so, does the lower bird density and diversity found in more intensive farms diminish that service? These are interesting research questions not only because of their practical importance but also because of their potential to contribute to a better understanding of trophic structures in terrestrial ecosystems.

Some of our experimental work in the coffee agroecosystem of southern Mexico has been aimed at answering the question of whether the food web involving both vertebrate and invertebrate predators corresponds to a competition or a trophic chain situation (two of the three alternatives pictured in Figure 5.1). In other words, are there three or more effective trophic levels in this food web? If the system corresponds to competition among the carnivores (vertebrate predators mainly eating the herbivores and not the invertebrate predators), the impact of insectivorous birds would be positive on the plants. However, if it corresponds to a trophic chain, with four trophic levels (vertebrate predators eating mostly invertebrate predators), the impact of the vertebrate predators on the plants would be negative.

The basic qualitative structure of the coffee agroecosystem in Mexico is as follows. Both invertebrate and vertebrate predators consume a group of arthropod herbivores, some of which are either pests or potential pests. The invertebrate predators include some beetles, lacewings, wasps, robber flies, ants and spiders, with spiders being the

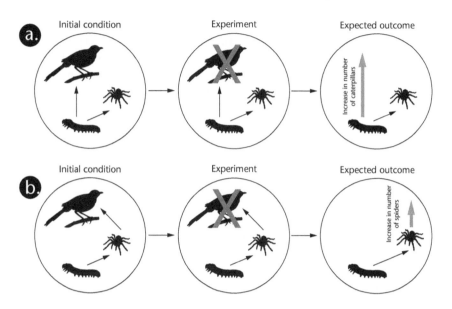

FIGURE 5.4 Expected initial response from the experimental removal of the vertebrate predators from the system. (a) If the basic structure of the system is carnivore competition, removing the vertebrate predators will initially release the herbivores from control since they are the most immediately affected by the birds. (b) If the basic structure of the system is a trophic chain, removing the vertebrate predators will initially release the spiders from control since they are the most immediately affected by the birds. Later on, we expect the spiders to increase in the case of competition (a), and the herbivores to decrease in the case of a trophic chain (b).

most obvious. The vertebrate predators include toads, lizards, bats and birds, with bats and birds being the most common. Thus, if we experimentally remove the effect of the birds and bats (vertebrate predators) from the system, we expect that the initial response will be an increase in herbivores if the system is structured as carnivore (predator) competition (Figure 5.4(a)), but an increase in spiders if the system is structured as a trophic chain (Figure 5.4(b)). While both alternatives suggest an ultimate increase in spiders and decrease in herbivores, the underlying structure suggests a time lag, enabling a potential experimental separation of the two if experimental results are followed over time. That is, if birds and bats are mainly consuming herbivores, experimental removal of these flying vertebrates would result initially in an increase in herbivores (as pictured in Figure 5.4(a)), followed later by an increase in spiders due to a release of competition (by birds and bats) and a plentiful supply of prey. However, if birds and bats are mainly consuming spiders, experimental removal of the flying vertebrates would result initially in an increase in spiders (as pictured in Figure 5.4(b)), followed later by a decrease in herbivores since the increase in spiders would increase predation pressure, although only after a lapse in time.

Working on four coffee farms in the Soconusco region of Chiapas, Mexico, we set up an experiment consisting of 32 large cages designed to exclude the birds (Figure 5.5). These "exclosures" were constructed of transparent monofilaments of nylon (5 centimeter mesh) fishing nets covering at least ten coffee plants. A total of 616 coffee plants were sampled, half of which were inside the exclosures and half outside.

The results of the experiment are presented in Table 5.1. The initial response to the manipulation (three months after initiation of the experiment) was a significant increase in spiders, but no increase in herbivores, precisely the signal expected from a trophic chain, a four-trophic level (including the plants) system (Figure 5.4(b)), as discussed above. However, as a check on this result, we examined the subsequent response of the system six months after initiation of the experiment. If the system was indeed behaving like a trophic chain, we would have expected a delayed response of decreasing herbivores due to increased predation pressure from spiders subsequent to the removal of birds. After six months, there was indeed a clear response of herbivores to the treatment. However, this response was not in the direction expected if the system was a four-level trophic chain. Rather than the expected decrease, the herbivores *increased* in the treatment that excluded vertebrate predators (Table 5.1), a strange result indeed. It appears that our formulation of the food web alternatives (Figures 5.2 and 5.4) must have been too simplified.

We attribute this surprising response to the presence of parasitoids, which significantly affect herbivores, but are consumed mainly by spiders, especially the web weavers,[42] and not extensively by birds or bats, which clearly prefer the larger arthropods.[43] Parasitoid numbers showed a significant decline when vertebrate predators were excluded both after three and six months (Table 5.1). Indeed we found a significant negative correlation between spiders and parasitoids.

These results strongly suggest that the spiders are consuming parasitoids, which are also natural enemies of the herbivores. In this intraguild predation case, the guild "arthropod predators" is effectively subdivided into spiders and parasitoids. The result is that when birds are excluded, the spiders increase in abundance and the parasitoids decrease in abundance, finally resulting in an increase in herbivores to the detriment

FIGURE 5.5 Researchers setting up cages designed to prevent birds from eating arthropods on coffee plants.

TABLE 5.1 Results of experiment in which birds were excluded from the system. Data for herbivores, spiders and parasitoids are percentages of all individual arthropods sampled that were found in the category (* is equivalent to statistical significance of less than 0.05).

	After three months			After six months		
	*🕷	*🐛	*🦟	*🕷	*🐛	*🦟
🐦	41	53	63	40	42	61
🐦✗	59	41	37	60	58	39

of the plant. In this experiment, we also recorded a significant increase in herbivory (damage to plants by herbivores), demonstrating that the impact of the vertebrate predators cascades all the way down to the plant. Thus, this study suggests that the food web in the coffee agroecosystem is organized as a five-level trophic chain (Figure 5.6); birds and bats reduce spiders, spiders capture large numbers of parasitoids, and parasitoids kill herbivores, which consume the coffee plant.

It has long been suspected that birds limit the populations of arthropods in forests and agroecosystems.[44] The implicit assumption of these and other exclosure studies is that these vertebrate predators serve important ecosystem functions by directly consuming herbivores. However, our study suggests that intraguild predation (omnivory) is implicated as well, and that the reduction of herbivores when birds and bats are removed could be, at least partially, the result of a trophic cascade with five trophic levels (vertebrate predators, spiders, parasitic wasps, herbivores and plants).

Thus, the food web in this coffee agroecosystem is characterized as a system that has five levels and behaves primarily as a trophic chain, but with important web-like connections. The originally postulated "arthropod carnivore guild" is complicated by omnivory,[45] with a structure that requires a larger number of "effective" trophic levels in its basic conceptualization. We suggest that the basic structure (vertebrate predators, arthropod predators and herbivores) is widespread in terrestrial food webs; however, intraguild dynamics (e.g., birds and bats preying on spiders and spiders preying on parasitoids) may significantly alter the expected dynamics of the web, including modifications of the effective arrangement of the trophic levels, as well as the ultimate impact on the primary producers, and consequently, in agroecosystems, on the yield of the crop in question.

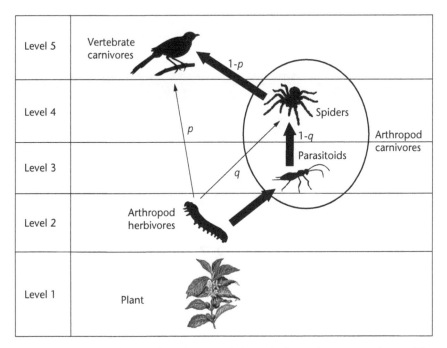

FIGURE 5.6 Revised food web as discovered through the exclosure experiment. The parameters *p* and *q* represent the proportional significance of each of the relevant energy channels (e.g., *p* is the proportion of the bird's food that is composed of herbivores). The experiments suggest that *p* and *q* are relatively small for the system under study.

Many previous studies have documented the effect of vertebrate predators, mainly birds and lizards, on spiders,[46] and of spiders on herbivores,[47] but few have documented a negative relationship between spiders and parasitoids. To our knowledge, the only other study documenting this relationship is that of Schoener and colleagues[48] on small Bahamian islands. The majority of the spiders captured in our study were web builders in the families *Theridiidae* (24 percent), *Tetragnatidae* (11 percent) and *Araneidae* (10.3 percent). Studies of prey catch of these families and other web spiders have shown that they capture significant numbers of parasitoids in their webs.[49] Therefore, it is a reasonable hypothesis that the trophic cascade suggested in our study could be found in other terrestrial ecosystems dominated by web spiders.

We might even go so far as to say that the trophic structure uncovered in this experiment may be very widespread. Those terrestrial food webs characterized by the four essential elements found in our study – vertebrate predators (mainly birds and bats, but possibly lizards as well), spiders, parasitoids and herbivorous arthropods[50] – may very well be generally characterized as the five-level system we describe here. With the exception of the ecosystems dominated by large grazing mammals[51] and those on small islands or agroecosystems with no vertebrate carnivores,[52] the general

limitations of the particular physical characteristics of the five elements (vertebrate predators, spiders, parasitoids, herbivores and plants) suggest our results could be quite widespread. If further study reveals p and q to be generally small (Figure 5.6), as they appear to be in the coffee example, perhaps revisiting the question of energy transfer[53] in terrestrial ecosystems would be in order.

Trait-mediated effects in food webs

What is trait mediation?

The sorts of trophic structures of the previous section, when involving three or more components, always contain the underlying structure of "indirect effects," a deceptively simple idea: basically, a friend of my enemy is my enemy, or an enemy of my enemy is my friend. So, for example, when a bird eats spiders, it basically has a positive effect on (is a friend of) the insects that the spiders would have eaten. In real ecosystems, these sorts of interactions become more complicated. For example, in Figure 5.6, the birds have a positive effect on the parasitoids (by eating spiders), the spiders have a positive effect on the herbivores (by eating the parasitoids) and the parasitoids have a positive effect on the plants (by eating the herbivores). However, the birds also have a positive effect on the plants (the birds eat the spiders, which eat the parasitoids, which eat the herbivores, which eat the plants – four cascading negative effects gives a positive effect). How to put all of these complicated indirect effects together is a challenge for ecological theory that Richard Levins attacked some time ago in an underappreciated methodology called loop analysis. Basically, the loop from (1) the birds having an effect on (2) the spiders, which have an effect on (3) the parasitoids, which have an effect on (4) the herbivores, which have an effect on (5) the birds, is a loop of effects from the birds back to the birds (in terms of loop analysis, this is a loop of length five). By combining loops of different lengths in a food web, one can frequently make statements about the stability of the web.[54] It is a complicated method, but then again, stability of food webs is a complicated issue.

Such complications are beyond the present text. However, there is another sort of indirect effect that we suggest may be even more important than the obvious ones that arise from these trophic connections. For example, in the simple case of a tri-trophic cascade, in which the birds eat the spiders and the spiders eat the caterpillars (Figure 5.4(b)), if we examine more closely the natural history of the species involved, we see how some evident complications could arise.[55] The small caterpillar lands in the spider web and, as it struggles to free itself, the spider detects the vibrations in the web, rushes out, captures and poisons it, and either eats it or wraps it in silk to be eaten later. Certainly from the caterpillar's point of view, any bird that eats those spiders is beneficial, and the beneficial effect is "indirect" in the sense that the bird, being an enemy of the spider, is a friend of the caterpillar. However, in the real world, something more complicated can happen. For instance, as the bird approaches the web, the spider senses the presence of this potential predator and retreats to the central funnel of its web, specifically constructed so as to hide from potential predators. However, if the spider is hiding in the funnel of the web, it is either unable to detect the caterpillar that gets stuck in the web, or is afraid to go out and attack it, and the caterpillar may very well struggle and escape from the web. In this case, the bird clearly has a positive effect on the caterpillar, but the effect occurs not through the process of consumption (the bird eating the spider) but through a change in the behavior of the spider. Suppose that the spider normally eats ten insects per day, but that if the birds are around continually harassing, perhaps only five, or even just one insect per day can be eaten. That is, the mere presence of the bird has changed the rate at which the spider can feed; it has changed a behavioral trait of the spider. Such changes are known as "trait-mediated indirect interactions" (Figure 5.7).

These kinds of interactions are now commonly recognized in ecology.[56] Indeed, trait-mediated effects seem to be more common in nature than we first imagined, sometimes completely overwhelming the density-mediated effects.[57] It is really quite a simple and intuitive idea, but it takes on some very complex forms in the real world and has some unusual consequences as its effects cascade through a complicated ecosystem.

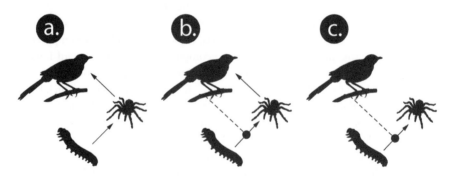

FIGURE 5.7 Illustration of the idea of trait-mediated interaction (arrows indicate positive effects, and dashed lines with closed circles indicate negative effects). (a) The generalized case of a tri-trophic chain in which energy goes from the caterpillar herbivore to the spider predator to the bird predator. The indirect effect here is the fact that the bird reduces the population density of the spider, which has an indirect positive effect on the herbivore. (b) If the bird acts to scare the spider such that its feeding efficiency is reduced, the bird effectively acts to reduce the rate of energy transfer from the population of the herbivore to the spider. There are thus two indirect effects in this case, one an effect mediated by the changes in density of the spiders due to predation by the bird and one due to the bird changing the "trait" of predation rate of the spiders. The trait-mediated indirect effect is indicated by the dashed line with closed circle (for negative effect) on the end. (c) Where the bird does not eat the spider at all, but has a negative effect through modifying its feeding trait. Where a system is located along the implied gradient that runs from (a) to (b) to (c) signifies whether density-mediated indirect effects or trait-mediated indirect effects are more important.

Conceptualizing trait-mediated effects as fundamental non-linearities

Recalling previous discussions of complexity and the importance of non-linear effects, it is worth noting that a trait-mediated indirect interaction implies an extra "non-linear" effect in the system (the rate at which the spider eats caterpillars is itself a function of the presence of birds).

A common trait-mediated interaction in nature involves the ubiquitous ant–hemipteran mutualism. Ants have been called the first farmers of the natural world because they frequently "cultivate" hemipterans. These are sap-sucking insects that pierce the phloem of plants and excrete a sugary solution called honeydew, which the ants harvest. In exchange for this energy-rich liquid, the ants protect the hemipterans by scaring away parasites and predators (Figure 5.8(b)). Ant–hemipteran interactions are abundant and widespread in arthropod food webs and frequently result in an indirect benefit to plants, because the increased activity of the ants on plants where hemipterans are being tended deters other herbivores that could otherwise cause more damage to the plant (more than the hemipterans – Figure 5.8(c)).[58] However, more relevant to the issue of trait-mediated interactions is the fact that the ants benefit hemipterans by protecting them against predators, and harassing or chasing away the potential predators frequently accomplishes this. For the protection to be effective, the ant does not need to kill and consume the natural enemy of the hemipteran. In other words, the ants affect the ability of the predators to prey on the hemipterans, thereby affecting the predation rate of predators on the hemipterans (Figure 5.8(b)). This is a classical example of a trait-mediated interaction that introduces a fundamental non-linearity into the system.

The complicated system of trait-mediated interactions associated with the Azteca ant

In Chapter 4, we described the spatial distribution of the ant *Azteca* on the shade coffee plantations of Chiapas, Mexico, and in particular, within the 45-hectare plot in *Finca* Irlanda, where we have conducted most of our work on this species. In that chapter, we also introduced some of the key players that interact with *Azteca*, forming a complex ecological network that is, at least partially, responsible for the self-organized aggregated pattern exhibited by the ant. Among the key interactions described in the previous chapter was the mutualism between *Azteca* and the green coffee scale, *C. viridis*. Although the ants nest within shade trees, they forage and tend scales on coffee plants nearby their nest. In this section, we will explore how a complex ecological network centered on the *Azteca*–green coffee scale mutualism is structured, the importance of trait-mediated interaction in this network, and how this network plays a role in the ecosystem service of pest control within the coffee agroecosystem.

The *Azteca*–*C. viridis* interaction is a classical mutualism in which the ant obtains energy directly from the honeydew produced by the scale insects and, while tending

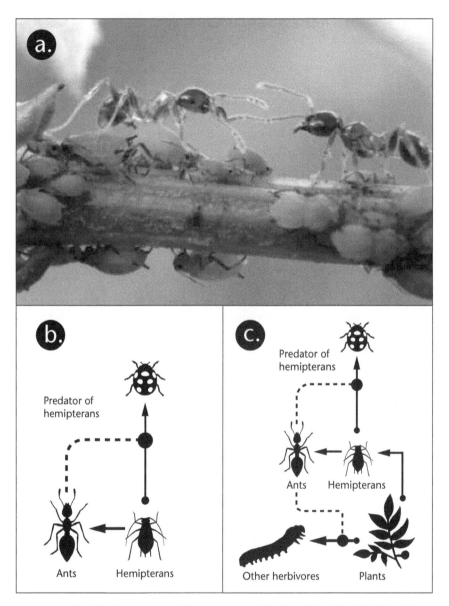

FIGURE 5.8 Ant–hemipteran mutualism as an example of a trait-mediated indirect interaction. (a) *Pheidole megacephala* tending a colony of aphids for honeydew. (b) Diagrammatic representation of the trait-mediated interaction involved in the ant–hemipteran mutualism. (c) The two trait-mediated interactions in which ants deter both the predators of the hemipterans and the other herbivores of the plant. Solid lines represent energy flow and direct interactions, dotted lines represent the trait-mediated interaction, arrows represent positive effects and solid circles represent negative effects. Frequently the effect of the ants in deterring the other herbivores is greater than the effect of the hemipterans on the plant.

Source: (a) Alex Wild

the scales, protect them from natural enemies (Figure 5.8(b)). Among the most important natural enemies in the system is the lady beetle *A. orbigera* (Figure 4.15(c)). This beetle is a voracious predator of the green coffee scale. An individual beetle can devour an average of 40 scales per day – that is, if the scales are not being tended by ants. When ants are tending the scales, the beetles have a hard time getting close to the scales since they are harassed and sometimes killed by the ants (Figure 5.9). This protective behavior of the ants represents the first trait-mediated interaction in the network. Similar to the diagram in Figure 5.8(b), the ants affect the rate at which the beetles can prey on the green coffee scales.

At this point, we return to another seemingly unrelated actor, the phorid fly (previously introduced in Chapter 4). This fly is a parasitoid of the ant and, in addition to directly affecting the ant through parasitism, it has an important trait-mediated indirect effect on the ant's ability to protect the scale insects. In a detailed experiment, Liere and Larson placed individuals of *A. orbigera*, the coccinellid beetle, on coffee plants with different levels of ant activity.[59] They then recorded the time the beetles were able to remain on the plant before the ants kicked them off. They found what appears to be a critical threshold of ant activity below which the beetles are not affected (Figure 5.10(a)). However, more importantly, they found that when phorid flies arrived, ant activity went down to that critical level (i.e., where ants had no effect on the presence of the beetles) (Figure 5.10(b)).

FIGURE 5.9 *Azteca* attacking an adult of the coccinellid beetle *A. orbigera.*

Source: Shinsuke Uno

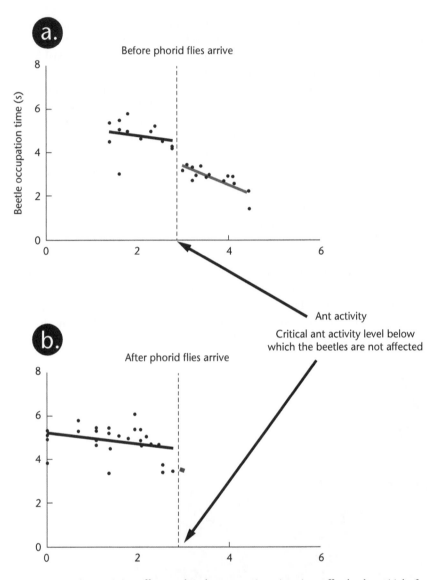

FIGURE 5.10 Ant activity effect on beetle occupation time in coffee bushes, (a) before and (b) after the arrival of phorid flies. Note that after phorid flies arrive, ant activity is effectively canceled, from the point of view of the beetles.

Source: Modified from Liere and Larson (2010), reprinted with permission

The phorid flies thus create the conditions for a trait-mediated cascade.[60] The ants impede the ability of the beetles to prey on the hemipterans (Figure 5.11(a)), and the phorid flies impede the ability of the ants to impede the ability of the beetles to prey on the hemipterans (Figure 5.11(b)). This trait-mediated effect of the phorids thus creates two levels of non-linearities in the system, since the rate of predation is changed by ant harassment which itself is changed by the presence of the phorid fly (Figure 5.11(b)).[61]

Yet there is another component of the system that complicates the situation dramatically. A curious set of adaptations in some lady beetles, including the one of concern here, enables them to consume hemipterans even when under the protection of ants.[62] In these cases, the beetle larvae have some form of protection against the aggressive behavior of the ants. The adaptive significance is obvious. The hemipterans reach high densities in areas where the ants are tending them and, therefore, the beetle larvae have an ample supply of food in these areas. Since the beetle larvae do not fly and thus have a restricted range, to emerge and develop in

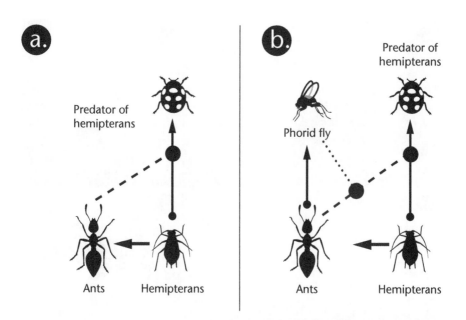

FIGURE 5.11 Ant–hemipteran mutualism affected by the phorid fly as an example of a trait-mediated cascade. (a) The basic trait-mediated indirect interaction of the ant protecting the hemipteran (same as Figure 5.8(b)). (b) The phorid fly parasitoid added to the system, whereby its effect on reducing ant activity has the effect of reducing the ant's ability to reduce the predation rate of the hemipteran predator. The dotted line indicates the second-level trait-mediated effect added by the phorid fly to interfere with the original first-level trait-mediated effect produced by the ant harassment of the predators, indicated by a dashed line.

patches with ants has obvious benefits. If females lay eggs where there are large concentrations of hemipterans, the resulting offspring will have an adequate supply of food; otherwise, food will always be scarce since the larvae can only move a limited distance in search of more food. For the lady beetle of concern here, *A. orbigera*, the protection for the larvae is an array of waxy filaments that the larvae develop as they grow (Figures 5.12(a) and (b)), effectively protecting them from the attacks of the ants (Figure 5.12(a)). Furthermore, in addition to being protected from ant attacks, the beetle larvae are inadvertently protected from their own parasitoids by the ants (while scaring away parasitoids of the scale insects, *Azteca* ants also scare away the parasitoids that attack the beetle larvae).[63] Thus, the presence of the ants insures an abundant supply of food and protection against natural enemies for the beetle.

However, even as the protective device contained by the larvae is obviously adaptive, it generates a problem. Mothers must be able to lay eggs where ants are actively patrolling (since that is where the concentrations of hemipterans are) but must place the eggs where ants cannot find them (unless the eggs themselves are

FIGURE 5.12 Photos of *A. orbigera*. (a) *Azteca* attacking *A. orbigera* but getting its mandibles filled with the sticky waxy filaments that cover the body of the larvae. (b) *A. orbigera* eating *C. viridis*. (c) Eggs and first instar larvae (white arrows) of *A. orbigera* on an old pupal case of the same species. (d) *A. orbigera* eggs hidden under *C. viridis*.

Source: © Heidi Liere

protected from the ants, which does not normally seem to be the case). Two of the main safe sites for depositing eggs in this system are in the old discarded waxy covering of pupae of the same beetle species (Figure 5.12(c)) or directly underneath the adult scale insects (Figure 5.12(d)). In both cases, the eggs are largely protected from the attacks of the ants. However, yet another problem arises for the beetle. How can the mother spend the time searching for safe sites to lay eggs in a place where the ants are constantly attacking her?

The answer to this question is quite obvious in light of the information presented in Figure 5.10 (also Figure 4.8) – wait for the attack of the phorid flies and take advantage of the reduction in ant activity to seek out safe sites for the eggs. However, the details of how this is accomplished could serve as inspiration for an espionage movie. Communication among insects is frequently conducted through chemicals called infochemicals. When the chemical produced is used for communication among members of the same species, it is called a pheromone. Ants, in particular, are bags of chemicals, all carrying some information crucial for the maintenance and reproduction of the colony. One of these chemicals is an alarm pheromone that is released when the ants perceive a problem and need to defend the colony. When this pheromone is released by one or a few ants, many individuals swarm out of the nest, running and attacking whatever comes close. Unfortunately for the ants, but not surprisingly for us, the phorid flies have evolved an ability to detect this chemical signal of their host. However, once attracted to the general area in which this pheromone has been released, the flies seem disoriented and unable to locate the individual ants they seek to attack. It turns out that phorid flies have poor acuity in eyesight and need movement to be able to locate individual ants on which they can oviposit. The ants, in turn, react to the presence of a phorid fly by becoming motionless.[64] They assume a defensive posture with their mandibles open and their thorax and head facing upward (Figure 5.13(a)). Clearly, this motionless state is adaptive in that it denies the phorid flies the movement they require to be able to attack. However, something curious happens to the other ants in the general vicinity. As can be seen in Figure 5.13(b), ants in the surrounding area also adopt that posture. This occurs over a distance of meters, which suggest that the ants under attack by phorid flies release a pheromone that traverses a relatively long distance (compared to the size of an ant) and that alerts other ants to the presence of the phorid flies. In other words, the ants seem to be releasing a pheromone that effectively says, "Look out, there is an attacking phorid in the neighborhood."

In effect, the ant's behavioral response to the presence of phorid flies creates a window of opportunity for the beetle adults to search for safe oviposition sites.[65] For all practical purposes, when the phorid flies are present, beetles behave as if there were no ants (Figure 5.10). Furthermore, the effect of the phorid flies lasts long enough to provide the female beetles with enough time to find high-quality oviposition sites. This basic natural history makes this system a great example of a trait-mediated cascade (Figure 5.11(b)).

However, the story does not end with the phorid flies tapping into the ants' communication system. Like any good espionage story, it continues with higher

FIGURE 5.13 Ant response to attack by the phorid fly parasitoid. (a) Photograph of an ant (not *Azteca*) in the typical defensive posture taken when phorids are present. (b) Almost all of the ants in the general vicinity of the phorid attack assume the same posture, indicating they are receiving some sort of signal from the particular ant under attack.

Source: authors

levels of sophistication in eavesdropping. Using a two-arm olfactometer that detects insects' preference for particular scents or volatile chemicals, Hsun-Yi Hsieh discovered that female beetles (but not males) are attracted to ants when under attack by phorid flies.[66] In other words, female beetles can eavesdrop on the ant "conversation" to find out when the ants are under attack by the phorid flies. Furthermore, testing mated and unmated females, she found that the attraction to ant pheromones is manifested only after mating and continues to increase for at least seven days after copulation. These results demonstrate that gravid female beetles are able to detect the phorid alert pheromone (or pheromones) and take advantage of the low ant activity that results from phorid attacks.

In the same study, Hsieh also found that both male and female beetles can detect and are attracted to the scent of the scale insects (or volatiles emitted by coffee leaves with scales). Thus, it appears that both male and female beetles find patches of high concentrations of scales using these chemicals. Since the larvae of the beetle have dramatically restricted movements and are attacked by several parasitoids, there is clear pressure for female beetles to oviposit in ant-tended areas, where high prey density and low risk of parasite attack are secured. However, the aggressive behavior of ants apparently renders female beetles incapable of ovipositing in these high-quality areas. Female beetles can avoid this problem by being able to eavesdrop on the ants to detect when the phorid flies are attacking the ants. This ability allows beetles to take advantage of the low ant activity periods to search for sites where their eggs can be hidden and protected against ant predation after ants resume their normal activity levels.

More generally, this system is an example of a complicated system of cascading trait-mediated indirect interactions facilitated by the ants' chemical communication system. The first trait-mediated indirect effect is the ants' interference with the ability of the female beetles to oviposit in sites with high scale abundance, due to the ants' mutualistic interaction with the scale insects. The second trait-mediated indirect effect is the phorid flies' interference with the first trait-mediated effect (i.e., with the ants' interference with beetle oviposition). Furthermore, preliminary studies suggest that these cascading effects can get even more complicated by facilitating the parasitoids that attack the beetle larvae. Experiments conducted by Hsieh show that the presence of the phorid flies facilitates beetle parasitoids by reducing ant activity. When the phorid flies are present, ants reduce their protection of scale insects and become less aggressive toward any parasitoids that come close to the scales. So when the phorid flies arrive, ants focus on defensive behavior and abandon their tending activities. This results in higher parasitization of the beetle larvae in the presence of phorid flies. Through those cascading trait-mediated indirect interactions (effects on effects on effects), the phorid flies indirectly facilitate the coccinellid beetles, but also the parasitoids of the beetles (as illustrated in Figure 5.14).

From a theoretical point of view, these results support the hypothesis of a trait-mediated cascade (Figure 5.11(b)). It is worth noting that considerably more interesting system dynamics result from this second-order trait-mediated cascade.

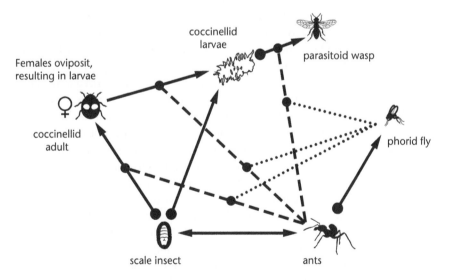

FIGURE 5.14 Diagram showing cascading trait-mediated interactions that result in phorid fly facilitation of a coccinellid beetle and its parasitoid. Arrows indicate positive effects, solid circles indicate negative effects, solid lines indicate direct interactions, dashed lines indicate first-order trait-mediated indirect interactions and dotted lines indicate second-order trait-mediated indirect interactions. Note how this figure is an extension of Figure 5.11(b).

This sort of unavoidable behavior resulting from a trait-mediated cascade underlines the importance of such theoretical structures and challenges approaches that remain mired in linear and quasi-linear assumptions, even as the absolute number of components increases. It is arguably the nature of the couplings that determine whether such complexity will emerge from a system, and not the number of variables considered. This observation places special importance on the search for other trait-mediated cascades in nature.

Trait-mediated indirect effects as coupling agents in food webs

In a previous section, we introduced the idea of conceptualizing trophic structures generally as coupled oscillators. For example, the trophic chain of Figure 5.1 can be conceptualized as a set of two coupled oscillators, birds and spiders representing one predator–prey oscillator, and spiders and caterpillars representing another. Here we tentatively explore the idea of coupling oscillators through trait-mediated indirect effects.[67]

As argued earlier, a potentially useful way to conceptualize food webs is to think about how two sets of consumer–resource oscillatory systems may be connected. For example, lady beetle predators of aphids would be one oscillator and flies that parasitize ants would be the second oscillator. The parasitoid flies consume the ants and the lady beetles consume the aphids; thus, each system is a consumer–resource oscillator. If no other interactions are occurring, then these two consumer–resource systems will be oscillating independently of each other. Now assume that the ants have a mutualistic interaction with the aphids (Figures 5.15(a) and (b)). The ant–aphid mutualism links the two oscillators, making predictions a much trickier proposition. Furthermore, in Figures 5.15(a) and (b), the ant–aphid mutualism is conceptualized as a "direct" mutualism, as it frequently is characterized in the literature. However, this is perhaps misleading,[68] since the benefit the ants provide for the aphids emerges from the ants' behavior in protecting the aphids from their natural enemies. Consequently, formulating the mutualism as direct is not really correct. The indirect effect of lady beetle adults being denied access to the aphids is an example of a trait-mediated effect and is more correctly formulated as an "effect on an effect" (Figure 5.15(c).), with the obvious consequence that the coupling of the two oscillators will generate more complicated behavior.

These trait-mediated couplings, well known in the experimental literature, are employed in ecosystem and community models less frequently than is perhaps merited.[69] The direct mutualist coupling (Figure 5.15(b)) itself produces complicated oscillations, first completely synchronized (i.e., C_1 and C_2 peak at close to the same time), but, with higher degrees of coupling, morphing into a complicated set of chaotic attractors, not unlike other forms of direct coupling.[70] However, modeling the process of mutualism as a combination of direct and trait-mediated indirect interactions, as in Figure 5.15(c), even though the process remains qualitatively similar (i.e., coupling through mutualism), reinforces the unstable aspects of the original coupling, and generates both in-phase coordination and anti-phase coordination,

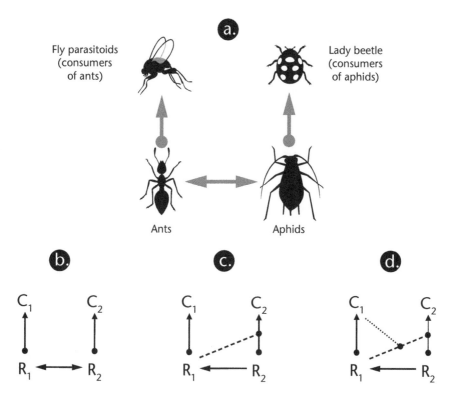

FIGURE 5.15 Illustrative example of trait-mediated cascade coupling (arrows indicate positive effects and circles indicate negative effects). Basic consumer resource oscillators are connected initially through direct mutualism. (a) The natural history of the example, in which a species of ant and its fly parasitoid form one oscillator and a species of aphid and its lady beetle predator form the other, where the ants and aphids are related through a mutualism. (b) Symbolic representation of the case in (a). (c) One direction of direct mutualism is replaced by an indirect trait-mediated coupling, representing the aphids "directly" providing honeydew to the ants and the ants interfering with the lady beetle predators, a trait-mediated effect. (d) The trait-mediated "cascade," in which the behavior of the parasitoids of the ants interferes with the trait-mediated effect of the ants against the natural enemies of the aphids.

depending on the strength of the coupling. This complex system becomes even more complex when we add another trait-mediated interaction (Figure 5.15(d)). For example, as we noted in the previous section, the fly parasitoids of the ants, in addition to consuming the ants directly, reduce their foraging activity. In that way, the flies affect the ability of the ants to affect the ability of the lady beetles to consume hemipterans. This is an example of a trait-mediated cascade.[71] From a theoretical perspective, these kinds of trait-mediated coupling of oscillatory systems can become intractable very quickly.[72]

Regardless of how difficult it is to understand the general behavior of these systems in which the oscillators are coupled through trait-mediated interactions, it is arguably important that we understand them better both from a theoretical and an empirical perspective, since they are common in both aquatic and terrestrial ecosystems. Indeed, in many ways, communities and ecosystems are collections of oscillators, more like a network of springs than a network of boxes and arrows, as they are often represented in textbooks.

In this chapter, we revealed a series of trophic and trait-mediated cascades that we have uncovered in the coffee agroecosystem. These density- and trait-mediated interactions proved to be more complicated than they seemed to be at first glance. From a theoretical perspective, these complex interactions can lead to very complex, frequently unpredictable behaviors. From a practical perspective, they can result in the loss of crops due to an unexpected pest outbreak, or to autonomous pest control by which potential pests remain steadily at low levels.

When farmers make observations of ecological interactions on their farms, they frequently see the direct density-mediated interactions, like a bird eating a caterpillar. Sometimes they can even piece together indirect interactions, like ants tending hemipterans that feed on the crop. It is far more difficult, sometimes even impossible, to "see" complex cascading and trait-mediated interactions of the sort described in this chapter. Don Walter Peters, the farmer who inspired us to conduct the bird exclosure experiment, is very knowledgeable about many direct ecological interactions that take place on his farm. He knows that leaf miners mine the coffee leaves, that the coffee berry borer bores into the coffee berries, that spiders and ants prey on the coffee berry borer, that the rust and other diseases attack his coffee plants. During his morning stroll through the farm, he can observe ecology in action. However, without ecological experimentation, he could not have predicted that patches of *Azteca* ants with scale insects are essential for the survival of the lady beetle that ultimately controls the scales on the rest of the farm. These kinds of ecological studies contribute to a deeper understanding of the complexity of nature, more generally, and the complexity and consequences of ecological interaction on the coffee farm.

Notes

1 Holmes (1990).
2 Greenberg et al. (2000); Van Bael et al. (2003, 2008); Perfecto et al. (2004); Philpott et al. (2004, 2008a, 2009); Van Bael and Brawn (2005); Borkhataria et al. (2006); Gunnarsson (2007); Kellermann et al. (2008); Whelan et al. (2008); Johnson et al. (2009, 2010); Mooney et al. (2010).
3 Kalka et al. (2008); Williams-Guillén et al. (2008).
4 Borkhataria et al. (2006).
5 Sekercioglu (2006); Mooney et al. (2010); Jedlicka et al. (2011).
6 Polis et al. (1989); Spiller and Schoener (1990); Gunnarsson (2007); Federico et al. (2008); Van Bael et al. (2008).
7 Polis and Holt (1992); Polis and Strong (1996); Holt and Polis (1997); Daugherty et al. (2007).

8 Johnson et al. (2010).
9 Perfecto et al. (2004).
10 Pimm and Lawton (1978).
11 Winemiller (1990); Hall and Raffaelli (1991); Polis and Holt (1992); Strong (1992); Rosenheim et al. (1995); Polis and Strong (1996); Finke and Denno (2002); Müller and Brodeur (2002); Arim and Marquet (2004); Duffy et al. (2007); Janssen et al. (2007).
12 Spiller and Schoener (1994); Holyoak and Sachdev (1998); Arim and Marquet (2004).
13 Pimm and Lawton (1978); Pimm (1982); Pimm and Rice (1987); Law and Blackford (1992); Fagan (1997); Holt and Polis (1997); McCann and Hastings (1997); McCann and Yodzis (1998); Mylius et al. (2001); Emmerson and Yearsley (2004); Tanabe and Namba (2005); Vandermeer (2006a); Namba et al. (2008).
14 Vandermeer (2006a).
15 Vandermeer (2006a).
16 For example, Holling (1973); Kirlinger (1986); Law and Blackford (1992); Hofbauer and Sigmund (1998).
17 McCann and Yodzis (1998).
18 If we allow T = top predator, P = other predator and H = herbivore, the standard equations, presuming a linear relationship between predation rate and food density, are

$$\frac{dT}{dt} = a_{TH}HT + a_{TP}TP - m_T T$$

$$\frac{dP}{dt} = a_{PH}HP - a_{TP}TP - m_P P$$

$$\frac{dH}{dt} = rH(1-H) - a_{TH}HT - a_{TP}TP$$

but usually ecologists relax the assumption that there is a linear relationship between predation rate and food density, and allow a saturating function define a_{ij}, such that

$$a_{TH} = \frac{\alpha_{TH}}{b_{TH} + H}$$

$$a_{TP} = \frac{\alpha_{TP}}{b_{TP} + P}$$

19 Vandermeer (2006a).
20 Vandermeer (2006a).
21 Hutchinson (1961).
22 Armstrong and McGehee (1980).
23 This idea was perhaps first articulated formally by Levins (1979), who showed that the non-linearity effectively made a single resource appear to be more than just a single resource, a point made even more explicitly by Meszéna et al. (2006).
24 Vandermeer (1993, 2004, 2006a); Huisman and Weissing (1999); Vandermeer et al. (2002a); Vandermeer and Pascual (2006); Benincà et al. (2009).
25 Hastings and Powell (1991).
26 Vandermeer (1993, 2006).
27 Elton (1927).
28 Lindeman (1942).
29 Hairston et al. (1960); Pimm and Lawton (1978); Post (2002).
30 Pascual and Dunne (2005).
31 Hairston et al. (1960).
32 Paine (1969); Pimm (1982); Persson (1999); Yodzis (2000).

33 Elton (1927); Pimm (1982); Carpenter and Kitchell (1996); Schmitz et al. (2000); Dunne et al. (2002).
34 Cohen et al. (1993); Emmerson and Raffaelli (2004); Carbone et al. (2007).
35 Polis et al. (1989); Schoener (1989); Cohen et al. (1993); Spiller and Schoener (1994); Holt and Polis (1997).
36 Cohen (1978); Hastings and Conrad (1979); Paine (1980); Pimm (1982); Briand and Cohen (1987); Cousins (1987); Schoener (1989); Post (2002).
37 Schoener (1989); Spiller and Schoener (1990); McCann et al. (1998); Yodzis (2000); Stouffer et al. (2007).
38 Spiller and Schoener (1990, 1994); Hairston and Hairston (1997); Oksanen and Oksanen (2000).
39 Montoya and Solé (2003).
40 Holmes et al. (1979); Marquis and Whelan (1994); Sanz (2001); Van Beal et al. (2003); Van Beal and Brawn (2005); Kalka et al. (2008).
41 Greenberg et al. (2000); Mols and Visser (2002); Hooks et al. (2003); Van Bael et al. (2003); Perfecto et al. (2004); Philpott et al. (2004); Borkhataria et al. (2006); Gunnarsson (2007); Kellermann et al. (2008); Whelan et al. (2008); Williams-Guillén et al. (2008); Johnson et al. (2009); Philpott et al. (2009); Johnson et al. (2010); Mooney et al. (2010).
42 Ibarra Núñez (2001).
43 Holmes et al. (1979); Marquis and Whelan (1994); Greenberg et al. (2000); Van Beal et al. (2003); Gruner (2004); Philpott et al. (2004).
44 Greenberg et al. (2000); Van Bael et al. (2003, 2008); Perfecto et al. (2004); Philpott et al. (2004, 2008a, 2009); Van Bael and Brawn (2005); Borkhataria et al. (2006); Gunnarsson (2007); Kellermann et al. (2008); Whelan et al. (2008); Johnson et al. (2009, 2010); Mooney et al. (2010).
45 Polis et al. (1989); Polis and Holt (1992); Spiller and Schoener (1994); Holt and Polis (1997); Hodge (1999).
46 Schoener and Spiller (1987); Spiller and Schoener (1990); Greenberg et al. (2000); Van Bael et al. (2003); Gruner (2004); Philpott et al. (2004).
47 Riechert and Lockley (1984); Wise (1995); Ovadia and Schmitz (2002); Schmitz and Sokol-Hessner (2002).
48 Schoener et al. (2002).
49 Völkl and Kraus (1996); Weisser and Völkl (1997); Ibarra-Nuñez (2001).
50 Memmott et al. (2000).
51 Völkl and Kraus (1996); Weisser and Völkl (1997).
52 Schoener (1989).
53 Hairston and Hairston (1997).
54 Levins (1974).
55 Riechert and Hedrick (1990).
56 Golubski and Abrams (2011); Kamran-Disfani and Golubski (2013).
57 Werner and Peacor (2003).
58 Styrsky and Eubanks (2007).
59 Liere and Larsen (2010).
60 Hsieh and Perfecto (2012).
61 Hsieh et al. (2012).
62 Vantaux et al. (2012).
63 Liere and Perfecto (2008).
64 Mathis et al. (2011).
65 Hsieh et al. (2012).
66 Hsieh et al. (2012).
67 The general idea of trophic webs as coupled oscillators is explored for very simple situations in Vandermeer (2006). Benincà et al. (2009) provide an example of teasing out subtle coupling effects from a real community.
68 Jha et al. (2012).

69 Some examples can be found in Abrams (1984, 1991, 1995); Bolker et al. (2003); Golubski and Abrams (2011).
70 Vandermeer (1996, 2006); Vandermeer et al. (2001); Golubski and Abrams (2011).
71 Liere and Larsen (2010).
72 Some interesting advances have been made with these sorts of complicated couplings by Golubski and Abrams (2011) and Kamran-Disfani and Golubski (2013).

6

INTERACTIONS ACROSS SPATIAL SCALES

Introduction

The Brazilian Pantanal is a most amazing place, presenting us with a spatial panorama of swamp, savannah, and forest (e.g., Figure 6.1). Organisms are constrained in this space. So, for example, if you wished to encounter a capybara (Figure 6.1(c)) in the Pantanal, a large marshland area in western Brazil, you would not go to any of the "islands" of terrestrial vegetation that dot the marshland (Figure 6.1(a)), but

FIGURE 6.1 Images from the Pantanal of Brazil. (a) An area of the Pantanal, near Campo Grande in southern Brazil. (b) The giant anteater, common in the upland areas of the Pantanal. (c) The capybara, the world's largest rodent, common in the marshy areas of the Pantanal.

FIGURE 6.2 Finer scales of the Pantanal. (a) Vista showing patches of open water and various kinds of marsh grasses forming non-random and non-uniform patchwork. (b) Close-up of the vegetation in one of the open water areas with various vegetation textures also forming a non-random pattern.

rather you would search the marshland itself. If you were interested instead in seeing a giant anteater (Figure 6.1(b)), you would concentrate your search in the upland patches. And then, at closer range, we see a similar patchy pattern in the vegetation (Figure 6.2), qualitatively repeated at even closer magnification. The more we study spatial patterns in nature, the more it looks like all ecosystems exhibit similar patterns at a variety of scales, or what is called fractal character.

In ecology, it is increasingly recognized that the spatial patterns of ecosystems are extremely important in the general structure of those ecosystems, as we already discussed in Chapter 4. Two of the most useful theoretical frameworks in ecology are based on the idea that the habitats of organisms are spatially structured (island bio-geography and metapopulation theories), and the phenomenon of self-organization emerging from ecological interactions has become standard fare as well (see Chapter 4).

Less attention has been given to the study of interactions among spatial patterns at various spatial scales, and serious studies of population dynamics in spatially self-organized systems are unfortunately scarce. One example of these kinds of studies is the case of metapopulations or source–sink populations that emerge from the spatial organization of the ant *Azteca* and the scale insect *C. viridis* on the coffee farms in southern Mexico, as discussed in Chapter 4. Yet we have observed in this same system that spatial patterns appear to be self-organized at a variety of distinct scales. The *Azteca* ants form patches at the scale of hectares, but other ants form patches at the scale of meters. And most importantly, these ants do indeed interact with one another, suggesting that studying interactions across such scales may reveal details that would otherwise be obscure.

The importance of this subject is not simply its potential contribution to general ecological understanding. It takes on special significance when we attempt to deal with the provisioning of ecosystem services, such as pest control. If pest control agents are limited to particular patch types, the origin of those patches is of practical importance.

In this chapter, we explore this idea with the system of *Azteca*, at a large scale, and a group of other ant species, at a smaller scale. Many of these species have been shown to be potentially excellent predators of some coffee pests. We have already discussed the self-organized spatial pattern of the *Azteca* ant in Chapter 4. We begin our discussion here with a description of the smaller-scale dynamics of the other ants.

Small-scale patterns in the ant community

We think we have a reasonably good idea of how the spatial pattern of *Azteca* nests gets formed, namely through a Turing-like process of nest expansion coupled with natural enemy control (Chapters 4 and 5). A few natural enemies – such as the fly parasitoid, the lady beetle predator and the fungal disease – may all be implicated in the process of pattern formation and simultaneously responsive to that pattern with their own population dynamics. All of this was introduced in Chapter 4, and further detailed in Chapter 5. Having discussed the dynamics of those large-scale patterns, we now turn to a smaller spatial scale. At this smaller scale, we also see evidence of spatial pattern among various other predators and the organisms with which they interact, along with important implications for their population dynamics and potential contributions to ecosystem services (as will be discussed in Chapter 7). The most evident of these are the ants, both those that forage on the vegetation and those that forage on the ground.

During a casual walk through a shade coffee plantation, one is prone to encounter many species of ants. In part, this is due to the fact that coffee is grown in the tropics, where species richness is high, but also it helps that shade coffee farms offer a forest-like environment in which many species can co-exist. To the untrained eye, it may seem like ants are everywhere, that there is no particular pattern in their distribution. However, a myrmecologist, able to distinguish among different species, may notice a non-random pattern in the way in which the various species are distributed. She may notice a patch of *Crematogaster*, a genus of ant that has a heart-shaped abdomen, followed by a patch of fire ants, followed by a patch of a species of *Pheidole*, the most diverse ant genus in the world. It is evident that these ants do not occur randomly on the farm, but rather that some sort of spatial pattern is evident, perhaps associated with some patchiness in the underlying vegetative structure or soil moisture conditions or other abiotic factors, but also perhaps associated with the biological interactions among the species, the self-organization of the ant system.

Indeed, our studies thus far strongly suggest that there is such "self-organization" at work here, although its specific nature seems more related to the ecological process of competition, rather than to effects from natural enemies, as it seems to be at the larger scale with *Azteca*. It appears that the large-scale pattern observed with *Azteca* is organized from the "top down."[1] That is, organisms above the ant in the energy transfer chain, such as their parasitoids, control its pattern. On the other hand, the smaller-scale patterns that we observe in the rest of the ant community seem organized from the "bottom up," either by organisms below the ants in the energy transfer chain (in other words, their food) or perhaps by the control of space (in other words, their nesting sites). One of the ecological processes involved in bottom-up control is competition – consumers seek the same or similar foods and thus compete with one another, or one species seeks to dominate a space sought by another species and the two species thus compete with one another for space. We thus begin this section with some comments about ecological competition and its role in the formation of spatial patterns.

Ecological competition and spatial pattern: the theory

Although writings on the subject of competition and spatial pattern are extensive and eclectic, it is nevertheless possible to make sense of this literature with a dichotomous classification. Some studies find the emergence of a mosaic of relatively discrete patches that retain a general qualitative structure for long periods of time.[2] This pattern is usually referred to as a fixed mosaic, and is frequently observed in communities of ants.[3] Other studies concentrate on the formation of so-called "moving spiral waves,"[4] based on the existence of "intransitive loops," the basic idea of the children's game of rock, scissors, paper. In ecological competition, if Species A beats Species B, and Species B beats Species C, and Species C in turn beats Species A, that cycle is referred to as an intransitive loop.[5] These two basic forms are illustrated in Figure 6.3. Although the difference between a mosaic pattern and a spiral pattern is visually striking, the actual detection of "spiralness" is difficult for any situation in the real world. However, that is not really the important part of this dichotomy. In the case of the spiral patterns (Figure 6.3(a)), the spatial positions of the species are variable in time, whereas in the fixed mosaic pattern (Figure 6.3(b)), the spatial positions of the species are constant in time. So the test of which of the two patterns better describes a natural situation is not so much whether they appear spiral or patchy, but rather the extent to which the pattern remains constant in time. If the underlying pattern is a spiral, we expect the pattern to change over time. If the underlying pattern is a mosaic, we expect it to be constant over time.

A related topic commonly treated in the ecological literature is the effect that spatial pattern has on the process of competition,[6] effectively the other side of the coin of competition creating spatial pattern. Naturally, there are many reasons that spatial pattern may exist, not just the self-organization described here. It may be that a particular spatial pattern in the distribution of an organism is driven by the type of vegetation or variation in microclimate, or any other type of factor exogenous to the ecological dynamics of the particular organism in question. Those spatial patterns, whether exogenous or endogenous, have long been thought to affect the way in which competition eventually plays out – which species go locally extinct and which survive, or whether species will coexist with one another for a long period of time.

Sessile organisms are likely to engage in strong competition. Only a single tree can occupy a particular site in a forest such that if the spatial model considers the scale of an individual tree as the relevant scale, competition must be strong by definition, since there is no possibility of two individuals coexisting at the very local

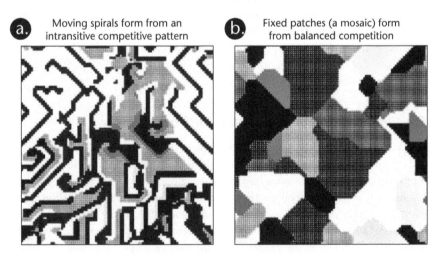

a. Moving spirals form from an intransitive competitive pattern

b. Fixed patches (a mosaic) form from balanced competition

FIGURE 6.3 Theoretical patterns generated by computer model, illustrating the two extreme forms of pattern generated by spatially explicit competition. (a) The spatial pattern that emerges from multiple intransitive loops creating moving spirals, a pattern probably difficult to actually detect in nature. (b) The fixed mosaic pattern that emerges from the situation in which competition is balanced between the competitive effect and competitive response. These patches remain constant over time.

Source: Adapted from Vandermeer and Yitbarek (2012)

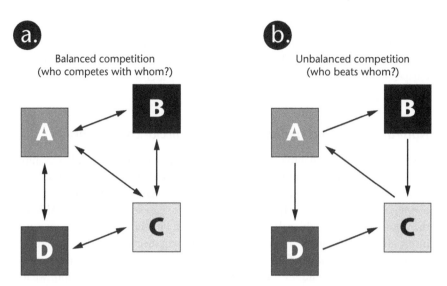

a. Balanced competition
(who competes with whom?)

b. Unbalanced competition
(who beats whom?)

FIGURE 6.4 The two extreme forms of competitive arrangements, where the arrowhead indicates "competitively dominates over." (a) Balanced competition in which it is not possible to say which member of a pair will win in competition. This is the general form that generates a fixed mosaic in space. (b) Unbalanced competition in which the winner in any paired match-up is strictly determined. This is the form that generates spiral waves (note that both A > B > C > A and A > D > C > A are intransitive loops).

level. In other words, no two trees can occupy the exact same site. Ants are considered sessile organisms because the relevant reproductive unit is the colony, not the individual ant, and they tend to be sessile or semi-sessile. For this reason, ants are likely to engage in strong competition, since no two ant colonies can occupy the exact same site.

The tendency of species assemblages to form patches in space, whether fixed mosaics or moving spirals, is a function of the balance rather than the intensity of competition. Competition is balanced when the effect of Species A on Species B is the same or similar to the effect of Species B on Species A, and unbalanced when the effect of Species A on Species B is different to the effect of Species B on Species A. This suggests a conceptual framework in which the competitive structure of a community falls on a gradient ranging from highly balanced to highly unbalanced. In this conceptualization, a strong fixed mosaic structure tends to form at the balanced extreme and the moving spiral waves tend to form at the unbalanced extreme (Figure 6.4).[7]

There are many complicating factors involved in the competition among ant species[8] and using this theoretical framework is fraught with the problem that it is likely too simplified to really account for many real-world patterns. Consequently, in addition to asking whether the small-scale spatial pattern emerges as self-organized through the process of interspecific competition, it is important to know something about the natural history of the species involved. In the following section, we inject a dose of realism by describing the natural history of a specific assemblage of ant species in two plots on a coffee farm in Chiapas, Mexico. We then link this local-level ant assemblage with the large-scale spatial pattern of *Azteca*, described in Chapter 4, and discuss the relevance of these spatial patterns for the biological control of several pest species in the system.

Natural history and spatial pattern: the special case of ants

Tropical ants tend to fall into two major groups, terrestrial and arboreal. For example, *Azteca*, the subject of the previous section on large-scale patterns, nests and forages arboreally, which is to say in the vegetation above the ground. By contrast, *Pheidole protensa* (discussed more fully below) nests and forages on the ground. Although there are exceptions to this rule,[9] it is generally useful to consider the ant community in any given area as either the terrestrial, ground-foraging community or the arboreal community. Most studies of ant communities focus on either the terrestrial (ground) or the arboreal communities.[10] Here we note how, in the case in question, we lose a great deal of information and insight if we so restrict our attention.

While most species nest and forage either on the ground or in the vegetation, some species forage both arboreally and terrestrially regardless of where they nest. It is a perhaps obvious but nevertheless important observation that this latter category connects the arboreal community to the terrestrial community. Thus, the focus on either the arboreal or the terrestrial community may leave out an important component of the whole structure since the connection, even if weak, makes the two seemingly independent communities part of the same metastructure.

In Chapter 4, we examined the spatial pattern of *Azteca*, whose spatial pattern develops over a relatively large area (i.e., hectares). Although *Azteca* is the most evident arboreal nester in the system, at least two other species are relatively common as arboreal nesters with large colonies (*Crematogaster* sp. and *Campanotus textor*[11]), whose spatial pattern may be similar in scale to that of *Azteca*, but who have thus far been studied only at a relatively small scale. Turning to the other ants in the system, those who form spatial patterns at a much smaller scale (square meters), there are 10–15 such species that commonly nest in tree cavities throughout the farm, both in the coffee trees themselves and the shade trees.[12] Additionally, at least one species, *Pheidole synanthropica,* is a common arboreal forager even though it nests on the ground, and, finally, some 30–40 species are known to be ground foraging and ground nesting. Clearly, it is a complicated matter to consider the spatial organization of this large ant assemblage.

Despite this enigmatic complexity, it seems that the general spatial patterns seen with these ants are a consequence of interspecific competition, not the Turing process described in the previous section on pattern formation in *Azteca*. We arrived at this conclusion based on our generalized natural history observations as well as a substantial literature suggesting that competition plays an important role in structuring ant assemblages at this scale.[13] Sampling ants on coffee farms, we have observed a similar spatial pattern that has been reported in other studies; various species tend to form patches that are relatively discrete, each species occupying its own space and the patches fitting together almost as if they were pieces in a jigsaw puzzle. This pattern is commonly referred to as a spatial mosaic and, as we noted above, is commonly reported in ants.[14] Our studies suggest that much of the spatial pattern is indeed a consequence of the underlying structure of competitive interactions among the various species,[15] although that structure is not simple, as previously discussed.

It is important to note that our elaboration of the spatial pattern in this section is limited to those species of ants that are attracted to baits. As was acknowledged some time ago,[16] any species assemblage is defined by the sampling methods (moths that come to light traps, visible/singing birds, bats that get trapped in mist nets, etc.). Ecologically, the ants we sampled are generalist ants that forage on the forest floor and coffee plants, and our baiting technique is just a convenient manner of sampling them. However, we do not pretend to be studying "the entire ant community," but only those generalist species that are expected to engage in competitive interactions. Indeed, from other studies at this site,[17] we have determined that our sampling method (with tunafish baits on the ground) generates a subset of approximately 30 percent of ants that can be sampled from a detailed search through the leaf litter. However, it is likely that the ants we do sample with the baiting technique are in competition with one another, since they are generalists and are attracted to the same bait (oil and tunafish).

The major players in small-scale structuring

For several years, we sampled the ground-foraging and arboreal ant community in two plots within a shade coffee farm in Mexico using the baiting technique. Through systematic sampling at these two sites, plus natural history observations, we pieced together what appear to be the major features of spatial community structure, quite distinct in the two sites, even though the background is similar, both in the physical environment and the relevant species pool. The vast majority of the sampling points were dominated by one of the six principal species that we took as the subjects of our spatial analysis: *P. synanthropica, Pheidole protensa, Solenopsis geminata, Azteca, Crematogaster* sp. and *Solenopsis picea*. The first three nest in the ground and the latter three are arboreal nesters. The foraging patterns are somewhat more complex.

Azteca, the subject of Chapter 4, is a very aggressive ant that forms large colonies, nesting in the shade trees of the coffee system and sometime establishing satellite nests on the coffee bushes. It actively tends hemipterans in both the shade trees and the coffee bushes – mainly the green coffee scale, *C. viridis*, in the coffee. It mainly stays in the trees and coffee bushes, but sometimes forages on the ground very near its nesting tree. It is actively attacked by at least four species of phorid fly parasitoids in the genus *Pseudacteon*, as discussed in Chapter 4. At a large spatial scale, it forms clusters of nests, with most likely all members of the same colony within a given cluster, which is to say the spatial extent of a single colony is likely very large (see Chapter 4). It is polygynous, with each nest having multiple queens, and budding sister colonies consisting of at least one queen and associated brood and workers.

FIGURE 6.5 Wax cast of the nest of *P. synanthropica*. The top of the nest is to the left.

It dominates trees and very rarely are any of the many other arboreal species in the system found on the same trees that have an *Azteca* colony. It is important in the present study because of its necessary restrictions in space, requiring a shade tree to position its nest, thus limiting its foraging range by a force independent of the competitive effects of the other species.

P. synanthropica has a medium-sized ground nest that is roughly cylindrical with a diameter of 10–15 centimeters and a length of 20–30 centimeters and with multiple laterally flattened chambers (Figure 6.5). This species is unusual in that it nests in the ground but forages equally readily on the ground or in the vegetation, frequently tending scale insects and other hemipterans. Second only to *Azteca*, it is the species that tends large numbers of *C. viridis*. It forms very large and aggressive swarms at baits and seems capable of excluding many other species, when its nest is close enough for effective recruitment. Although it is a very common species, it escaped the attention of taxonomists until 2010.[18]

P. protensa is a minute, slow-moving, ground-nesting and ground-foraging ant. The structure of its nest is enigmatic, but presumably is in the form of small subterranean nests. Multiple entrances are common – in three 0.5 × 1.0 meter survey areas, we encountered 7, 8 and 13 nest entrances, some of which were clearly entrances for the same colony, while others were for distinct colonies judging by aggressive encounters with individuals originating from distinct entrance holes. As is common for this genus, there are two morphological forms: very large head capsule ("majors," in the lexicon of ant biologists) and very small head capsule ("minors"). Majors and minors seem always to occur together, and no evident division of activities between majors and minors occurs.

S. geminata is the infamous tropical fire ant, known to Neotropical travelers for its habit of crawling up one's leg surreptitiously and stinging *en masse*. It is a ground nester, creating a central chamber with radiating covered tunnels, presumably a defensive maneuver against the phorid flies that regularly attack it,[19] not the same species that attack *Azteca*. Such tunnels also make it frequently victim to the

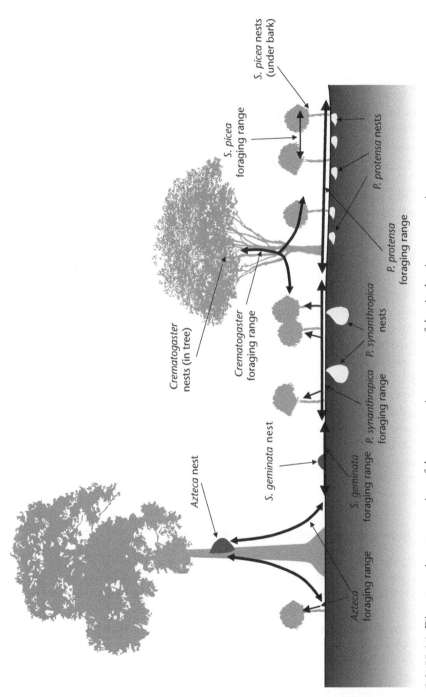

FIGURE 6.6 Diagrammatic representation of the community structure of the six dominant ant species.

Source: Modified from Perfecto and Vandermeer (2013)

underground raiding army ant, *Labidus coecus*.[20] Its dome-shaped central chamber presumably collects heat and it is a common sight in more open areas. It is especially tolerant of high temperatures.[21] It is extremely aggressive at baits, normally displacing many of the other species in this system. On the other hand, in at least one instance, its long-term competitive dominance has been questioned based on its relatively lower ability to discover resources compared to a species of *Pheidole*.[22]

The genus *Crematogaster* is notoriously difficult taxonomically and consequently we are not certain of the formal identification of the species so we just refer to it as *Crematogaster* sp. It nests arboreally and forages mainly arboreally, although it is occasionally encountered foraging on the ground. Several other species in the genus tend to form loosely connected hubs that together form a megacolony, which we suspect is the case here also.

S. picea is a minute black ant that is commonly encountered foraging arboreally. Its nests are seemingly loosely organized under the flaking bark on the trunks of trees, including the coffee bushes on which it is regularly found. We have only occasionally encountered it foraging on the ground, although it has been reported to occur in samples of leaf litter.[23] It is known to form very large colonies where it is difficult to delimit borders of the colonies.

The structure of the community of this collection of six species is qualitatively obvious. *P. synanthropica* easily gains a competitive advantage through its excessively broad foraging range. *Azteca* evidently excludes all other ants near its nest but is restricted in its foraging area by the need for a relatively large tree in which to nest, thus creating an automatic spatial constraint on the system. Neither *Crematogaster* nor *S. picea* are able to form successful nests in areas where *P. synanthropica* dominates. *P. protensa* appears to exclude *P. synanthropica* on the ground, thus indirectly facilitating the arboreally foraging *Crematogaster* sp. and *S. picea*. *S. geminata* seems able to dominate all ground sites when the physical conditions are appropriate (local light gaps), although its relationship to its phorid parasitoids is likely to alter its competitive advantage,[24] and the voracious army ant *L. coecus* undoubtedly plays a major role as one of its predators.[25] The general qualitative structure of the community of these six locally abundant species is shown in Figure 6.6.[26]

The nature of the small-scale spatial pattern

During the years 2009–2011, we sampled ground and arboreal ants in two small plots (approximately 50 × 50 meters) to get some insights regarding their spatial pattern, the mechanism of pattern formation and the potential implications for pest control. The sampling was conducted using a 4 × 4 meter grid in which tunafish baits were located at each point. We noted that the spatial pattern formed by these species was different for each of the sites, both in terms of generalized appearance and dynamic stability. The terrestrial foraging ants at Site I do not seem to maintain a fixed mosaic over time (Figure 6.7). By contrast, the terrestrial foraging ants at Site II seem to maintain a fixed mosaic with the two predominant species *P. synanthropica* and *P. protensa* (Figure 6.7). The activity of *Azteca* has an important effect at both

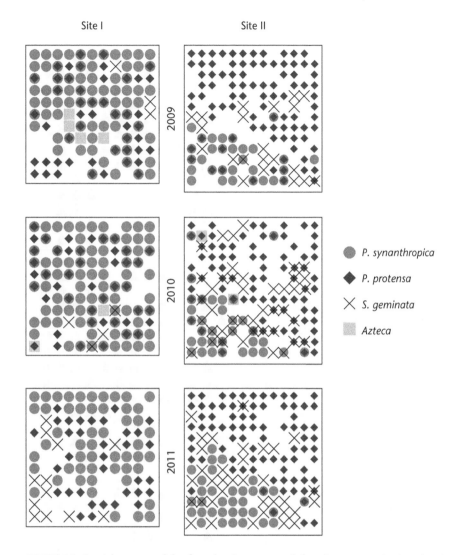

FIGURE 6.7 Spatial pattern of the four dominant ground-foraging ant species in a 4 × 4 meter grid.

Source: Modified from Perfecto and Vandermeer (2013)

sites, but it is restricted to a small area around its nesting tree and is not well captured at the coarse scale that the sampling was conducted.

The arboreal ants at the same two sites (Figure 6.8) reflect, to some extent, the pattern of the terrestrial foragers, but with significant exceptions. One of the dominant ground foragers, *P. synanthropica*, is overwhelmingly dominant at Site I, but occupies only the same area as it does on the ground in Site II. This is not surprising given the natural history of this species. It nests on the ground and forages both

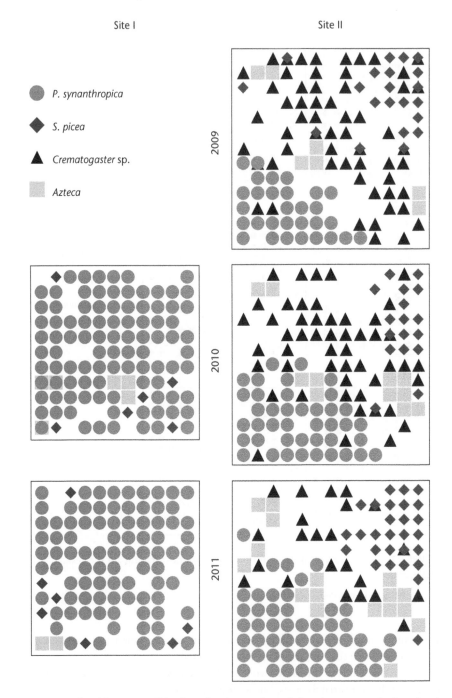

FIGURE 6.8 Spatial pattern of the four dominant arboreal-foraging ant species in a 4 × 4 meter grid. Sampling in 2009 was not spatially explicit for Site I.

Source: Modified from Perfecto and Vandermeer (2013)

terrestrially and in the coffee bushes. It seems that it restricts other species that forage arboreally, yet that ability to restrict other species is clearly determined by its nesting position. In Site II, this same species is restricted to only a corner of the plot, leaving the rest of the plot open to foraging from two other arboreal dominants, *Crematogaster* sp., and *S. picea*. It is notable that *Crematogaster* sp. has never been found at Site I, yet it is quite abundant at Site II, always in approximately the same area of the plot. As in the case of the terrestrial foragers, at Site II the three dominant arboreal-foraging species appear to form a fixed mosaic, the three "patches" retaining their relative position in space over the three-year period. On the other hand, because of the overwhelming dominance of *P. synanthropica* at Site I, the question of year-to-year constancy is rather difficult to answer.

Comparing Figures 6.7 and 6.8, it is evident that there is a connection between the arboreal foragers and the terrestrial foragers. Most important is the role of *P. synanthropica* as the only one of the dominant species that forages extensively both terrestrially and arboreally. *P. synanthropica* dominates the arboreal foraging areas at both sites, but, as is clear in Site II, its arboreal foraging area is largely determined by where it is nesting on the ground.

The spatial pattern of this community of six dominant ant species is understandable in terms of competitive interactions, taking into account the idiosyncrasies of the nesting and foraging habits of each species. What is perhaps most interesting is the dramatic difference between the two sites, despite the apparent similarity of background habitat conditions. The complete absence of *Crematogaster* sp. from Site I is one of the most obvious differences between the two sites, although some other subtle differences in community structure also exist (e.g., the presence in relatively high abundance of some other species).[27] Precisely how these differences could map into the observed differences in the two spatial patterns is not evident. However, it seems reasonable to suggest that *P. synanthropica* is able to dominate arboreally mainly because it has its nests located nearby, and being somehow restricted in those nesting sites at Site II opens up the foraging areas to *Crematogaster* sp. and *S. picea* in the coffee bushes. However, why there is a clear division between foraging areas of *P. protensa* and *P. synanthropica* at Site II but not at Site I is not evident.

An alternative explanation for the pattern is that *P. synanthropica* is unable to compete successfully with *P. protensa*, something that would seem unlikely to anyone having observed these two species together. Yet, as we have shown in other circumstances, an apparently dominant or aggressive species may lose out to a less obviously dominant or aggressive one, especially if the less dominant one is an especially good forager.[28] In the case of *P. protensa*, its abundant nests in the areas that it dominates suggest that it can dominate most resources that fall to the forest floor simply because of the density of foragers from so many nests. However, when *P. synanthropica* has access to arboreal resources, like scale insects, those extra resources may give them a competitive edge so they can out-compete, or at least coexist with, *P. protensa*. Under this scenario, the accidental existence of *Crematogaster* sp. in the one site may limit the ability of *P. synanthropica* and thus make it vulnerable to the effects of competition from *P. protensa*.

What is interesting is not only that the two sites are different in their spatial patterns, but that the pattern changes dramatically from year to year at one site but not at the other. This corresponds, qualitatively, to the theoretical results discussed earlier,[29] in which at one extreme of competitive organization, very rigid spatial mosaics are formed that remain constant over time, while at the other extreme, the spatial pattern is an ever-changing patchwork (Figure 6.3). The competitive organization relates to the degree to which intransitive competitive loops might be involved in the competitive process (Figure 6.4). When competitive intransitivities are likely, the mosaic that is formed is constantly changing its spatial form, as contrasted to the state when intransitivities are unlikely, in which case fixed mosaics are most likely. It is thus tempting to suggest that the particular complex of species in Site I contains intransitive competition whereas that of Site II does not. Precisely where these intransitivities might lie is completely unknown at the present time.

Interaction of the two spatial patterns and consequences for biological control

Ants as predators of coffee pests

It is now well established that several species of ant are major predators in the coffee system, most importantly having a demonstrable effect on the coffee berry borer (*Hypothenemus hampei* Ferrari), one of the major concerns of coffee producers the world over (Figure 6.9).[30] Of the species occurring at our site, we know unequivocally that *Azteca, P. synanthropica, P. protensa, Wasmannia auropunctata, Pseudomyrmex*

FIGURE 6.9 *Ganaptogenys sulcata* preying on the coffee berry borer on a Colombian coffee plantation.

Source: Francisco Posada and Moises Vélez of Cinecafé

simplex, Pseudomyrmex ejectus, Pseudomyrmex PSW–53 and *Procryptocerus hylaeus* are direct predators on the coffee berry borer. However, the mode of predation is different for each of the species. *P. synanthropica* and *Azteca* are effective predators as the coffee berry borer arrives at the coffee bush, preventing it from entering the berry in the first place.[31] However, other narrower species (e.g., *P. simplex* and *P. ejectus*) can enter the entrance hole made by the borers and attack them within the coffee seed, but not when the coffee plant is being patrolled by *Azteca*.[32] When fruits containing the borer fall to the ground, *P. protensa* is an effective predator, entering the fruits to predate.[33] Thus, the spatial pattern, which results in local dominance of particular species, is bound to have an effect on the general efficiency of the biological control of this and probably other pest species on coffee plantations. Furthermore, the predatory effect of the ants is only part of the overall interaction network that, we argue, controls pests in the system, as we discuss more fully in Chapter 7.[34]

One of the other obvious potential pests in the system is the green coffee scale insect, *C. viridis*. This one is particularly enigmatic, since it is mutualistically associated with the dominant *Azteca* ant. Yet, as we discuss in great detail in Chapter 7, the spatial pattern formed by this ant is critical to the maintenance of the pest control function in the system as a whole. The important point for the present discussion is that the spatial pattern formed by *Azteca* at the very large scale contributes in an important way to the overall ecosystem function of pest control. Thus, we are forced to conceptualize the system in at least two distinct spatial scales, the large-scale pattern formed by *Azteca* and its associates (as discussed in Chapter 4), and the smaller-scale pattern formed by competing ants both on the ground and in the vegetation. Both of these spatial patterns influence pest control in the coffee agroecosystem.

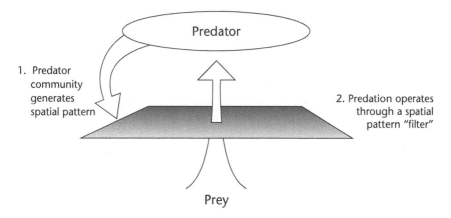

FIGURE 6.10   Basic structure of the interaction between predator and prey causing (1) spatial pattern, and (2) spatial pattern filtering of the predator–prey interaction.

The dialectics of predation and spatial structure

Much has been written connecting the dynamics of predation to spatial structure.[35] What many of these approaches have in common, although perhaps not explicitly stated as such, is the dialectical interaction between spatial structure and the process of predation. On the one hand, predation, by acting as a density-dependent mortality factor, can create the spatial structure itself.[36] On the other hand, the spatial structure affects predation,[37] which, in turn, affects the spatial structure, and so on and so forth.[38] We can effectively view the predation system emerging through a filter of spatial organization, while the filter of spatial organization itself is created by the dynamics of the predation system. We diagram the basic idea in Figure 6.10.

In our particular system, a key aspect of the relationship between spatial pattern and the pest control system is the keystone mutualism between *Azteca* and the scale insect *C. viridis* previously discussed in Chapter 5. This mutualism forms the center-piece for both the ongoing biological interactions and the resultant spatial structure, forming clusters of nests of *Azteca* at the large scale (macro scale) and partially determining the spatial structure of the other species (discussed in the previous section) at a smaller scale (meso scale). These patterns, at two spatial scales, are thus mediated by the keystone mutualism.

At both scales, we conceive of the spatial pattern acting as a filter for the dynamic processes in the network (e.g., predation), and the dynamic processes in the network acting as a constructive force of the spatial pattern, as depicted in Figure 6.10 for a simple predator–prey interaction. When the dynamic processes operate at both spatial scales, we see a dialectical connection between them. For the particular system we study, we illustrate this complicated two-scale system in Figure 6.11, which is an extension of the illustration in Figure 6.10. In this case, the interaction web that includes the predator is responsible for the creation of the spatial pattern "filter", and the dynamic processes associated with the many ecological interactions are themselves realized through that filter. In the system that we have studied exten-sively, there are two distinct spatial scales, a macro scale (from what we have deduced from the 45-hectare permanent plot on a traditional shade coffee farm), and a meso scale (deduced from studies in plots of less than one hectare, like those in Figures 6.7 and 6.8), and two distinct interaction networks, intimately connected to one another through the keystone ant, *Azteca* (Figure 6.11).

The macro-scale component of the system is centered on the keystone ant *Azteca* and its mutualist partner the scale insect *C. viridis*, along with one of the key natural enemies of the scale, the lady beetle *A. orbigera*. The dynamics of space with this system were described in the previous chapters. The Turing mechanism reported to exist is dependent on not only an activator (the tendency of the ants to move to nearby trees and form satellite nests), but also an inhibitor. The phorid parasitoid[39] that attacks *Azteca* is certainly a possible contributor to the inhibition process. However, several other possibilities exist, including the predatory coccinelid beetle (*A. orbigera*) as well as the fungal pathogen (*L. lecanii*) of the ant's mutualist associate, the green coffee scale. Thus, the large-scale dynamics may in the end involve a

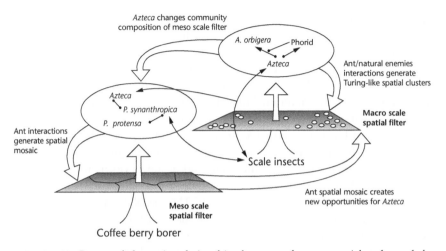

Coffee berry borer

FIGURE 6.11 Proposed dynamic relationships between the two spatial scales and the predator–pest complex in the coffee agroecosystem. The macro-scale filter refers to the spatial distribution of *Azteca* nests (small white circles), which was deduced from our 45-hectare permanent plot (Vandermeer et al. (2008)). The meso-scale filter refers to the patchwork mosaic of different ant species, which was deduced from plots of less than one hectare. At the macro scale, the predator is *A. orbigera* (the coccinellid beetle), which is embedded in the *A. orbigera–Azteca*–phorid complex (negative interactions symbolized by closed circles and positive interactions with arrowheads) that results in the control of *C. viridis* (the scale insect). This complex also generates the spatial pattern at the macro scale (patches of *Azteca*). At the meso scale, the predators are represented by the *Azteca–P. synanthropica–P. protensa* complex (negative interactions symbolized by closed circles). The patchwork mosaic of these species will determine the control of the coffee berry borer. This patchwork of the ant community will influence the formation of the macro-scale spatial filter by providing new opportunities for *Azteca* to establish new colonies.

variety of repressive factors operating in subtly different ways, but always in the context of a spatial filter, itself caused by the complicated interactions of the players involved.[40] These dynamic interactions create a "Turing-like" structure in which *Azteca* nests continue occupying new space, but become susceptible to natural enemy attack with increasing local nest density – an activator coupled with a repressor in the sense of the basic Turing mechanism.[41] A pattern of clustered ant nests results, both in our models and in the field, at this macro scale, as explained in Chapter 4.

In the meso-scale component of the system, the keystone ant is embedded in the ant community, which constitutes a major predatory pressure on the coffee berry borer.[42] Some of the non-*Azteca* ants also tend scale insects (e.g., *P. synanthropica*) and others do not (e.g., *P. protensa*), while some of the ants actively prey on the berry borer and others do not. Furthermore, some ants mainly forage on the ground (e.g., *P. protensa*), others mainly in the trees (e.g., *Azteca*), and a few (e.g., *P. synanthropica*) forage both on the ground and in the trees. The result is a complex pattern of competitive interactions among the ant species resulting in the formation of a spatial mosaic (as described in the previous section), in which relatively large patches are formed by dominant species, and those patches of various species form a sort of jigsaw puzzle pattern in space. Whether or not there is effective predation by ants on the coffee berry borer depends on the nature of that spatial mosaic.[43] Thus, predation on the coffee berry borer depends on the nature of the meso-scale pattern (the ant mosaic), which in turn is formed by the interactions among the ant species.

As illustrated in Figures 6.10 and 6.11, our most general conceptual framework consists of spatial structure, at two scales, acting as a filter for a herbivore-anchored interaction web. The herbivores, in this case, are the scale insects and the coffee berry borers. The complexity of this framework is most evident at two points of contact: first, where the interaction web creates the spatial pattern, and second, where the spatial pattern affects the interaction web. This spatial complexity adds to our understanding of this system, incorporating the dual spatial component to the already discussed trait-mediated indirect interactions and trait-mediated cascades (Chapter 5), which, we argue, are all important elements in the understanding of autonomous pest control, one of the subjects of the next chapter.

Notes

1 Vandermeer et al. (2008).
2 Durrett and Levin (1994, 1998); Bascompte and Solé (1995); Adler et al. (2007).
3 Majer et al. (1993); Blüthgen and Stork (2007); Dejean et al. (1997).
4 Boerlijst and Hogeweg (1995); Johnson and Seinen (2002); Edwards and Schreiber (2010); Rohani et al. (1997).
5 Vandermeer and Yitbarek (2012); Vandermeer (2013).
6 See for example Neuhauser and Pacala (1999).
7 Vandermeer and Yitbarek (2012); Vandermeer (2013).
8 Yitbarek et al. (2011).
9 Blüthgen and Feldhaar (2010); Dornhaus and Powell (2010); Perfecto and Vandermeer (2013).
10 See various summaries in Lach et al. (2010).
11 Philpott et al. (2006a); Philpott (2010).
12 Philpott et al. (2006a); Livingston and Philpott (2010); Philpott (2010).
13 Gotelli and Ellison (2002); Parr and Gibb (2010); Cerda et al. (2013).
14 Majer et al. (1993); Blüthgen and Stork (2007); Yitbarek et al. (2011).
15 Perfecto and Vandermeer (2011); Vandermeer and Yitbarek (2012).
16 Fisher (1956).
17 Armbrecht and Perfecto (2003).
18 Longino (2009).
19 Feener and Brown (1992); Morrison et al. (2000).
20 Perfecto (1992).
21 Torres (1984).
22 Perfecto (1994).
23 Longino (2009).
24 Morrison (1999); LeBrun and Feener (2002, 2007); Tillberg et al. (2007); Feener et al. (2008).
25 Perfecto (1992).
26 Perfecto and Vandermeer (2013).
27 Perfecto and Vandermeer (2013).
28 Perfecto (1991, 1994); Perfecto and Vandermeer (1996).
29 Vandermeer and Yitbarek (2012).

30 Vélez et al. (2000, 2001, 2003); Bustillo et al. (2002); Varón Devia (2002); Perfecto and Vandermeer (2006); Armbrecht and Gallego (2007); Larsen and Philpott (2010); Philpott et al. (2012); Jímenez-Soto et al. (2013).
31 Jímenez-Soto et al. (2013).
32 Larsen and Philpott (2010).
33 Perfecto and Vandermeer (2013).
34 Vandermeer et al. (2010a).
35 Kareiva et al. (1990); Hassell et al. (1991); Comins et al. (1992); Pascual (1993); Pascual and Levin (1999a, 1999b); Alonso et al. (2002); McCann et al. (2005).
36 Cousins (1980).
37 Gilinsky (1984); de Roos et al. (1998).
38 Liere et al. (2012).
39 Vandermeer et al. (2008).
40 Vandermeer et al. (2008, 2010a); Jackson et al. (2009); Liere and Larson (2010); Liere et al. (2012); Hsieh and Perfecto (2012); Philpott et al. (2012).
41 Turing (1952); Vandermeer et al. (2008, 2010a).
42 Armbrecht and Gallego (2007); Jaramillo et al. (2006); Bustillo (2002); Vélez et al. (2003); Perfecto and Vandermeer (2006, 2013); Larsen and Philpott (2010).
43 Perfecto and Vandermeer (2014).

7

BIODIVERSITY AND ECOSYSTEM SERVICES

Introduction: the nature of ecosystem services

The notion that ecosystems provide services that are of utility to *Homo sapiens* is certainly an ancient idea. In Ancient Greece, Epicurus (341 BCE–270 BCE) maintained that the environment established fixed limits on what we might extract for our general happiness, and in the later part of the nineteenth century, Engels warned us about the unintended consequences, or what today would be translated into the losses of ecosystem services, of the unwise manipulation of nature, in his famous quote:

> Let us not, however, flatter ourselves overmuch on account of our human conquest over nature. For each such conquest takes its revenge on us. Each of them, it is true, has in the first place the consequences on which we counted, but in the second and third places it has quite different, unforeseen effects which only too often cancel out the first. The people who, in Mesopotamia, Greece, Asia Minor, and elsewhere, destroyed the forests to obtain cultivable land, never dreamed that they were laying the basis for the present devastated condition of these countries, by removing along with the forests the collecting centers and reservoirs of moisture. When, on the southern slopes of the mountains, the Italians of the Alps used up the pine forests so carefully cherished on the northern slopes, they had no inkling that by doing so they were . . . thereby depriving their mountain springs of water for the greater part of the year, with the effect that these would be able to pour still more furious flood torrents on the plains during the rainy seasons. Those who spread the potato in Europe were not aware that they were at the same time spreading the disease of scrofula. Thus at every step we are reminded that we by no means rule over nature like a conqueror over a foreign people, like someone standing outside nature – but that we, with flesh, blood, and brain,

belong to nature, and exist in its midst, and that all our mastery of it consists in the fact that we have the advantage over all other beings of being able to know and correctly apply its laws.[1]

Although the idea is certainly not new, the specific phrase "ecosystem services" was adapted from the term "environmental services" more recently. Harold Mooney and Paul Ehrlich[2] attribute the origin of the phrase to George Perkins Marsh in the late nineteenth century and a variety of nature writers in the mid-twentieth century (e.g., Fairfield Osborn, William Vogt and Aldo Leopold)[3], and cite 1970 as the date of the first published use of the phrase in the book *Man's Impact on the Global Environment* by the Study of Critical Environmental Problems (SCEP).[4] In what seems to us to be an attempt to popularize something that had been evident to Plato, Aristotle and Epicurus, the United Nations Millennium Ecosystem Assessment (2005) took what appeared to be the easiest route to get folks to take notice: talk about the money in their pocket. If people are willing to pay, either directly or through their taxes, for a variety of "services," such as getting their hair cut or fire fighting, why not use the ecosystem as the metaphorical barber or firefighter? We must admit that there is something a bit unseemly about the need to convince people to *not* destroy the place they live because it provides them with stuff. However, the term is so well entrenched in contemporary environmental discourse that challenging it would be Sisyphean. So we reluctantly adopt this crude terminology.

In adopting this terminology, however, we cannot simply ignore some contradictory issues that emerge from its use. On the one hand, it is evident that the wetlands that covered the southern coast of Louisiana did indeed "function to" partially protect or "provided the ecosystem service of" partially protecting New Orleans from Hurricane Katrina. Likewise, many other obvious functions or services could be mentioned. Yet, other examples may be less evident. Is the build-up of dry leaves in Australia's eucalyptus forests a service? Since eucalyptus forests normally need fire to survive over the long haul, the dry matter does serve to maintain that ecosystem, but the houses built on the edge of the forest can be less described as being "served" by the fires that sometimes burn through entire towns in Australia. An ecosystem function or service that is frequently cited in the literature is the cycling of nutrients. Yet, what is it that is served? Will faster recycling of nutrients be a good thing for ecosystems, for humans? The fact of the matter is that nutrients indeed do cycle – it is a principle of biogeochemistry. Whether they "should" cycle either quickly or slowly is a function of a human decision about what is needed or wanted.

Here we find a critical definitional point. A service is provided for some need and a function does something to keep either the system running or one of its "products" available. These are effectively normative issues. And when speaking of natural ecosystems, natural forests, savannahs, wetlands etc., it is not at all clear what the normative desire is, nor which might be the best in the case of competing normative ideologies. Was it generally a good thing that fires were controlled by the Anglo-Australians, interrupting the fire management systems the Australian Aboriginals used? Was the service provided by the unburned vegetation better in some sense than the service provided by the fire disclimax regimes managed by the Aboriginals? These are questions that cannot be answered in a non-normative fashion. Do the forests that invaded the tundra following the retreat of the last glacier in Michigan provide a service to the ecosystem? The question itself makes no sense unless some preference for tundra or forest is established, and nature knows no such preferences.

It is quite a different issue when talking about human-managed systems – there is, by definition, normative content. We can, with no philosophical ambiguities, speak of the function or service provided by an ecosystem or some component thereof. Bees may serve to pollinate crops, parasitic wasps may serve to control pests, trees may function as moderators of temperature and carbon sinks, birds may serve to attract touristic adventurers. And since this book is about a human-managed system, the deep and perhaps obscure normative issues associated with ecosystem function and service do not really emerge. Normative is not a criticism; it is, well, the norm. However, as it is frequently the case in issues related to human affairs, what is good for some people is bad for others. Ecosystem services are not an exception. For example, the parrots so loved by well-off tourists from the Global North act as pests of the mango trees so enjoyed by the local farmers of the Global South – one person receiving a service, another receiving a disservice. Sometimes the same person experiences these trade-offs. For example, on coffee farms the shade trees serve many functions, like moderating microclimate, capturing carbon, suppressing weeds, providing organic matter and nitrogen to the soil, providing habitat for biodiversity, etc., but, at very high densities, they may reduce yield.

The Millennium Ecosystem Assessment grouped ecosystem services into four categories: *provisioning*, such as the production of food, fuel and fiber; *regulating*, such as the control of climate and disease; *supporting*, such as nutrient cycles; and *cultural*, such as spiritual and recreational benefits. It is clear that the coffee agroecosystem merits analysis within each of these categories, among multiple services or functions. However, in keeping with the focus on the research in which we have been personally involved, we restrict our discussion to the ecosystem services of pest management, climate change mitigation and pollination.

Pest management

Our approach

In the past 20 years of research on coffee farms in Mexico and Central America, we have encountered at least 30 different caterpillar species that eat coffee leaves (Figure 7.1). None of them, so far, has become a pest. Of the 850 insect species that have been reported to eat coffee worldwide, approximately 200 are Neotropical, yet only a relative handful have become significant pests.[5]

That there are so many insects that feast on the coffee plant is perhaps not so surprising. After all, coffee is a perennial plant that currently has a worldwide distribution. But why do so few of them become pests? The fact that coffee contains caffeine, a chemical that protects the plant from herbivores, is part of the answer. The

FIGURE 7.1 A sampling of lepidoptera that eat coffee leaves, but never become pests (so far).

production of caffeine is almost certainly an evolutionary response of the coffee plant to the pressure of herbivory. However, as always happens in nature, herbivores evolve detoxifying mechanisms to be able to deal with the plant's strategic defenses. This is known as the "Red Queen hypothesis." Just as the Red Queen explained to Alice (in *Alice in Wonderland*) that you have to keep running just to stay in the same place, the plants evolve chemicals to protect themselves from herbivores, and then the herbivores evolve mechanisms to detoxify the chemicals that the plants evolved, and so on and so forth. So at any point in time, we witness not only what has evolved, but also what is evolving.

Given this process of evolution, and the high number of herbivores reported to eat coffee, why are more insects not reported as significant pests of coffee? The answer may be found in another one of our casual observations while working on coffee farms. We frequently bring caterpillars into the lab to rear them so as to see what the adult butterfly or moth looks like, as most naturalists interested in insects do as a matter of course. In the case of almost all of the species we have encountered, when we bring them into the lab they turn out to contain parasitoids or pathogens that kill them before they reach maturity (Figure 7.2). It seems that while there are plenty of potential pests waiting in the wings, they are held in abeyance by an equally impressive array of natural enemies.

We frequently recall the study by one of our colleagues, Helda Morales, in traditional maize farming in the highlands of Guatemala.[6] She began with the idea of surveying traditional maize farmers about their pests and the methods they use to control them, aiming eventually to conduct experiments to see if, in the context of Western scientific experimentation, their methods made ecological sense. She began her farmer interviews by asking the question: "What kind of pests do you have on your farm?" And their responses were shocking. Every farmer said the same thing: "Pests? We don't have any pests." This was not something that a young ecologist beginning her career wanted to hear about one of her first planned experiments –

FIGURE 7.2 A parasitoid wasp attacking a potential pest caterpillar, on a shade coffee farm in southern Mexico. (a) The wasp crawling on the surface of the caterpillar. (b) The wasp ovipositing into the body of the caterpillar. (c) Wasp cocoons (several weeks later) emerged from the body of the caterpillar.

no pests basically equaled nothing to study. However, Morales reflected and modified her questionnaire; "What kinds of insects eat your maize?" she asked instead. To that question, she received a long list of insects that, since they eat maize, could be pests, but apparently were not. She then compared that list to a list of insect pests in maize in Central America. Here she encountered many species occurring in both lists. So the same insect species that were maize pests according to agricultural entomologists were non-pests according to traditional Mayan farmers.

In her interviews, Morales followed up with questions about why certain species were *not* pests and here she got all sorts of answers. Most farmers clearly understood the basics of biological control and tied the diversity of natural enemies to the failure of these species to reach pest status. Others focused on the nutritional status of the maize plant itself and its ability to resist the herbivores trying to chew on it. Others focused on something that a modern ecologist might refer to as the "health" of the farm. The point is that these farmers knew what they were doing. They were not "responding" to a problem on their farm; they were managing their farms such that those problems would not emerge in the first place. They were not trying to cure a disease; they were trying to prevent the disease from getting there in the first place.

At about the same time that Helda Morales was learning from the indigenous farmers of Guatemala about managing farms to prevent pest outbreaks, four insect ecologists and a pest management specialist published an important perspective piece in the *Proceedings of the National Academy of Science*, calling on a "total system approach to sustainable pest management." They wrote:

> A fundamental shift to a total system approach for crop protection is urgently needed to resolve escalating economic and environmental consequences of combating agricultural pests. Pest management strategies have long been dominated by quests for "silver bullet" products to control pest outbreaks. However, managing undesired variables in ecosystems is similar to that for other systems, including the human body and social orders. Experience in these fields substantiates the fact that therapeutic interventions into any system are effective only for short-term relief because these externalities are soon "neutralized" by countermoves within the system. Long-term resolutions can be achieved only by restructuring and managing these systems in ways that maximize the array of "built-in" preventive strengths, with therapeutic tactics serving strictly as backups to these natural regulators.[7]

Here we see a convergence of ideas from Western scientists and indigenous farmers, something that might serve as guiding principles for agroecologists interested in pest control. In the context of ecosystem services, if maintaining the farm relatively free of pests is a goal (hardly a debatable point), then the forces that do so could be thought of as providing that service. In this sense, pest control is an ecosystem service. However, this service does not automatically occur. Like other services (whether ecosystem types or others), it needs to be cultivated and managed. In order to effectively provide this ecological service, it is imperative that the ecology of the system be understood. And, since it is essential to understand not only how to control pests once they become pests, but also how to prevent insect herbivores from becoming pests in the first place, the essential role of ecological research in the endeavor is obvious.

This necessary ecological research must happen in a broad intellectual sphere, since we are dealing with something that is both complex and underdeveloped as a science. In such a situation, sources of insight are bound to be diverse, sometimes arising in unexpected places and contexts. We recall a visit we made several years ago to a shaded cacao farm in northern Brazil. Much like coffee, cacao is frequently cultivated under a forest-like canopy. We noticed that this farmer had tied long strings from one of the shade trees to the surrounding cacao trees. Curious, we enquired as to why. He went on to describe in great detail that there was a kind of ant that nested in the shade trees and was a voracious predator of the main cacao pest, a bug in the family Myridae. He also noticed that these ants avoided walking on the ground and would forage on the cacao trees only when the shade-tree vegetation where their nest was located touched the cacao vegetation. Putting his ingenuity to work, he designed a system of radiating strings from the trees where the ants had their nest to the cacao plants. In a sense, he was helping the ants provide the ecosystem service of pest control. Looking closely, we discovered that the ant the farmer was talking about was *Azteca*, the ant we have been studying in the coffee agroecosystem of Mexico and a keystone species in a complex web of pest control in coffee.

This short vignette underscores a fundamental principle that has deep historical roots. As noted by Sir Albert Howard when he was sent to India by Queen Victoria to teach Indian farmers the wonders of the new agricultural technologies that were emerging from the advances of the Industrial Revolution, "by 1910 I had learnt how to grow healthy crops, practically free from disease, without the slightest help from . . . all the . . . expensive paraphernalia of the modern experiment station."[8] He specifically acknowledged the traditional Indian farmers as his teachers. This early foray into the world of farmer-to-farmer and farmer-to-scientist dialectics has been repeatedly rediscovered. As Richard Levins has noted, the combination of the profound knowledge of the farmer with the broad theory of the scientist is likely the best way not only to develop useful technology for promoting ecosystem services, but also of discovering new scientific principles of ecology more generally.

Much has been written about the details of pests of coffee and ways to control them,[9] including literature reviews for specific types of pests such as nematodes[10] and the coffee berry borer.[11] Rather than reviewing the literature again, we focus on the results of our research on *Finca* Irlanda to illustrate the concept of autonomous pest control, the type of pest control that emerges from ecological interactions within a diverse agroecosystem. Although our group has worked on many ecological aspects of the coffee agroecosystem, an emphasis of our work has been on trying to understand ecological interactions that result in reduced pest problems. Reflecting on the research process itself, what we present in the following sections is a summary of what we have been able to discover regarding coffee herbivores and their natural enemies. As in any ecological research project, we began with a simple observation

– either by ourselves, from the literature or from the farmers – and then started the inquiry process.

Vertebrate insectivores

Recall the anecdote from Chapter 5. We were walking on the farm with Don Walter Peters, the owner of the farm at the time, when we ran into a shade tree that was being completely devoured by caterpillars. He told us he was not worried because the migrant birds were on their way and would take care of the caterpillars – and they did. As a passionate birder, Don Walter had been observing the comings and goings of migrant birds through his farm for years. He had noticed that the caterpillars that could potentially damage a significant number of the *Inga* trees that provide shade for the coffee appear just a few weeks before the migratory birds arrive, and soon after that the birds begin to feast on the big juicy caterpillars. This is biological control in action.

Inspired by the efficiency with which the birds eliminated this potential pest, we set up a large "bird" exclusion experiment to investigate the effects of birds on coffee pests. We put "bird" in quotation marks because, as noted in Chapter 5 and discussed later in this chapter, the exclusion netting also kept out other organisms, mainly bats. The detailed results of the exclusion experiment are explained and discussed in Chapter 5. Briefly, what we found was that the overall effects of birds occurs through complex interactions rather than the direct consumption of herbivore pests. It appears that birds have the main negative effect on orb-weaving spiders, which in turn negatively affect the parasitic wasps in the system.[12] Thus, the spiders, by negatively affecting the parasitoids, have a general "negative function" for the agroecosystem, but the birds, through a complex trophic cascade, have a positive function (see Figure 5.6).

Yet the functional significance of birds may be even more complicated. Some birds have a density-dependent response to resources, like the migratory birds that control the caterpillars on Don Walter's shade trees. In other words, when a resource becomes suddenly abundant, some bird species respond by concentrating their foraging on that resource, sometimes referred to as "prey switching".[13] That led us to think that if a particular insect herbivore increases its population and is on its way to becoming a pest, a number of birds could potentially shift foraging strategies and focus instead on that herbivore, preventing it from reaching outbreak levels. In other words, birds that under normal conditions would have only minimal effects on pest control, under other conditions could play an important role in pest control. In general this phenomenon, when couched in terms of biodiversity function, is called the "insurance hypothesis,"[14] so called since the collection of species (i.e., the bio-diversity) may not function as biological control under normal circumstances, but may be called in for this critical service at special times, when conditions change.

In seeking to examine the insurance hypothesis of biodiversity function, we simu-lated a pest outbreak by placing caterpillars at higher than normal densities on coffee plants inside and outside bird exclosures on two types of farms: a low-statured/

monocultural shaded plantation with low bird diversity and abundance, and *Finca Irlanda*, which is a high-statured/diverse shaded plantation with high bird diversity and abundance. We then counted the number of caterpillars left on the coffee plants to determine if birds had an effect on controlling the caterpillars and if the effect was different between the high-shade and the low-shade coffee plantations. Although the number of caterpillars declined in all four treatments (some may have fallen from the plants and others may have been eaten by invertebrate predators such as spiders and predatory wasps), caterpillars in the no-exclosure high-shade treatment declined by 75 percent in 24 hours, which is significantly more than in the other three treatments, which declined by an average of little less than 50 percent in the same amount of time (Figure 7.3). These results suggest two things: first, that birds (and/or bats) can potentially prevent a pest outbreak, therefore acting as an insurance against pest outbreaks; and second, that this ecosystem service is lost in the low-shade system. As is evident in Figure 7.3, we found no difference in the number of caterpillars remaining on plants inside and outside the exclosures on the low-shade farm. We speculate that the birds that were able to discover the caterpillars and prey on them were absent from the low-shade farm, which indeed had less diversity and abundance of birds and bats. This is precisely what the insurance hypothesis says – the more biodiversity, the more likely it is that there will be a species that will be able to take care of a potential outbreak.

Our work (both the artificial outbreak experiment described above and the large exclosure experiment described in Chapter 5) suggests that birds and/or bats can reduce coffee herbivores either through direct consumption, when the prey is abundant, or through cascading effects such as those described in Chapter 5.

The emphasis on birds as potential biological control agents has long been a popular topic in ecology. Most of the earlier experiments were conducted in the temperate zone and concluded that birds had little influence on pest control.[15] However, since then, much evidence has accumulated suggesting that birds can strongly influence the populations of their arthropod prey.[16] The pioneering work of Russ Greenberg,[17] from the Migratory Bird Center in Washington DC, inspired a wave of studies investigating the role of birds on coffee plantations.[18] After all these studies, there is now strong evidence that birds do indeed provide the ecosystem service of pest control to coffee farms and can help improve yields and revenues for coffee farmers. Furthermore, meta-analyses comparing studies of agroforestry systems, including coffee, have concluded that the top-down effects of birds are stronger when migratory birds are present and that a higher diversity of predators results in a higher removal of herbivores.[19] In particular, the number of species of birds seems to be the main factor determining whether or not birds in general will be effective in regulating arthropods.[20] However, it is important to point out that a few species had a disproportionate effect on arthropod removal, suggesting a "sampling effect"[21] – that is, a higher probability that a particularly effective species will be responsible for the observed effects when biodiversity is larger.

Finally, we must acknowledge the potential role of bats in pest control ecosystem services in coffee. Almost all the evidence for "bird" effects comes from exclosure

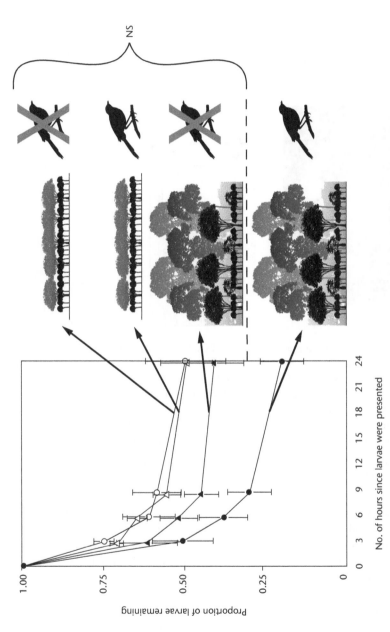

FIGURE 7.3 Results of a pest outbreak experiment. Caterpillar larvae were introduced on artificial leaves attached to coffee bushes under four different treatments: exclosures (preventing bats and birds from accessing the coffee) in high- and low-shade areas, and controls in high- and low-shade areas. In the low-shade area (open symbols), the artificially introduced larvae declined at significantly lower rates than in the high-shade area (closed symbols), and in the absence of birds and bats the decline was significantly less rapid than with these vertebrate insectivores, but only in the high-shade area. Thus, birds and/or bats seem to offer something of an insurance policy against pest outbreaks, but only on the high-shade farms. (NS = not significantly different from each other)

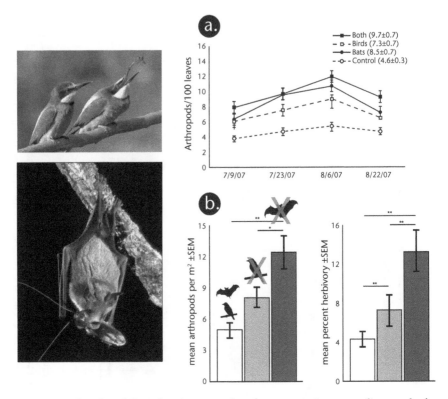

FIGURE 7.4 Results of diurnal and nocturnal exclosure experiments to disentangle the effect of birds and bats on arthropods. (a) All treatments (bird exclosure, bat exclosure and bird and bat exclosure) had significantly more arthropods than the control for all sampling periods, indicating that birds and bats have an effect on reducing the number of arthropods in coffee plants. (b) The exclosure of bats alone resulted in significantly higher numbers of arthropods and significantly higher levels of herbivory than the exclosure of birds alone or the exclosure of both bats and birds.

Sources: (top photo) Pierre Dalous (2012); (bottom photo) Bat Conservation International/Merlin D. Tuttle; (a) from Kalka et al. (2008); (b) from Williams-Guillén et al. (2008)

experiments that, as discussed in Chapter 5, also exclude bats since the exclosures are left day and night for at least several days, if not months. It took two clever mammal ecologists, Williams-Guillén and Kalka, to disentangle the effect of birds and bats on the arthropod community. By establishing exclosures that were up only during the day, only during the night, and all day, and by comparing the results with controls in which bats and birds could access the plants all the time, they were able to demonstrate that the effect of bats and birds on arthropods is additive (Figure 7.4).[22] Furthermore, by conducting the experiment during both dry and wet seasons, Williams-Guillén was able to demonstrate that in the wet season, when the migratory birds are absent, bats have a stronger effect on arthropods than birds. More recent work by Karp and Daily in coffee agroforests[23] and Maas in cacao agroforests[24] further demonstrates the ecosystem services of these vertebrate insectivores in agroforestry systems. However, this work suggests that we are just beginning to understand the trophic and trade-mediated interactions that may be involved, and that the relationship between biodiversity and ecosystem services is not a straight linear relationship.

Ants as predators

In our own work, we began with an emphasis on ants as predators, since they are well-known generalist predators in agricultural systems and likely to be interacting with many other organisms.[25] On the one hand, it is clearly the case that ants in general are capable of controlling several pests in coffee,[26] and that particular species of ants can be effective at controlling specific pests, for example, the coffee berry borer[27] and the coffee leaf miner.[28] Yet when considering the ant fauna as a whole, Philpott and colleagues[29] were unable to demonstrate any effect of ant removal on the herbivore trophic level. These divergent results may reflect the complexity of interactions – some positive and some negative, some mutually beneficial and some detrimental, some direct and some trait-mediated – that occur within the arthropod community in the coffee agroecosystem. The net result of those interactions could be neutral for the herbivore under some conditions or dramatically positive or negative under others. What follows is a discussion of our attempts to unravel the complexities of the ant-mediated ecosystem service of pest control on coffee farms.

We start with *Hypotheneus hampeii*, the coffee berry borer (henceforth CBB), because it is currently the most important insect pest in coffee worldwide, with an estimated economic impact of US $500 million.[30] The CBB is a small beetle that burrows into the coffee seed, destroying much of the seed itself, but also creating opportunities for fungal and bacterial entrance into the seed (Figure 7.5). Since it attacks the seed directly, it is usually regarded as an especially important pest. The origin of this pest remains unclear, but it was first reported in Gabon in 1901[31] and has spread throughout the coffee-producing world.

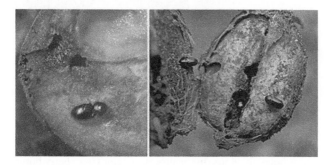

FIGURE 7.5 (a) The CBB and (b) the damage it does.

Source: (a) Daniel Karp

The CBB has proven to be a particularly difficult pest to control. Attempts to do so with chemicals have largely failed (it lives inside of the seed, so spraying only kills those individuals that happen to be flying outside of the fruit at the time). Several parasitoid wasps have been identified that attack the berry borer, and the famous fungus *Bauvaria* spp. is being investigated as a potential bio control agent. Although a great deal of research on control of this pest has been produced in the past decade, it remains a significant pest, of concern among coffee producers worldwide.

It is now well established that several species of ants are major predators in the coffee system, most importantly having a demonstrable effect on the CBB.[32] An example is the large solitary forager *Gnaptogenys* sp., that sits and waits for a beetle to emerge from its opening (Figure 6.9). The invasive little fire ant, *Wasmannia auropunctata*, enters the entrance hole made by the beetle and preys on both adults and larvae (Figure 7.6(a)). The ant *Azteca* successfully deters the borer from entering fruits. To get to the seed, an individual beetle requires approximately an hour to burrow its way through the fruit. If it attempts to do so in an area being patrolled by *Azteca* ants, it most frequently is unsuccessful as the *Azteca* grabs its abdomen and pulls it from the hole (Figure 7.6(b)).[33] Although the ant frequently simply drops the beetle to the ground, its harassment has the effect of reducing the damage caused by the beetle to the coffee beans.[34]

In a field study, Armbrecht and Gallego[35] discovered that ants could reduce the population of borers by 27 percent – certainly not the 100 percent a farmer might wish, but a significant amount of predation nevertheless. Furthermore, they found that 21 different species of ground-foraging ants were capable of preying on the CBB. Bustillo and colleagues[36] report seven different genera of ants preying on the borer by entering the fruit through the hole made by the borer, and Gonthier and colleagues[37] report on experiments with eight different species of ants, six of which successfully deter beetles from entering the berries in an experimental situation.

FIGURE 7.6 Ant predation on the CBB. (a) *W. auropunctata* carrying a berry borer that it probably removed from its entrance hole. The insert illustrates that the size of the ant permits it to enter the entrance holes. (b) *Azteca* in the process of removing a berry borer that is beginning to bore a hole into the berry.

Source: (a) (insert) Alex Wild

Of the species on *Finca* Irlanda, we know unequivocally that *Azteca, P. synanthropica, P. protensa, W. auropunctata, P. simplex, P. ejectus, Pseudomyrmex PSW-53,* and *Procryptocerus hylaeus* are direct predators on the CBB. However, as noted above, the mode of predation is different for each of the species. *P. synanthropica* and *Azteca* are effective predators at the point where the CBB arrives at the coffee bush, preventing it from entering the berry in the first place. Other arboreal species are smaller and/or more narrow-bodied (e.g., *P. simplex* and *P. ejectus*) and can enter the entrance hole made by the borers and attack them within the coffee fruit.

As far as the CBB is concerned, it faces as much of a threat from predators on the ground, since coffee berries that fall to the ground during harvest time contain the next generation of CBB. When fruits containing the CBB fall to the ground, ants can be effective predators, entering the fruits to predate or attacking them when they exit the seeds. Thus, the activity of ground-foraging ants is of potential importance for the overall control of this pest. As described in Chapter 6, on *Finca* Irlanda some areas are dominated by two species of *Pheidole* ants, *P. synanthropica*, a relatively large ant, and *P. protensa*, an extremely small ant. Both forage extensively on the ground. After careful measurement of the head capsule, it was clear that the larger species would have difficulty entering a coffee berry through the hole made by the borer, while the size of the smaller species would allow such penetration easily. Indeed, observations of both species revealed that *P. protensa* easily entered and exited through the damage hole while *P. synanthropica* was unable to do so, even when it appeared to be trying.

These two species exist in a spatial mosaic (Figure 6.7), which is to say that they form a patchwork in space where one species tends to dominate an area of several square meters while the other dominates in other patches of similar size. This pattern of distribution in space provided us with the opportunity to study the predatory effect of these ants on the CBB, by placing fruits infected with borers on the ground in areas known to be dominated by *P. synanthropica* (the large species) or areas known to be dominated by *P. protensa* (the small species). We found a very significant difference between the two species one week after placing the coffee berries on the ground, but that difference declined somewhat after another week, presumably because either *P. protensa* or some other species was eventually able to find the berries in patches dominated by *P. synanthropica* (Figure 7.7).

Thus, the spatial pattern of the ant assemblage, which results in the local dominance of particular species (see Chapter 6), is bound to have an effect on the general efficiency of the ants acting as predators of this and probably other pest species on coffee plantations. Some patches may have ants that prevent the beetle from entering the fruit in the first place (e.g., patches dominated by *Azteca* or *P. synanthropica*), while in other patches the beetle will have easy access to enter the fruit, but be subject to predation by ants entering the fruit while the berries are still attached to the plant (e.g., patches dominated by *W. auropunctata* or *P. simplex*), yet in other patches there may be no ant species that acts as a predator. Finally, the CBB inside the berries that fall into the ground may be predated by small ground-foraging species (e.g., *P. protensa* and *W. auropunctata*) or they may encounter predation when the

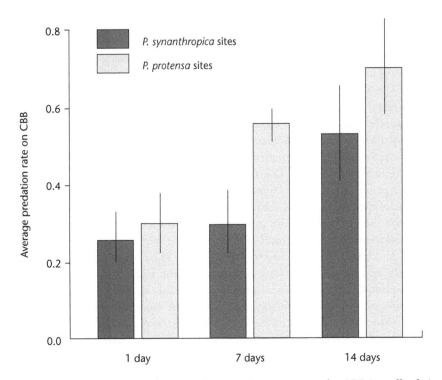

FIGURE 7.7 Predation rate of *P. synanthropica* and *P. protensa* on the CBB in coffee fruits placed on the ground.

next generation of adult beetles attempt to leave the berry to colonize new sites (e.g., *P. synanthropica*). The specific composition of an ant patch matters.

However, the differences among ants in terms of predation on the CBB goes beyond their size and ability (or inability) to enter through the CBB entrance hole or whether or not they forage in the trees or on the ground. In a detailed behavioral study of *Azteca* and *P. synanthropica*, Jiménez-Soto and colleagues[38] filmed the ants interacting with the CBB and were able to demonstrate that these two species, while foraging in the trees and not being able to penetrate the entrance holes of the CBB, exhibit very different behavior when encountering a live CBB. While both ants showed a high rate of detecting and seizing the CBB on a coffee plant, once the borer was captured, *P. synanthropica* proceeded to take the borer back to its nest in half of its encounters, while *Azteca* simply threw the borer off the plant almost 80 percent of the time.

Furthermore, there are other complications that mediate predation. For example, Pardee and Philpott,[39] further exploring the effect of the phorid parasitoids (see Chapter 5), compared predation rates of *Azteca* on the CBB both with and without phorid fly parasitoids. Their results demonstrate that the phorids cause the ants to reduce foraging activity so as to evade attack from the parasites, resulting in less active predation on the CBB and thus more coffee berries successfully attacked by the beetle, another trait-mediated indirect effect of the phorid flies (Figure 7.8). Essentially, the phorid parasitoids neutralize the effect that the ants have on the CBB.

Philpott and colleagues[40] followed up on these results, asking the question whether the trait-mediated effect that reduced the efficacy of predation on the CBB by the *Azteca* ants might be offset by increased species diversity of ants in the system. They basically repeated the experiments of Figure 7.8, but this time with two other species of predatory ants in the system, *P. simplex* and *P. hylaeus,* two species that nest in the hollow twigs of the coffee plants. The question was whether the presence of other predators would cancel out the trait-mediated effect that so strongly reduced the predatory effect of *Azteca* on the CBB. Their results showed very clearly that when the phorids were present, the loss of biological control activity of *Azteca* was replaced by the activity of the other species (Figure 7.9). In other words, the

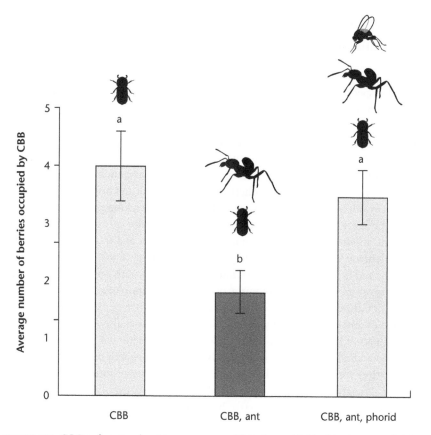

FIGURE 7.8 CBB infestation level in treatments: CBB alone, CBB with *Azteca*, and CBB with *Azteca* and phorid flies (*Pseudacteon lascinosus*). Error bars represent standard error and letters denote significant differences.

Source: Modified from Pardee and Philpott (2011)

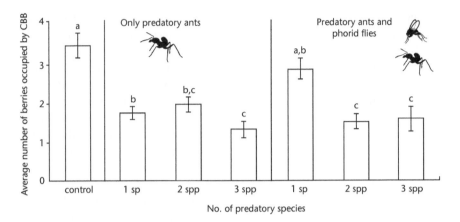

FIGURE 7.9 Effects of species diversity of predatory ants on the efficiency of biological control of the CBB. The presence of other species of ants in the system, other predators, rescues the biological control force of the predatory system when the phorids are present.

Source: Adapted from Philpott et al. (2012)

increased species diversity of predators had the expected effect of effectively rescuing the biocontrol function of ants in general.

From this accumulated series of observations and experiments with a variety of ant species, we see that the effects of ants on the CBB, although generally negative, cannot be easily characterized for a number of reasons. First, the mode of predation is distinct for different species, with some species attacking adults on the vegetation before they go inside the berry, others attacking the adult and brood inside the berry while it is still attached to the plant, and yet others attacking the adult and brood inside the berry after it drops to the ground. Second, the ants sometimes interfere with one another, sometimes complement one another and sometimes respond to other elements in the ecosystem (e.g., the phorid flies). Thus, understanding ant behavior and the multiple modes of interactions of ants with one another and with other organisms is important for evaluating the contribution of ants to the ecosystem service of CBB control on coffee farms. Furthermore, since one of the most obvious features of ants is their patchy distribution in space, spatial concerns must be an integral part of the analysis. As we saw earlier, the efficiency of the predation of *P. synanthropica* and *P. protensa* in attacking the CBB is dependent on the mosaic spatial pattern they form, presumably through their competitive interactions with one another. Understanding, at a very general level, this process of spatial patterning is thus related to the search for effective biological control.

Azteca *and the pest control complex*

We started the section on the ecosystem service of pest control by describing our approach, which is one of embracing complexity and avoiding the "one pest, one control agent" paradigm. In other words, rather than looking for magic bullets, whether chemical or biological, we emphasize the internal strength of agroecosystems that contain high biodiversity. And, although the term "internal strength" may evoke mystical forces of dubious origins with little scientific grounding, nothing could be further from the truth. We use the phrase "internal strength" to refer to a network of ecological interactions with feedback mechanisms that result in pest regulation without external intervention – in other words, autonomous pest control as envisioned by Lewis and colleagues as cited at the beginning of this chapter.

In addition to the CBB, there are three other organisms that are considered pests of coffee on the farm where most of our research has been done, two of which we have extensively studied. These are the green coffee scale and the coffee rust disease. In this section, we examine the ecological network involving the ant *Azteca*, as discussed previously in Chapters 4, 5 and 6, and present this network as an example of autonomous pest control in the coffee agroecosystem.

The green coffee scale and the myrmecophylous beetle

The green coffee scale (*C. viridis*), which is tended by *Azteca* (Figure 7.10), is similar to the scales almost everyone is familiar with on house plants (sometimes the same species). It is a sporadic pest on a wide host range, including coffee and citrus trees.[41] It reaches high densities only when associated with ants and is usually regarded as "pesky" and a nuisance, but not really a pest. However, it occasionally gains significant pest status.[42]

As discussed in Chapter 4, if we take a bird's-eye view of the coffee farm, these ant nests are not randomly distributed over the farm, but rather occur in distinct clusters (Figure 4.7). This clustered pattern of *Azteca* results in a similar clustered pattern of the green coffee scale since they only reach high levels when the ants tend them. As described in Chapter 5, the lady beetle, *A. orbigera*, is a major predator on this pest and we judge it to be the main reason that the coffee scale outside *Azteca* patches (i.e., on the rest of the farm) rarely reaches a population density at which it would be considered a pest in this system. The adult beetles are not able to survive very well within the clusters of ant nests because the ants harass and sometimes kill them (Figure 5.10). However, the beetle larvae are protected against ant predation because they are covered by waxy filaments (Figure 5.12(a) and 5.12(b)). Furthermore, the *Azteca* ants also inadvertently protect the beetle larvae from their own parasitoids because they harass any parasitoids that come close to the scales. From this series of ecological interactions emerges a curious situation for the coccinellid beetle. As larvae, they do well within ant clusters because they have plenty of food (scales tended by ants) and are in an enemy-free environment. However, as adults, they get harassed by ants within ant clusters, but do well outside ant clusters, where they survive by eating (and controlling) the scales that disperse across the farm. From the point of view of pest control, this is an interesting situation since the control of scale insects at the level of the entire plantation depends on allowing the existence of patches with high density of the pests, for the successful production

FIGURE 7.10 The green coffee scale (*C. viridis*) being tended by the *Azteca* ants.

of the next generation of adult beetles. It is clear that the clusters of ant nests create the conditions for the beetle to persist and act as the main biological control agent of the scale insects on the rest of the farm.

The coffee rust disease

The history of the coffee rust disease is a famous horror story.[43] Sri Lanka (previously Ceylon) was occupied by various European powers, which, as part of classical colonial strategy, needed to establish a local revenue-generating scheme. Coffee was an obvious choice and it became the gold of the land. The Dutch, and later the British, got into the coffee business in a big way and rapidly evolved an economic model totally dependent on this export commodity. Then disaster struck in the form of a rust disease that appeared to have come out of nowhere (it is thought to have originated in Africa, the original home of coffee). That rust was the classical coffee rust, *Hemileia vastatrix*, and was so devastating that after only a few years coffee had to be completely abandoned across the entire island. That is why, today, one hears of Ceylon tea, not Sri Lankan coffee, and is at least part of the cause of English tea-drinking habits.

The next part of the rust story is especially interesting. Since it was so devastating in Sri Lanka, and later in Java and Sumatra too, the rush to bring coffee to the Americas in the late nineteenth century was accompanied by extreme caution not to bring the rust disease along with it.[44] Despite such caution, the disease was recorded in Brazil in 1970. Shortly thereafter, it arrived in Central America and the

Caribbean, and with it much concern over its potential devastation. Phytosanitary methods of various types, based on dubious sources, were introduced, including reducing the shade covering the coffee plants and planting resistant varieties. In the end, the worries were largely unfounded. Even on remote coffee farms that had not adopted any of the suggested phytosanitary methods nor used the new resistant coffee varieties, the rust became (and remained, until recently) a nuisance, but not the disaster that had been feared. In the light of a very real history of having caused disasters earlier in south-east Asia, the non-devastating persistence of the coffee rust in Central America is something that cries out for an explanation,[45] a point to which we return momentarily.

An unusual outbreak of the coffee rust disease occurred during the closing months of 2012. This epidemic was especially severe, with some reports anticipating as much as a 40–50 percent reduction in yield over a region that extended from southern Mexico to Colombia.[46] For example, our regular monitoring of a small plot in southern Mexico reflects the severity of the 2012/2013 epidemic, where in one test plot over 60 percent of the plants experienced more than 80 percent defoliation and almost 9 percent of the plants died. Similar patterns were reported anecdotally from Mexico to Columbia. For example, in Costa Rica, where losses of up to 50 percent were anticipated, special legislation was enacted in January 2013 authorizing relief payments to affected farmers.[47] We return to a discussion of this particular epidemic below.

In an interesting twist of ecological events, it turns out that the ant *Azteca* is also related to the biological control of the coffee rust. As described in the previous section on the green coffee scale and the myrmecophylous beetle, the clusters of *Azteca* nests generate the background habitat for source populations of the green coffee scale. However, they also harbor an important natural enemy of both the scale insects and the coffee rust, the white halo fungus, *L. lecanii* (Figure 7.11(a)). It is mostly known as an entomopathogen occasionally promoted as biological control for various insect pests,[48] but also known to attack other fungi, in particular the

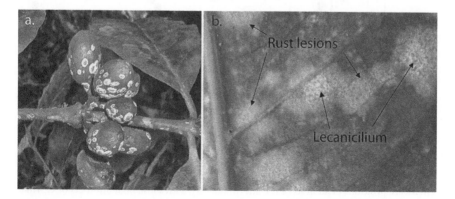

FIGURE 7.11 The white halo fungus disease (caused by the fungus *L. lecanii*) attacking (a) the green coffee scale and (b) the spores of the coffee rust.

coffee rust. This fungus infects the scale insects regularly, yet only becomes epizootic (i.e., epidemic) when the scale insects are locally very abundant (Figure 7.11(a)), which happens only when they are under the protection of the ants. It is then within these ant clusters that *L. lecanii* proliferates and provides spores that disperse from

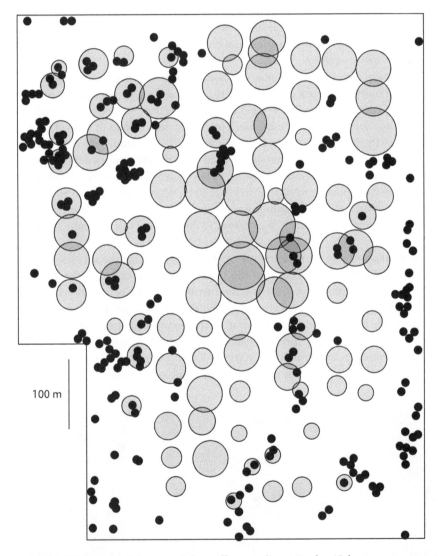

FIGURE 7.12 Survey of *Azteca* and the coffee rust disease in the 45-hectare permanent plot. Gray shaded bubbles are proportional to incidence of coffee rust disease, and black circles are positions of all nests of the ant (frequently with substantial numbers of *C. viridis* surrounding them).

Source: Modified from Vandermeer et al. (2009)

these nuclei to attack the coffee rust (Figure 7.11(b)). Studies of rust incidence demonstrate that the epizootic presence of *L. lecanii* within an ant cluster has a negative influence on the rusts for a radius of up to 15 meters.

Since the green coffee scale is part of the mutualistic association with the *Azteca* ant, it can be expected to reflect the large-scale distribution of the ant. That is, when ant nests are very common in a general area, we expect the green coffee scale to also be common. These areas where the scale insects are common are in turn expected to form foci where spores of the white halo fungus are especially abundant. Thus, if the white halo fungus exerts at least partial control over the coffee rust, we would expect to find a spatial correlation between ants and the rust (patches of ants make patches of scales, which make patches of white halo fungus, which reduces rust). In Figure 7.12, we show the distribution of coffee rust infection in the permanent 45-hectare plot, along with field estimates of the coffee rust disease intensity, during a normal year (i.e., moderate attack of the coffee rust). The notable point is that the rust is negatively correlated with the distribution of *Azteca*, and more formal statistical tests verify this general impression.[49]

While such general pattern is evident at the large spatial scale (Figure 7.12), the dynamics of the disease play out at a smaller scale.[50] It is reasonable to assume that spores from active epidemics on the green coffee scale could directly attack *H. vastatrix*. However, it is also the case that the soil serves as an environmental reservoir of viable propagules of *L. lecanii*,[51] and these propagules can be translocated from the soil onto the coffee plant via rain splash.[52] Therefore, it is also conceivable that spores of *L. lecanii* accumulate in the soil during one wet season and attack the rust when it emerges from dormancy during the subsequent wet season.[53] All of this implies that at least some of the disease dynamics of this system operate at a relatively small scale, even though some of its important background conditions are determined at a large scale.

In a two-year census (2009 and 2010) of both *L. lecanii* and *H. vastatrix,* the two species were concentrated in the lower half of a census plot in the September 2009 survey, but by 2010 the center of the *H. vastatrix* infection had very clearly moved to the upper region of the plot, and the rust was largely absent from the plants that had been heavily infected the previous year (Figure 7.13). This pattern is precisely what would be expected if the white halo fungus acting as a disease for the coffee rust has spatial dynamics operating at this relatively small scale.

Despite the apparent subtlety and variability of the controlling effect of *L. lecanii*, its regulatory effect on *H. vastatrix* may be substantial. The magnitude of the role the *L. lecanii* may play in preventing outbreaks of *H. vastatrix* depends on the details of the population dynamics of a complex web of interactions between multiple species, and these interactions themselves likely vary substantially over space and time. Therefore, while evidence has been accumulating that *L. lecanii* has a negative effect on the prevalence of *H. vastatrix* under field conditions, quantitatively assessing this effect remains difficult. Given its widespread distribution throughout coffee farms that maintain shade trees within which *Azteca* can nest, however, there is a high potential for *L. lecanii* to play an important regulatory role.

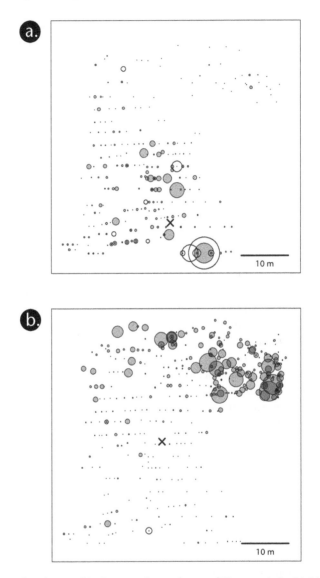

FIGURE 7.13 Abundance of *L. lecanii* and prevalence of *H. vastatrix* in (a) 2009 and (b) 2010. Diameters of open circles are proportional to the estimated number of *L. lecanii*-infected *C. viridis* on coffee plants, with the largest circle corresponding to the largest local infection. Note that the locations of the centers of the epidemics are influenced by fungal concentrations that are too small to see clearly at this scale. Dark gray circles are proportional to the number of leaves per coffee plant with lesions of *H. vastatrix*, with the largest circles corresponding to the largest observed number of lesioned leaves (254). Crosses mark the centers of the *L. lecanii* concentrations. Note that the locations of the centers of the epizootics are influenced by fungal concentrations that are too small to see clearly at this scale.

Source: Data from Jackson et al. (2012)

The coffee leaf miner

The coffee leaf-mining moth, *Leucoptera coffeella*, has not been a major pest in Mexico, although in other areas (e.g., Puerto Rico) it is regarded as one of the main pest problems in coffee production (Figure 7.14). It is also growing in its pest status on the farm where we work in Mexico. This moth is commonly associated with certain management options – for example, the reduction or elimination of shade trees.[54] It is also likely a secondary (or resurging) pest resulting from the use of pesticides.[55] This species too has not yet been devastating, despite the clear potential for it to be so, and indeed its current importance as a major pest in other areas of the world.[56] Surprisingly, we find little evidence that *Azteca* acts as a predator of the coffee leaf miner, but several of the other arboreal ant species have been clearly implicated in the predation of this potential pest.[57] Since coffee plants that are well patrolled by *Azteca* have few or none of the other ant species, its effect on the coffee leaf miner is potentially positive through indirect interactions with the other ant species that do prey on this pest. On the other hand, the aggressivity of *Azteca* could also deter the oviposition by the adult leaf-mining moth, lowering its incidence in coffee plants that are patrolled by *Azteca* ants. Unfortunately, very few studies have examined the interaction between *Azteca* and the coffee leaf miner to reach any solid conclusions.

The pest control complex I

The ecological interactions centered on *Azteca* are clearly indirectly responsible for at least the partial control of the green coffee scale and the coffee rust. Not only that,

FIGURE 7.14 Outbreak of the coffee leaf miner (*L. coffeella*) in Puerto Rico in 2013.

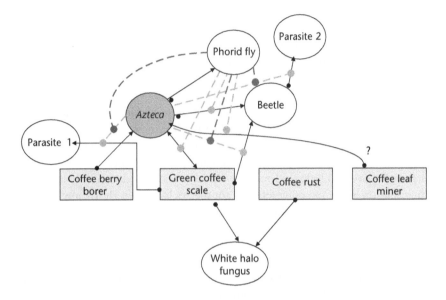

FIGURE 7.15 The ecological network involved in autonomous biological control, centered on the *Azteca* ant. Solid lines represent direct interactions, thick dashed lines represent trait-mediated indirect interactions of the first order and fine dashed lines represent trait-mediated indirect interactions of the second order. Circles represent negative interactions and arrows represent positive interactions. Pest species are inside squares. All the interactions included in this network have been studied and documented through field observations, surveys or manipulative experiments.

but (as was discussed earlier in this chapter) *Azteca* is also a predator of the CBB, especially at the stage when the borer is searching for berries or is in the process of boring into the fruit. Furthermore, because of its abundance and aggressivity on plants where it is tending scales, this species also has the potential to deter other potential pests such as the coffee leaf miner, which can inflict considerable damage under certain circumstances.[58] Therefore, here we have a situation in which an ant that occurs naturally on shade coffee plantations, through its complex indirect and trait-mediated interactions with other organisms, creates the conditions for the biological control of three and perhaps four of the main potential pests in coffee worldwide (Figure 7.15).

Connecting Azteca *with the other ant predators*

As complex as the *Azteca* system may seem (Figure 7.15), it is further complicated by connections with the other members of the assemblage of ant predators. As described in detail in Chapter 6, ants interact with each other, usually through competition or direct interference. Furthermore, there are subtle connections of various species in the network that extend to yet other guilds and form potentially important elements in the ecosystem service of pest control.

Nesting and foraging almost exclusively arboreally, the *Azteca* ants interact most strongly with other arboreal ants. One of the most locally abundant ant species that forages on coffee is *P. synanthropica*. This species nests on the ground but forages both in the coffee trees and on the ground. *P. synanthropica* also tends scales and is frequently found with groups of 50–100 scales on coffee plants. Its dynamic connection with *Azteca* is critical. When *Azteca* is expanding into a new area, it usually displaces *P. synanthropica* from those small groups of scales, using them as an inoculum for their own colonies of scales. Therefore, *P. synanthropica* seems to indirectly facilitate *Azteca* by maintaining small groups of scale insects, although both species compete strongly against each other and are hardly ever found foraging within the same coffee plant. Furthermore, *P. synanthropica* seems to exert competitive pressure on another group of ants, those that nest in hollow twigs. One of the predominant species of this latter guild is *P. simplex*, although about ten other species have similar habits.[59] *P. simplex* is a small ant, and a predator feeding on small insects, including the green coffee scale and the CBB. Coffee plants that are heavily patrolled by *P. synanthropica* often lack twig-nesting ants, presumably because of competitive pressure from *P. synanthropica* and similar ants (we refer to this group of ants as the *P. synanthropica* group). Species of the *P. synanthropica* group tend to dominate large areas (e.g., a circle of 15-meter radius).

More importantly, *P. synanthropica* is not able to withstand the competitive pressure of another common ground-nesting ant, a small species with a high nest density, *P. protensa*. It is unknown what tips the balance in favor of which species of *Pheidole*, but *P. protensa* is virtually incapable of foraging arboreally, and thus is unable to provide the local scale abundance that the *Azteca* ants need for their own nest expansion. So, where *P. protensa* dominates on the ground, *P. synanthropica* is unavailable to tend the scales, thus linking in a very indirect way the ground-foraging ant community with the nest cluster pattern formation of the *Azteca* ants. The precise importance of this particular function is difficult to evaluate completely, but given the well-known concept that very small ecological interactions can have very large ecological consequences, this component of the system is likely to be extremely important.

So the overall ant story can be tentatively summarized as follows:

1   The *Azteca* ants are extremely dominant where they occur, driving several general pest control processes (Figure 7.15), but they only occur on about 3–10 percent of the farm.

2   The ground-nesting arboreal-foraging ants (the *P. synanthropica* group) tend sucking insects, especially the green coffee scale, and prey upon whatever small arthropods they come across on coffee plants where *Azteca* is not dominant.

3   The ground-nesting ground-foraging ants (the *P. protensa* group, of which *P. protensa* itself is overwhelmingly dominant) strongly compete for nest sites and generally engage in strong competition with one another and with the *P. synanthropica* group.

4   The arboreal-nesting (mainly twig-nesting), arboreal-foraging ants (the *Pseudomyrmex* group) compete for both nest sites and food with one another and for food resources with the *P. synanthropica* group.

The guild structure connecting the various groups of ants is illustrated in Figure 7.16.
    This relatively complicated guild structure of approximately 50–70 species of ants is important for five reasons:

1   As indicated previously, the *P. synanthropica* group effectively maintains a residual population of scale insects that the *Azteca* ants seemingly need when the time comes to move their nest. As we have already indicated, moving their nests is a key element in the spatial structure formation, which, in turn, is important for maintenance of at least two key natural enemies in the system (the coccinellid beetles and the white halo fungus).
2   The ants in the *Pseudomyrmex* group are known predators of the leaf miner;[60] the *Azteca* ants, however, seem not to be as efficient, although they are generalist predators and do indeed prey on the leaf miner at least occasionally.[61]

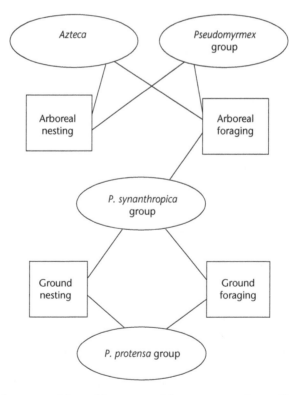

FIGURE 7.16 Summary of the guild structure of the ant community on *Finca* Irlanda. All groups contain species that are potential predators of the CBB and the coffee leaf miner. Several are mutualists, or potential mutualists, with the green coffee scale.

3   The *Azteca* ants[62] and many of the ants in the *Pseudomyrmex* group are predators of the CBB.

4   The ants in the *P. protensa* group may be important predators of the CBB in the old coffee berries that fall to the ground and provide refuge for the CBB during times when the plant does not have any berries.[63]

5   Although the keystone ant *Azteca* does not seem to affect the coffee leaf miner, the other ants in the system, especially the *Pseudomyrmex* group, have a potentially important effect.[64]

The pest control complex II

Connecting these other ant guilds with the *Azteca* network (Figure 7.15), and adding the information presented earlier on the vertebrate predators and the spiders, a more complete picture of the mechanisms of autonomous pest control emerges (Figure 7.17). The diversity of the system, especially the shade trees, plays a role in maintaining the pest control function of the *Azteca* network, not only because *Azteca* needs trees for nesting, but also because biodiversity provides insurance against the system's failure. For example, the phorid flies that attack *Azteca* can exert a dramatic indirect effect on their predatory ability, but the presence of the twig-nesting ants (the *Pseudomyrmex* group) insures that pest control is effective (Figure 7.9).

It is this complex network of ecological interactions that provides the "internal strength" of the system that helps control the main four potential pests in this coffee agroecosystem. It is what we call autonomous pest control. When the coffee agroecosystem becomes intensified or converted into a coffee monoculture, it will lose the *Azteca* ant, since the species needs trees to build its nest. Because of all the interactions involving *Azteca* (Figure 7.17), we consider it a keystone species. We propose that eliminating this species from the system, which would happen if the shade trees are eliminated, would result in outbreaks of the four pest species that are embedded in this ecological network.

The existence of such an ecosystem service emanating from ecological complexity is particularly important in the coffee agroecosystem, not only because coffee is so important in international trade and supports millions of small farmers worldwide, but also because its shaded nature has been intensively studied as a component in creating high-quality matrices in fragmented habitats for the purpose of biodiversity conservation (see Chapter 3). It is now well documented that a variety of taxa find refuge on shade coffee farms, sometimes at diversity levels approaching local natural habitats. Demonstrating the ecosystem service function of this biodiversity adds to our understanding of the importance of this agroecosystem.

The farm on which most of this work was accomplished had been an organic farm in production for almost 100 years (a more complete description is presented in Chapter 8). The organisms involved in the interaction web are well known to be associated with coffee, but their point of origin is not always known. The rust disease almost certainly comes from Africa, the lady beetle is known throughout the

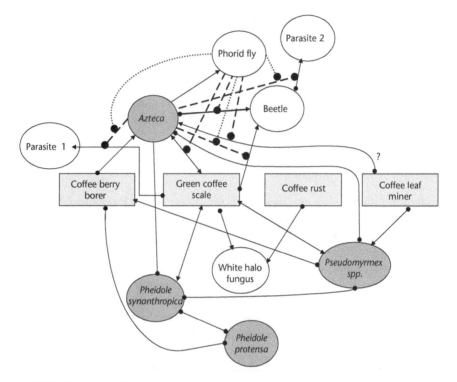

FIGURE 7.17 The ecological network centered on *Azteca* with the addition of the other agents discussed thus far, as related to the four main coffee pests. Solid lines represent direct interactions, thick dashed lines represent trait-mediated indirect interactions of the first order and fine dashed lines represent trait-mediated indirect interactions of the second order. Circles represent negative interactions and arrows represent positive interactions. Pest species are inside squares. Shaded ovals indicate species groups. All the interactions included in this network have been studied and documented through field observations, surveys or manipulative experiments. Details of complex interactions were presented in Chapters 5 and 6.

Neotropics, the white halo fungus is common throughout the tropics and most of the ants appear to be native to southern Mexico. Although almost all the work described in this chapter was done on this farm, our view is that, even if they are not obvious after a cursory study, interaction webs of this sort are common in agro-ecosystems in general. We just need to search them out.

Mitigating impacts of climate change

Evidence is now overwhelming that climate extremes are becoming more exaggerated. Along with elevated average global temperature, extreme climate events have become more frequent – an increased frequency of Category 4 and 5 hurricanes,[65] and an increased frequency of El Niño Southern Oscillation (ENSO) events,[66] among others. Furthermore, the double specter of more frequent droughts and more frequent flooding is now commonly found in models.[67]

As we have argued elsewhere,[68] agriculture – especially industrial agriculture – is one of the major culprits in this persistent and growing problem. However, it is also the case that agriculture is likely to be especially vulnerable to expected changes in climate, particularly for small-scale farmers who rely on rain-fed agriculture for their livelihoods. Because many crops are sensitive to changes in temperature and precipitation, and often have a narrow threshold for success,[69] changes in temperature and precipitation associated with extreme climate events are likely to affect both production and sustainability. Such sensitivities may be crucial in the tropics, where most agriculture is rain-fed and where climate change has a potentially large influence on productivity.[70]

Although the long-term, slow-changing climate events, such as gradually rising temperature or gradually changing precipitation rates, are of concern, farmers may be able to adapt by changing crops and varieties. It is the extreme climate events that come on rapidly and occur over a very short span of time that are perhaps more threatening to agriculture. Examples of such extreme events that are often related to crop damage and loss are the maximum temperatures on a single day or maximum one-day precipitation totals. If these extreme events coincide with an important step of the crop developmental cycle, production can be dramatically reduced. For example, eight-hour heat pulses that were applied to wheat during anthesis, to mimic elevated daytime temperatures, resulted in a decrease in harvest due to damaged flower development.[71] This suggests that an increased maximum temperature on a single day of flowering could affect production.

That agriculture is threatened by expected climate change is now beyond debate, and most climate scientists are not optimistic about the potential to reverse course. Indeed, even if CO_2 production were to cease completely today, the climate would continue to change for many years to come.[72] Consequently, the need for "adaptation" has entered the discourse and must be considered a major coping mechanism – perhaps at least as important as mitigation, remediation and restoration.[73] The Intergovernmental Panel on Climate Change (IPCC) synthesis report of 2007 states that climate change will have a major impact on food and water resources and suggests that adaptive measures must be developed. Accepting this necessity and focusing on agriculture, we must ask what "adaptation" would look like.

The search for potential agricultural adaptations to climate change has been broad in scope, but generally has focused on more technified management and human intervention – for example, genetic modification of crops,[74] changes in the location of production,[75] or the development of models for climate forecasting.[76] Less common is a recognition that management practices could contribute significantly to the arsenal of options available in pursuit of rational adaptation. Some evidence is beginning to emerge regarding the extent to which agroecological practices can offer resistance to the impacts of extreme climate events.[77] Furthermore, evidence of agroecological resistance in some agricultural systems appears to be a result of obvious microhabitat modifications.[78]

In a large-scale experiment that included 880 paired agricultural plots, Eric Holt-Giménez provided an example of how differences in management practices may create resilience to climate extremes. Holt-Giménez compared the response of small-scale farms under Sustainable Land Management (SLM) to conventional farms in Nicaragua after Hurricane Mitch, an extreme climate event that hit the country in October 1998. SLM includes a variety of soil conservation, agroecological and agroforestry practices and generally avoids external inputs such as pesticides and fertilizers.[79] On the other hand, conventional farms combined a variety of traditional and "semi-technified" practices, including the use of external inputs. None of the farms had irrigation or used heavy machinery. Evaluations performed after the hurricane found that despite high environmental variability among paired sites, SLM farms had, in general, more topsoil and higher field moisture measures, more vegetation within the system and lower economic losses than the conventional farms.

Although no such large-scale studies exist for coffee landscapes, Philpott and colleagues[80] provided evidence of the protective effect of shade tree cover to hurricane disturbance in the Soconusco region of southern Mexico subsequent to the devastating effect of Hurricane Stan, which hit the region in October 2005 just at the onset of the harvesting season. The hurricane dropped more than 500 millimeters of water in just a few days, creating massive landslides and floods, damaging roads, destroying bridges and flooding entire communities. It is estimated that up to 50 percent of the coffee harvest was lost that year due to the impact of the hurricane.[81] A large amount of the coffee in that region is grown in the mountains at altitudes between 500 and 2,000 meters, and most farms are located in rugged terrain with high slopes. In these fragile mountainous ecosystems, soil protection is essential and consequently most farmers maintain shade tree cover on their coffee farms. However, many farms in the region are "shaded monocultures" because their shade tree cover consists of one or a few tree species, usually of a single genus like *Inga*, at low densities (Figure 2.12). Philpott and colleagues found that landslides were more common on more intensified coffee plantations (i.e., those with few shade trees) than in the high-shade, more traditional, polycultural systems, corroborating other studies showing that landslides increase as the vegetation crown cover decreases.[82] In these fragile mountainous regions, even under normal climatic conditions, protection of the soil with vegetation is a priority. Shade trees within coffee agroforests have been shown to provide soil protection by maintaining a leaf litter cover, either through natural leaf fall or prunings, reducing raindrops' impact on the soil, improving soil structure and increasing water infiltration,[83] all of which lead to reduced soil erosion through run-off. These results suggest that the tree cover used on shade coffee farms provides some resistance to storm events, which translates into lower vulnerability and higher long-term sustainability under a climate change scenario.[84]

These two examples suggest that intensive agricultural systems may have lower resistance, lower resilience and higher vulnerability to extreme climate events, potentially affecting the long-term sustainability of crop production under global climate change.[85] In the case of the coffee agroecosystem, the shade trees planted with the coffee appear to serve an important ecological function by buffering the impacts of severe storms and reducing the farm's vulnerability to climate change. Therefore, it is important to further explore the idea of farming intensity and agroecological resistance in terms of understanding how management intensity may contribute to or mitigate farm vulnerability. Here, we present further evidence of this phenomenon, demonstrating the underlying mechanisms of agroecological resistance to climate change using the specific system on which this book is focused, the coffee agroforestry system.

Traditionally, coffee has been grown under the diverse canopy of shade trees, as we noted in Chapters 1 and 2. However, the gradual deforestation of shade trees within coffee agriculture has recently occurred in pursuit of larger production values. For example, over the last 40 years, intensification of coffee farming in Central America has resulted in the loss of more than 50 percent of the tree cover on coffee farms,[86] with native tree species being the most affected. Active deforestation is based on the assumption that shade trees compete for sunlight, nutrients and water with the coffee plants, thus lowering coffee yields. To be sure, there is some empirical support for such an assumption,[87] although as of now, the only study that looks at yield as a function of shade on actual functioning small-scale farms finds a pattern suggestive of a hump-shaped curve with highest production under 40–50 percent canopy cover.[88]

Although extreme climate fluctuations are already taking place, very little research has been conducted to better understand the sensitivity of coffee to changes in temperature and precipitation rates. Moreover, there have been very few initiatives to develop adaptive agricultural methods to protect coffee agriculture from climate change.[89] Because coffee has been traditionally grown as a rain-fed crop, and small-scale farmers have little access to external sources of water in many coffee-growing regions, a better understanding of coffee management in response to climate variability is necessary. The trend toward less shaded coffee systems, in combination with increasing climate variability, may become detrimental to production, since coffee is dependent on precipitation for its water supply.

Although coffee's specific requirements for water in different agricultural settings have not been quantified, it is well known that coffee production is dependent on the seasonal precipitation cycle of the tropics.[90] Water availability from precipitation may affect several key functions in the coffee plant. First, an extended period of drought is required for the flower buds to form.[91] The flower buds then open simultaneously in response to a sporadic dry-season rain shower, and are receptive to pollination for 48 hours after opening.[92] The beginning of the rainy season signals a period of rapid leaf flush and rapid fruit growth, which continues throughout the season.[93] Water availability has also been found to affect the maintenance of maximum photosynthetic rates,[94] fruit set levels,[95] and fruit size.[96]

Coffee is thus, among other things, vulnerable to both the quantity and timing of precipitation. Coffee plants are susceptible to plant stress and damage during the dry season and especially during times of drought, which can occur at important points of flowering and fruit development. With increasing climate variability, crops may be subjected to increasingly erratic precipitation events, leading to increased water stress within the system. Also, extreme events, such as ENSO, can cause extended droughts during the dry season at important developmental stages and put a large stress on the plants, thus reducing crop production dramatically.[97] Previous ENSO years have shown a decrease in production of 40–80 percent in southern Mexico, leaving many small producers impoverished and without other means to

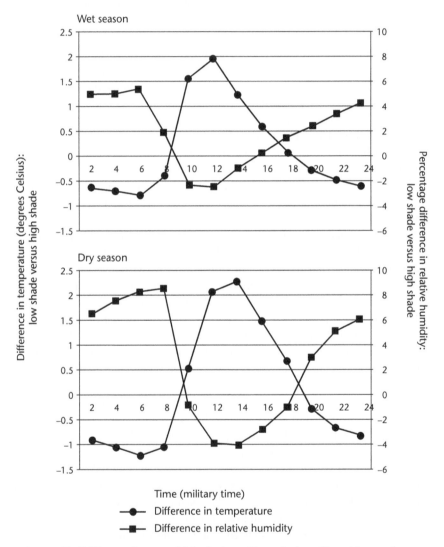

FIGURE 7.18 Difference between high-shade and low-shade coffee with regard to temperature and relative humidity during the course of a day in wet and dry seasons. The high-shade system is warmer at night and cooler during the day, and has lower relative humidity during the evening and higher relative humidity during the day, compared to low-shade coffee.

Source: Modified from Lin (2007)

earn income.[98] In Mexico, approximately 3.5 million people depend on coffee production;[99] therefore, large-scale extreme climate events that hit coffee-growing regions can affect many small producers and threaten the livelihoods of millions of people.

Coffee plants are quite sensitive to changes in microclimate. Arabica coffee, for example, has an optimum temperature range between 18 °C and 21 °C,[100] and temperatures above 24 °C decrease the net photosynthetic rate approaching zero at 34 °C.[101] Temperature can also affect fruit quality by accelerating development and ripening.[102] Shade trees have been shown to create a better microclimate for coffee plants by buffering temperature and humidity fluctuations.[103] In a study that compared areas of low shade-tree density with areas of high shade-tree density, Brenda Lin showed that the presence of increased shade cover reduced temperature at the most important point of the day when plants were most heat stressed.[104] The presence of shade also maintained higher temperatures in the evening when plants were most cold stressed (Figure 7.18).

Shade trees provide protection from other climate features as well, such as wind and storm events, which can defoliate coffee trees and decrease yields through flower detachment and premature fruit drop.[105] Reduction of wind speed passing over the crop may also be beneficial because it could potentially lead to less water loss through transpiration and soil evaporation.[106] The Lin study showed that greater levels of shade also increased water availability due to a decrease in evapotranspiration from the coffee layer and evaporation from the soil layer (Figure 7.19).

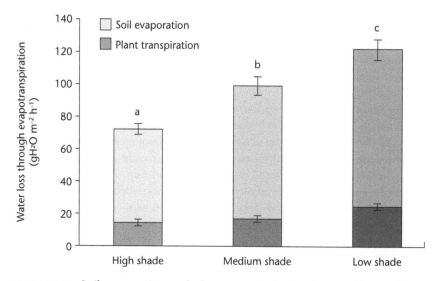

FIGURE 7.19 Soil evaporation and plant transpiration in three coffee production systems.

Source: Modified from Lin (2007)

Taken together, these studies suggest that the shade coffee agroecosystem may have a moderating effect on the expected consequences of global climate change by providing agroecological resistance to extreme climate events and by buffering micro- and macroclimatic conditions within the agroecosystem. This regulating ecosystem service is in no way exclusive to the coffee system and is not necessarily linked to biodiversity. The key for these particular ecosystem functions is the overstory layer. Indeed, as has been noted repeatedly, agroforestry in general should be seen as a tool in global efforts to mitigate global climate change.[107]

Pollination services

Bees and coffee yield

Kamal Bawa of the University of Massachusetts estimated that animals pollinate 89–99 percent of all flowering plants in tropical lowland rainforests.[108] It has been estimated that 70 percent of world crops experience increased size, quality or stability because of pollinating activities,[109] affecting at least 35 percent of the global food supply.[110] Furthermore, pollination contributes to the stability of food prices, food security, food diversity and human nutrition,[111] and is estimated to be worth US$200 billion worldwide.[112] Despite its general importance as a key ecosystem service, we are only just beginning to understand how anthropogenic land-use changes affect wild and domesticated pollinators.

Although crop pollinators include a wide array of animals, bees are usually regarded as the most important and most effective, at least with regard to important agricultural plants.[113] Bees are abundant and diverse, with an estimated 20,000–30,000 species worldwide,[114] living in a functionally diverse set of ecological niches, responding to both nesting and foraging resources at a broad range of spatial scales.[115] Most importantly, recent studies have demonstrated that crops, including coffee, experience higher or more stabilized fruit set in habitats with greater diversity of native bees.[116]

Coffee landscapes during flowering season have inspired poets and songwriters in Latin America. Entire landscapes become covered with millions of aromatic white flowers that provide a feast to native and non-native pollinators alike (Figure 7.20(a)). Pollination experiments have demonstrated that fruit weight and shape as well as fruit set and yield all increase with bee-mediated pollination.[117] Although some of these studies emphasize the role of the domesticated honeybee,[118] others highlight the importance of a diverse and abundant native bee community[119] (Figure 7.20(b)). In one study, it was estimated that pollination services provided by two forest fragments adjacent to the farm contributed an average of US$62,000, 7 percent of the total farm income, to a large (1,065-hectare) Costa Rican coffee plantation.[120] In a much larger study that included 21 Indonesian coffee farms, a 78 percent increase in yield with a fourfold increase in bee density was reported, associated with nearby forests.[121] In economic terms, this translates into an increase from US$6 per hectare for 20 bee visits per coffee plant, to US$55.1 per hectare for 80 bee visits per plant.

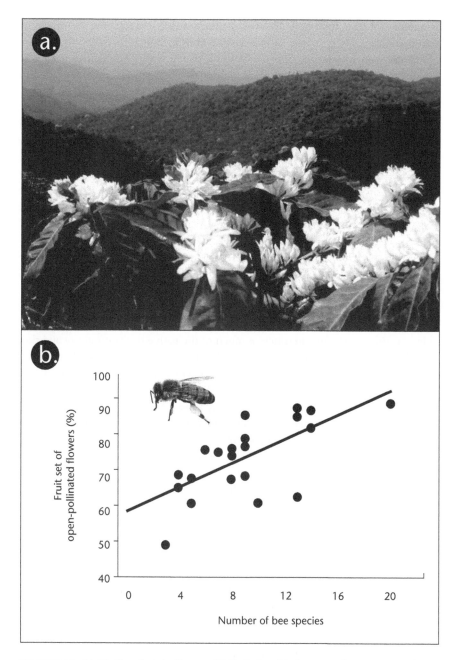

FIGURE 7.20 (a) Coffee plant in flower. (b) Relationship between bee species richness and coffee fruit set of Arabica coffee for 24 farms in Central Sulawesi, Indonesia.

Sources: (a) Shalene Jha; (b) modified from Klein et al. (2003b)

Regardless of how they are calculated, pollination services to coffee farms can be significant and can contribute substantially to farm income.

Focusing on the details, one might think that pollination would not be that important for the coffee industry since most of the high-quality coffee produced in the world is the self-compatible species, *Coffea arabica* (Arabica coffee), while the lower-quality (but higher caffeine content) species, *Coffea canephora* (syn. *Coffea robusta*, Robusta coffee), is an obligate outcrosser – in other words, pollen must come from a different individual than the ovum that is being fertilized. The fact that cross-pollination has an effect on coffee production[122] should thus not be surprising with Robusta coffee. However, this effect has also been found for the self-fertile Arabica coffee.[123] How could pollinators cause this repeatedly observed phenomenon in species that do not rely on outcrossing? Alexandra-Maria Klein, working with Ingolf-Steffan Dewenter and Teja Tscharntke from the Agroecology group in Gottingham, Germany, suggested that *C. arabica* may be an amphicarpic plant, with some flowers needing cross-pollination and others able to set fruit after spontaneous self-pollination.[124] If this were the case, then promoting a higher number and diversity of pollinators on coffee farms would be a good management strategy regardless of the species of coffee on the farm.

In the case of the coffee agroecosystem, it has long been suspected that the more traditional way of production, with diverse shade trees and minimal use of insecticides, encourages the maintenance of much of the native bee community and thus would be expected to contribute to the pollination ecosystem service. The process of intensification is expected to affect this service in a similar way that it affects any other service related generally to biodiversity. For example, decreasing the availability of flowering understory plants can negatively impact bee diversity, and can differentially affect the abundance of distinct bee functional groups.[125] Although in one study solitary bees responded positively to light intensity,[126] reducing the shade in a coffee system clearly has an effect on social native bees, as illustrated in Table 7.1, which translates into different visitation rates (Figure 7.21).

It is clear from these and other studies[127] that coffee vegetation management impacts bee community composition and foraging response. For example, one study found that the way in which farmers manage their coffee farms was more important for bee abundance and richness than the existence of native habitat in the landscape.[128] In this study, the most critical habitat variables for predicting bee abundance were tree species richness and percentage of canopy cover within the farms. Furthermore, bee species richness was most correlated with coffee bush density (a negative correlation) and flowering tree species richness (a positive correlation), all variables associated with the agroecosystem itself, rather than with larger landscape effects. Overall bee community composition did not vary substantially based on a gradient of forest cover in the nearby landscape.

Nonetheless, there is growing concern in the literature as to the relative importance of local- versus landscape-level factors for biodiversity and ecological processes more generally.[129] In pollination ecology, this translates into whether local or landscape factors are more important for bee density, diversity, community composition, floral visitation, pollination services and stability. Although several studies find that

TABLE 7.1 Bee species present in high-shade coffee (HSC) and low-shade coffee (LSC) agroforestry habitats and their functional group (FG), classified as an Africanized honeybee (AHB), a native solitary bee (NSOL) or a native social bee (NSOC).

Species	Family	LSC	HSC	FG
Apis mielifera, scutellata	Apidae	X	X	AHB
Ceratina eximia	Apidae		X	NSOL
Ceratinia ignara	Apidae		X	NSOL
Ceratinia sp. 1	Apidae		X	NSOL
Eulaema cingulata	Apidae	X	X	NSOC
Melipona beecheii	Apidae		X	NSOC
Nanotrigona testaceicornis	Apidae	X	X	NSOC
Plebia sp. 1	Apidae	X	X	NSOC
Scaptotrogona mexicana	Apidae		X	NSOC
Trigona fulviventris	Apidae	X	X	NSOC
Xylocopa tabaniformis, tabaniformis	Apidae	X	X	NSOL
Augochlora aurifera	Halictidae		X	NSOC
Augochlora nigrocyanea	Halictidae		X	NSOC
Dialictus sp. 1	Halictidae	X		NSOC
Dialictus sp. 2	Halictidae		X	NSOC
Halictus hesperus	Halictidae		X	NSOC
Halictus sp. 1	Halictidae	X		NSOC

Source: Reprinted with permission from Jha and Vandermeer (2009)

within-farm habitat management is more predictive of bee abundance and richness than landscape-level forest cover,[130] other literature reviews have concluded that bees are strongly affected by landscape-level factors, whether amount of or distance from natural habitat.[131] For example, several studies comparing coffee farms close and far from forests have documented higher overall bee richness, social bee richness, overall visitation rates, pollen deposition rates[132] and yield[133] in the sites near forests versus those further away from forests. Another study from India found that the size of adjoining forest fragments, rather than distance, positively influenced pollinator visitation to coffee flowers.[134] The difference between these findings could be explained by either distinct bee composition in the different study sites or the scale of measurement of the landscape-level effects.[135]

Interactions between pollinators and other organisms

When considering pollination ecology on coffee farms, one should also consider ecological interactions between pollinators and other organisms. Unfortunately, studies of ecological interactions between pollinators and other organisms in the coffee agroecosystem are almost non-existent. The notable example is a study by Philpott and colleagues.[136] Examining the relationship between ants and pollinators on coffee plantations in Mexico, the team eliminated ants from some flowering branches of coffee plants, leaving others with their normally foraging ants, and

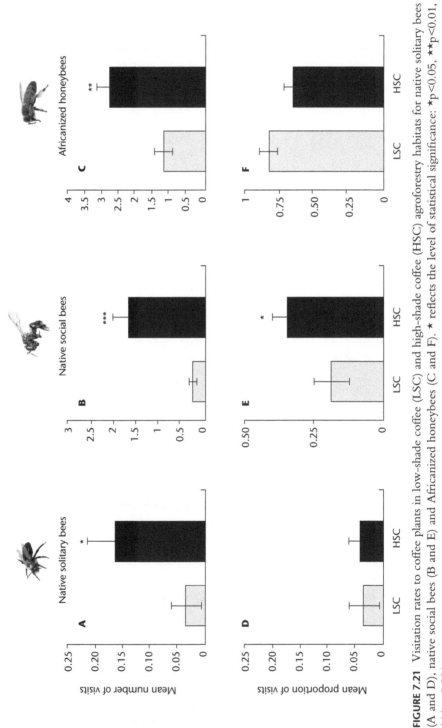

FIGURE 7.21 Visitation rates to coffee plants in low-shade coffee (LSC) and high-shade coffee (HSC) agroforestry habitats for native solitary bees (A and D), native social bees (B and E) and Africanized honeybees (C and F). ★ reflects the level of statistical significance: ★p<0.05, ★★p<0.01, ★★★p<0.001.

Source: Jha and Vandermeer (2009)

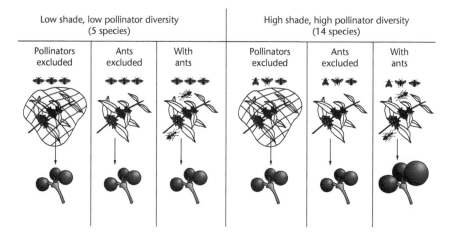

FIGURE 7.22 Results of the experiments of Philpott et al. (2006b). In the high-shade treatment, there was a greater diversity of pollinators. Excluding pollinators had no effect on yield when compared to the case of ants only being excluded in either habitat. However, in the high-diversity situation (high shade), when both ants and pollinators had access to the flowers, the average fruit size was significantly larger.

covered some flowering branches with bags, such that pollinators could not reach them at all. The experiment was conducted on both a high-shade and a low-shade coffee farm. They were expecting that the bagged branches would produce fewer fruits, since it had been shown before that pollinators increase fruit set and yield of coffee, as discussed earlier. The results of their experiment are shown in Figure 7.22.

These results are surprising given the basic biology involved. First, the team did not find the positive effect of pollinators on fruit production, as was expected based on previous published studies. Second, and more interestingly, they found a positive effect of ants on fruit weight, but only in the high-shade system. The explanation for this second result seems to be that the ants, by harassing the potential pollinators, cause them to move around more than they usually would, thus transferring more pollen over longer distances and providing a greater degree of outcrossing, which generally leads to greater average weight of fruit.[137] The fact that this result was obtained only on the high-shade farm can be explained by the fact that on the low-shade farm, flower visitations are dominated by the domesticated honeybee, *Apis mielifera*, an aggressive bee that is less likely to be disturbed by the much smaller ants. While the same domesticated honeybee is present on the high-shade farm, native social bees, which tend to be smaller, dominate flower visitations on that farm.[138] It seems that the disturbance action of the ants benefited fruit production on the high-shade farm by increasing pollinator movement and therefore pollination events.

Conclusion

Our studies on coffee agroecosystems have repeatedly shown that shade and more diverse coffee systems provide a higher degree of pest control, climatic resistance and pollination services. A recent global meta-analysis that examined the changes in biodiversity and ecosystem services with the intensification of coffee and cacao systems concluded that both biodiversity and ecosystem services declined with the intensification of these two agroecosystems.[139] Although much of the research that has been conducted into coffee agroforestry systems concerns their benefit for the conservation of biodiversity, evidence keeps accumulating of the benefits of these systems on productivity and long-term sustainability as well. The intensification of coffee farming may lead to increases in yield but, with the loss of ecosystem services like pest control, climatic resistance and pollination, the higher yields may very well be short-lived.

Notes

1 Engels (1883).
2 Mooney and Ehrlich (1997).
3 Marsh (1884); Osborn (1948); Vogt (1948); Leopold (1949).
4 Study of Critical Environmental Problems (1970).
5 Waller et al. (2007).
6 Morales and Perfecto (2000).
7 Lewis et al. (1997).
8 Howard (1943).
9 Le Pelley (1973); Waller et al. (2007).
10 Kumar (1988).
11 Damon (2000); Jaramillo et al. (2006); Vega et al. (2009).
12 Ibarra (2001).
13 Oaten and Murdoch (1975).
14 Yachi and Loreau (1999); Loreau et al. (2003).
15 Holmes et al. (1979).
16 Sekercioglu (2006); Van Bael et al. (2008).
17 Greenberg et al. (2000).
18 Perfecto et al. (2004); Philpott et al. (2004); Borkhataria et al. (2006); Kellerman et al. (2008); Van Bael et al. (2008); Johnson et al. (2009, 2010); Philpott et al. (2009); Railsback and Johnson (2011).
19 Van Bael et al. (2008); Philpott et al. (2009).
20 Philpott et al. (2009).
21 Huston (1997); Wardle (1999).
22 Kalka et al. (2008); Williams-Guillén et al. (2008).
23 Karp et al. (2013); Karp and Daily (2014).
24 Maas et al. (2013).
25 Way and Khoo (1992); Perfecto and Castiñieras (1998).
26 Philpott and Armbrecht (2006).
27 Velez et al. (2000); Bustillo et al. (2002); Varón (2002); Perfecto and Vandermeer (2006); Armbrecht and Gallego (2007); Larsen and Philpott (2010); Philpott et al. (2012); Jiménez Soto et al. (2013); Perfecto and Vandermeer (in press).
28 de la Mora et al. (2008).
29 Philpott et al. (2008a).
30 Durham (2004).

31 Le Pelley (1973).
32 Velez et al. (2000); Bustillo et al. (2002); Varón (2002); Perfecto and Vandermeer (2006); Armbrecht and Gallego (2007); Larsen and Philpott (2010); Philpott et al. (2012); Jiménez Soto et al. (2013); Perfecto and Vandermeer (in press).
33 Jiménez Soto et al. (2013).
34 Vandermeer and Perfecto (2006); Gonthier et al. (2013); Jiménez-Soto et al. (2013).
35 Armbrecht and Gallego (2007).
36 Bustillo et al. (2002).
37 Gonthier et al. (2013).
38 Jiménez-Soto et al. (2013).
39 Pardee and Philpott (2011).
40 Philpott et al. (2012).
41 Bess (1958).
42 Young (1982).
43 McCook (2006).
44 Fulton (1984).
45 Avelino et al. (2004).
46 Cressey (2013); International Coffee Organization (2013).
47 Inside Costa Rica (2013).
48 Roditakis et al. (2000).
49 Vandermeer et al. (2009); Jackson et al. (2012a).
50 Jackson et al. (2012a).
51 Meyling and Eilenberg (2007).
52 Jackson et al. (2012b).
53 Waller (1982).
54 Lomeli-Flores et al. (2009).
55 Staver et al. (2001); Fragoso et al. (2002).
56 Borkhataria et al. (2006).
57 de la Mora et al. (2008).
58 Fragoso et al. (2002); Lomeli-Flores et al. (2009).
59 Philpott et al. (2006a); Livingston and Philpott (2010).
60 de la Mora et al. (2008).
61 Lomeli-Flores et al. (2009).
62 Perfecto and Vandermeer (2006); Larsen and Philpott (2010).
63 Armbrecht and Perfecto (2003); Perfecto and Vandermeer (in press).
64 de la Mora et al. (2008).
65 Hoyos et al. (2006).
66 Dunbar et al. (1994).
67 Whetton et al. (1993); Fowler and Hennessy (1995).
68 Lin et al. (2011).
69 Gregory and Ingram (2000); Mendoza et al. (1997); Oram (1989).
70 Slingo et al. (2005).
71 Wollenweber et al. (2003).
72 Houghton et al. (2001); Solomon et al. (2007).
73 Berry et al. (2006); Parry et al. (2007).
74 Orindi and Ochieng (2005).
75 Assad et al. (2004).
76 Hansen (2005).
77 Holt-Giménez (2002); Philpott et al. (2008c).
78 Kiepe and Rao (1994).
79 Holt-Giménez (2002).
80 Philpott et al. (2008c).
81 Perez (2005).
82 Megahan (1978).
83 Beer et al. (1998).

 84 Starkel (1972).
 85 Gliessman (1998).
 86 Rice (1999).
 87 Fournier (1988).
 88 Soto Pinto et al. (2000).
 89 Adams et al. (2003).
 90 Carr (2001).
 91 Magalhaes and Angelocci (1976).
 92 Cannell (1983).
 93 Montoya and Sylvain (1962); Huxley and Ismail (1969); Cannell (1983).
 94 Nuñes et al. (1968).
 95 Ward (1882); Mayne (1935).
 96 Wormer (1964); Cannell (1976, 1985).
 97 Salinas-Zavala et al. (2002).
 98 Castro Soto (1998).
 99 Calo and Wise (2005).
100 Alegre (1959).
101 Cannell (1976).
102 Camargo (1985).
103 Beer et al. (1998).
104 Lin (2007).
105 Stigter et al. (2002).
106 Schroeder (1951); Teare et al. (1973).
107 Bäckstrand and Lövbrand (2006).
108 Bawa (1990).
109 Ricketts et al. (2008).
110 Klein et al. (2007).
111 Steffan-Dewenter et al. (2005).
112 Kearns et al. (1998).
113 Roubik (1995); Klein et al. (2007).
114 Michener (2000).
115 Steffan-Dewenter et al. (2002); Klein et al. (2008); Jha and Vandermeer (2009, 2010).
116 Kremen et al. (2002); Klein et al. (2007).
117 Manrique and Thimann (2002); Roubik (2002); Klein et al. (2003c); Ricketts et al. (2004); Olschewski et al. (2006); Veddeler et al. (2008).
118 Manrique and Thimann (2002); Roubik (2002).
119 Klein et al. (2003b); Badano and Vergara (2011).
120 Ricketts et al. (2004).
121 Veddeler et al. (2008).
122 De Marco and Coelho (2004); Ricketts et al. (2004); Klein et al. (2003c); Krishnan et al. (2012).
123 Klein et al. (2003a, 2003b).
124 Klein et al. (2003b).
125 Klein et al. (2008); Jha and Vandermeer (2009).
126 Klein et al. (2003b).
127 Klein et al. (2003b, 2007, 2008).
128 Jha and Vandermeer (2010).
129 Tscharntke et al. (2005); Rundlöf et al. (2008); Batáry et al. (2011); Tscharntke et al. (2012b).
130 Jha and Vandermeer (2010); Donaldson et al. (2002); Cane et al. (2006); Brosi et al. (2008).
131 Kremen et al. (2004); Klein et al. (2008); Ricketts et al. (2008); Batáry et al. (2011).
132 Klein et al. (2003b); Ricketts (2004); Munyuli (2011).
133 De Marco and Coelho (2004); Ricketts et al. (2004).
134 Krishnan et al. (2012).

135 Tscharntke et al. (2012b).
136 Philpott et al. (2006b).
137 Roubik (2002); Klein et al. (2003b).
138 Jha and Vandermeer (2009).
139 De Beenhouwer et al. (2013).

8

COFFEE, THE AGROECOLOGICAL LANDSCAPE AND FARMERS' LIVELIHOODS

The interpenetration of farmers' and biodiversity issues

A narrative emerges from the sum total of all previous chapters intersecting with the real lives of coffee farmers. From the question of biodiversity conservation (Chapters 2 and 3) to the question of ecological details such as Turing patterns and trait-mediated cascades (Chapters 4 to 6) to the application of these concepts in the context of ecosystem services (Chapter 7), it all comes together in conceptualizing the whole narrative as a dialectical interaction between the needs, desires and actions of the farmers who produce the coffee and the ecological forces with which they must work. Most important is the fact that the nature within which they must work is located in the tropics, which is also where most of the world's biodiversity is located. So the formal dialectical contradictions can be seen as the suite of "farmers' issues" interpenetrating the suite of "biodiversity issues." This conceptualization has taken on a variety of forms historically, perhaps driven too heavily by those whose focus is on the biodiversity rather than the livelihood of the farmer. Part of the intention of our research (and thus of this book) is to refocus the issue as one of the complex interconnections between the two parts of this dialectic. The proper way to look at the issue, we argue, is as a complex system involving the interpenetration of both farmers' issues and biodiversity issues.[1]

To focus our analysis, it is useful to begin with something of a historical review, first from the point of view of scholars who have been mainly concerned with the conservation of biodiversity and second from the point of view of scholars who have been mainly concerned with farmers' livelihoods and the agricultural enterprise. While our focus will be quite general, we will repeatedly return to the coffee agro-ecosystem as our ongoing example. After this historical review, we will revisit the relevant issues within the framework of the "matrix quality" point of view, recycling much of the detailed information presented in previous chapters, but within a more sociopolitical contextual framework.

The historical trajectory of biodiversity conservation in tropical lands

The key biodiversity versus agriculture debates (SLOSS, FT, LSLS)

The fact that most biodiversity is located in the tropics and that most natural habitat in the tropics has given way to agriculture, much of which has been for luxury crops for export, caused great concern to European travelers from the nineteenth century on. Especially in the late 1950s and 1960s, much was written calling attention to a looming disaster. If we are a single species on a planet on which all species seem to be interdependent, does it not seem rather foolish to be so unconcerned with the loss of the bulk of those species, especially in the tropics, given that most of the world's biodiversity is located there? It was an obvious point to make, and certainly a correct and disturbing one.

At that time, however, the only obvious experience in taking some form of action was the National Park system of the US, which was being replicated in other parts of the developed world. Almost without discussion, those addressing the question of biodiversity conservation in the tropics adopted this framework and the focus became one of deciding where to put national parks and how big they should be. On the one hand, it was assumed that large predators, such as lions and jaguars, needed very large areas to form successful populations, which would imply that preserved areas should be as large as possible. On the other hand, it was noted that many less charismatic species perhaps had very small ranges and that promoting a single large park could very well condemn them to extinction. Thus was born the "Single Large Or Several Small" (SLOSS) debate.[2] Suppose that conservation will be permitted in a fixed area; would it be better to put all that area into a single large preserve in one corner of the landscape, or would it be better to divide the area into many smaller preserves spread widely across that landscape?

Partly inspired by the theory of island biogeography, it was argued that the generalized relationship between area and number of species suggested that the larger the area of preserved natural vegetation, the more species to be expected in the preserved area. However, the theory itself said nothing about the arrangement of a group of preserved areas, and a debate emerged as to whether a Single Large Or Several Small preserved areas would be better for conserving species. The SLOSS debate, which was never really resolved, centered on the discrepancy between the actual distribution of species, which may be quite disjointed geographically, and the known relationship between extinction and habitat size. Some species are distributed in one contiguous geographic area while others are distributed in a patchy form across a broad landscape, as illustrated for a single example in Figure 8.1.

If a species is more-or-less randomly distributed across its overall range, a single large protected area could suffice for proper conservation protection. However, if the distribution is sparse, with relatively small subpopulations scattered throughout the larger area, or if genetic variability is spatially structured, a single large reserve is likely to miss some of the species or varieties that have restricted distributions. In

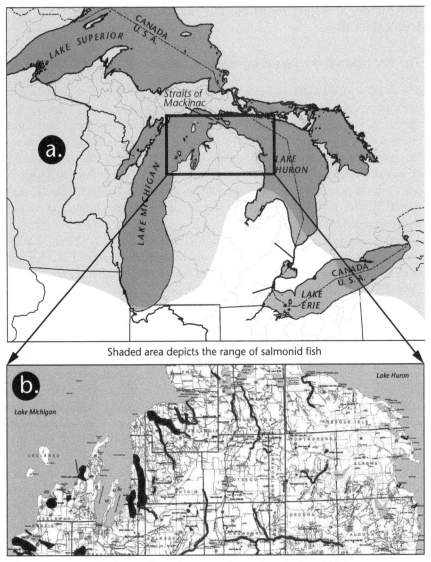

Black areas show the actual distribution of salmonid fish

FIGURE 8.1 Two depictions of the inland distribution of salmonid fish in the state of Michigan. (a) Classical distribution map showing the overall range of the family, in which the shaded area represents the range. (b) Map based on the Michigan Department of Natural Resources showing lakes and rivers in which salmonids actually occur (in black shading) – a small fraction of the potential aquatic habitat in the area, and an even smaller fraction of the overall geographic extent of the state. Establishing one large protected area in the middle of the map in (b) might very well miss all of the rivers and streams that contain the fish, whereas a large number of small refuges will likely contain at least portions of habitat where the fish occur.

the case of the Glanville fritillary butterfly, for example (Figure 8.2), it would be impossible to establish a single large contiguous nature reserve that could contain all the genetic variation known to exist (as depicted in Figure 8.2(b)). However, a collection of small reserves could clearly sample all of those populations and the fundamental metapopulation structure that this species is known to have could be maintained.[3] Later in this chapter, we revisit this debate and argue that, with the proper landscape perspective, the "Several Small" option usually makes more sense in the modern world.

However, the most critical error in the SLOSS debate was an error of omission. It was tacitly assumed that the people who lived in those areas that were to be designated as parks (also those who lived nearby) were irrelevant to the decision-making process. This gave rise to the next generation of conservation schemes that intended to combine parks or conservation areas with the concerns of the humans that lived there, or made their living in the to-be-preserved area, or in some way used the area. Thus, it was conceived that there be some sort of "core" area in which nature (mainly defined, naively, as not including *Homo sapiens*) would proceed unimpeded by humans, surrounded by a "buffer" area in which limited human activity would be permitted, but carefully controlled. Variations on this theme were abundant, but the basic idea was to integrate the development needs of humans with the national park idea into what collectively became known as an Integrated Conservation and Development Project (ICDP).[4] Such projects are still active, although their effects seem rather limited.[5]

A more recent development is based on historical precedent. With the arrival of Europeans in eastern North America, a process of deforestation ensued as a consequence of expanding agriculture. However, soon thereafter, the process of industrialization began and, soon after that, forests returned.[6] The dynamics that drove this process are evident at a broad qualitative level – wealth from agriculture drives local industrialization that, in turn, acts as a magnet for labor, which depopulates the countryside, leaving natural ecological succession to take over with the eventual transition to forest again. While this general view has many complications that drive local ecological and sociopolitical dynamics, as an overview of eastern North American forest history it seems historically accurate, and has been referred to as the "forest transition model" (the FT model).[7] Similar processes have been described for some European countries, the rural US south and most importantly, given its tropical location, Puerto Rico.[8] Based on this and other examples, it has been proposed that the FT model could be a framework for understanding tropical forest dynamics in general and might even be useful in negotiating a conservation agenda.

The Puerto Rican example is particularly interesting, not only for its tropical location, but because much of the "recuperated" forest was actually the consequence of abandoning shade coffee farms. Indeed, the FT model, if generally true, would apply in force when coffee agroecosystems were the basis of the agriculture that was being abandoned. As a point of fact, many of Puerto Rico's new forests are evidently old abandoned shade coffee plantations (Figure 8.3), and indeed may be stranded in

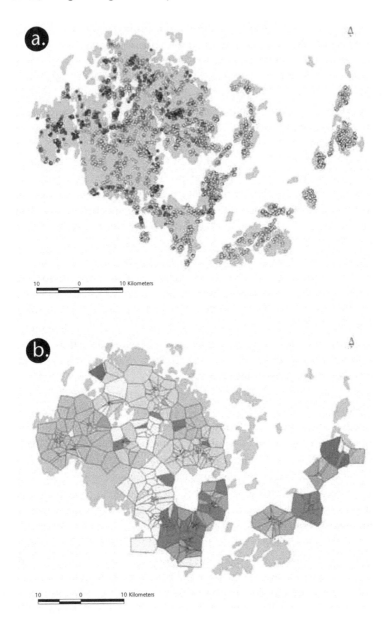

FIGURE 8.2 Habitat and genetic distribution of the Glanville fritillary butterfly on the Aland islands (Finland). (a) Distribution of the food plant, and hence the available habitat, of the larvae of the butterfly (darker areas indicate more dense populations). (b) Genetically homogeneous groups (same shades of grey represent genetic homogeneity). Clearly, any single large reserve will protect only a small number of genetic types, whereas a large number of reserves are more likely to conserve the known metapopulation structure and all of the genetic varieties in the general population.

Source: Modified from Hanski (2011)

FIGURE 8.3 Second-growth forest in an area previously occupied by a shade coffee farm. Larger trees are probably remnants from when the farm was operative. Smaller pole-size trees are mainly the pioneering species, *Guarea guidonia*.

an intermediate, non-pristine state for some time to come.[9] Unfortunately, the FT model seems laden with assumptions that are rarely met in nature. It certainly does not apply in general over all agroecosystems over all of the tropics, although it could be argued that, were it to apply anywhere, the coffee agroecosystem would be the place.

Although the FT model is usually proffered in an informal qualitative sense, there is an underlying quantitative logic on which it depends. Ultimately, it rests on two quantitative assertions and a seemingly logical conclusion. The two assertions are:

1   A given population density requires a certain land base to enable food production adequate to survival of the whole population.
2   The amount of food required to support that population, divided by current per-area productivity, equals the land area necessary for agricultural production.

The evident conclusion that stems from these two assertions is that the total land area minus the area necessary for food production is what could be available for conservation. The expectation from this conclusion is that the area for conservation will be maximized, either purposefully or through underlying dynamic processes, whenever both of two distinct assumptions are met: first, an assumption about rural–urban migration and second, an assumption about productivity.

The "rural–urban migration" assumption stipulates that with the reduction in rural population, more land will be available for conservation (fewer rural people will mean less use of land for agriculture, and thus natural regeneration of forest or other natural habitat). The "productivity" assumption stipulates that if per-area production can be increased, the land used for agriculture will be reduced, and consequently more land will be available for conservation (the same number of people needing food, but higher productivity, and thus less land needed for agriculture).

So rural people – those engaged in the act of food production – will move to the city in proportion to the labor efficiency of agricultural production; assuming that since they are not "needed" so as to satisfy the overall food needs of the society, they will not remain on the land, but will instead move to the city. Thus, as agricultural technology continues to improve efficiency, rural people will tend to move to the city and the land available for conservation will increase.

While it is certainly possible for the FT model to operate in this way, it is not in any way quantitatively assured that it actually will.[10] Does it necessarily follow that if the sons of Farmer Jones move to the city, his farm will reduce its production area? Does it necessarily follow that if Farmer Lin adopts new technology so as to double yield on her farm that Farmer Rivera will abandon his farm and move to the city? Indeed, as we elaborate in more detail below, the idea that improving the efficiency of agricultural production will cause people to leave agriculture strikes many economists as odd; indeed, quite the opposite might be expected i.e., more people being attracted to agriculture as it theoretically becomes easier (more efficient). In the end, theoretically, the issue is indeterminate. It thus makes sense to ask to what extent real-world data suggest that recent tropical situations replay the experience of the previous examples that had given conservationists such hope (e.g., Puerto Rico or New England).

In a detailed report on several case studies across the tropics, it was noted that sometimes the FT model appears to operate, sometimes not.[11] The study identified, as we mentioned above, an underlying contradiction in the basic ideas of the FT model:

1   "The belief that technological progress in agriculture reduces pressure on forests by allowing farmers to produce the same amount of food in a smaller area has become almost an article of faith in development and environmental circles."[12]
2   "Basic economic theory suggests that technological progress makes agriculture more profitable and gives farmers an incentive to expand production onto additional land."[13]

This contradiction suggests that whether the predictions of the FT model are true or not depends to a great extent on specific sociopolitical and ecological circumstances.

In a more extensive work, the same authors edited a series of chapters that included 17 in-depth case studies from Latin America, Africa and Asia.[14] Their conclusion from all these studies was that the issue of intensification of agriculture and its relationship to deforestation is complex and, effectively, that agricultural policy could be modified in such a way as to promote forest-preservative policies rather than policies that, however unintentionally, actually promote more deforestation with "improved" agricultural technologies. Examining these case studies, it is difficult to avoid the conclusion that, for the most part, conventional agricultural technological improvement results in more deforestation, not less. Of the 17 case studies presented, in 12 of them there was a clear indication that technological change had an effect on deforestation. Of those 12, nine showed increasing deforestation as a result of intensification (three of the nine suggested it could go either way, depending on circumstances), and only three suggested a clear decrease in deforestation with intensification. All cases were treated with the complex analysis they deserved, and seemed to negate quite effectively and completely the assumption that increases in agricultural technology automatically lead to land sparing. For example, in the Atlantic Coast of Costa Rica, increases in productivity of cattle land tended to increase deforestation, while increases in productivity of maize had the opposite effect.[15] In northern Zambia, changes in agricultural technology seemed to have the initial effect of decreasing deforestation, but later, as populations expanded either through local population growth or migration, the effect was the reverse.[16] To be sure, a few studies have shown support for the FT model,[17] but the great majority have shown either no effect or increased deforestation as a consequence of either agricultural intensification or decrease in the rural population.[18]

A variant of the FT model has emerged in recent years.[19] In this model, it is noted that with recent research showing that certain kinds of agriculture conserve more biodiversity than others, there is a push toward more "wildlife-friendly" farming. This idea of "sharing" agricultural land with wildlife is contrasted to the idea of "sparing" land to be untouched by human agents through the intensification of land already in agricultural use. The idea is that if more food could be produced on a smaller land base, more land could be devoted to pure conservation. It should be noted that the land-sparing idea has been around for quite some time and it was first proposed by Norman Borlaug, the father of the Green Revolution, and hence has also been called the "Borlaug Hypothesis."[20] We have already covered the technical side of this model in Chapter 3, and much of the detailed analysis of the FT model is directly applicable to this newer form of basically the same argument.

Critics of the land-sparing/land-sharing (LSLS) framework note that the model is based on a set of sometimes explicit, sometimes only implicit, assumptions.[21] First among the frequently implicit assumptions is the need to produce more food globally, at least in the near future. The problem with this assumption is that the production of calories in the world today already can accommodate the future population expected to stabilize at about nine billion sometime in this century. Today, agriculture provides more than 2,700 calories per person per day, a level considered sufficient to insure that the world's population is adequately fed.[22] Despite this, there are close to a billion people who are chronically undernourished.[23] The idea that a deficit in the amount of food produced is responsible for hunger in the world does not stand up to scrutiny. Yet there is great concern about food quantity throughout the world, a concern born mainly from the observation that so many people go hungry. Many people naively assume that *Homo sapiens* is a sufficiently moral species that it would never countenance a social organization in which an overabundance of an essential resource was available, yet withheld from a significant number of people. Refusing to accept that such a world exists, they are forced to conclude that starving people must mean that not enough food is available. Yet the data do not in any way support that position.

Even when we consider a future world with nine billion people, the argument that we need to increase food production by 70 percent[24] to satisfy the future demand for food ignores the inefficiencies of the industrial food system as well as alternatives that could increase food availability without intensifying agriculture. For example, it has been estimated that 30–50 percent of all food produced is wasted somewhere along the food chain.[25] Modest changes in human diets can also increase food availability without the need to increase yields. In 2008, 35–40 percent of the world's cereal production and 33 percent of all cropland area went to feed livestock. The annual caloric loss associated with feeding cereals to animals instead of using the cereal to feed people is equivalent to the caloric need of 3.5 billion people, even after accounting for the energy value of the meat. Some may argue that this reduction in meat production is unrealistic. However, a UNEP report estimates that even moderate changes in diets, which stabilize meat consumption to 2000 levels of 37.4 kilograms per capita in 2050 would free up enough cereal to feed 1.2 billion people.[26] In the end, we suggest that increasing the total amount of food in the world while maintaining the current system of distribution will do nothing to curb hunger; providing the poor with the means to produce or purchase the food already available will.

Second among the assumptions that go into the land-sparing/land-sharing debate is the fact that the negative secondary environmental consequences – the collateral damage, so to speak – of the industrial agricultural system are tacitly ignored. Just how much more of the world's oceans are we ready to sacrifice to dead zones? How many more aquifers will we permit to be poisoned with pesticides? How much more soil will we allow to be washed away? While agroecological techniques are also implicated in some negative environmental consequences, these are trivial compared to the massive consequences generated by the industrial system.

Taking the overall health of the human population and ecosystem into account, if we acknowledge that "feeding the people" also implies providing the people with long-term health, the negative collateral consequences of the industrial system must also be taken into account. People fed on high-fat/high-sugar/high-salt diets may be saturated with calories, but unhealthy (the industrial "system" considered in its entirety includes that diet as part of its definition). People who used to consume fish or other seafood products as part of their normal diet, but can no longer get them because of anoxic aquatic deserts, may be filled with calories from maize, but unhappy. People who consume food laden with endocrine-disrupting pesticide residues may have full stomachs, but face higher cancer risks.

The third assumption is that the negative secondary social consequences of the industrial system are minimized. As documented in many studies,[27] intensification along industrial lines can lead to further erosion of natural areas. For example, if a development plan promotes easy availability of chemical fertilizers in one area, it is most natural that hopeful migrants will thus be encouraged to come to that area and, if only natural areas remain uncultivated, there will be a tendency to claim those areas for agriculture. While such consequences could also be imagined under an agroecological framework (e.g., excessive nitrogen run-off is excessive whether it comes from the Haber–Bosch process or *Rhizobium*), previous evidence suggests that the industrial system is far more likely to engender such patterns.

The fourth assumption is that increased per-area productivity is always to be encouraged. Although this proposition sometimes seems obviously reasonable, it is not evident when viewed in a larger historical and social perspective. In many areas of agriculture today, the most pressing problem is overproduction, and has been so for many years. Dealing with this problem, frequently under the generalization of supply management, has demanded a great deal of analytical thought over the past century.[28] That Brazil, in the early twentieth century, purchased and subsequently burned almost 50 percent of its national coffee harvest[29] is only an extreme form. The famous coffee cartel of the Cold War, instituted as a part of the West's struggle against international Communism, is another. Even today, subsidy structures for maize production in the US allow giant grain companies to purchase maize at a price considerably below the cost of production, insuring that grain exports are competitive with even peasant producers in the Global South. While the desire to increase production by an individual farmer may be universal, the fact that overall overproduction depresses commodity prices and ends up negatively affecting all farmers is not lost on farmers, even if some economists may have trouble fully appreciating it.

Other critiques of the land-sparing option can be found.[30] It might be worth noting that the argument actually reverts to the original naivety of the isolated parks philosophy of the 1960s and loses whatever realism had been incorporated into conservation programs with the ICDP movement. With its implicit assumption that anything not in a preserved area is "sea," which is to say it is completely uninhabitable, and land "spared" for conservation needs to be free of that pesky species *Homo sapiens*, the early years of preservationism seem to have made a comeback. The LSLS framework effectively presumes that it is perfectly acceptable to intensify agriculture to the point that no biodiversity can survive in it, as long as productivity keeps going up. This assumption is, to say the least, contrary to the bulk of recent literature on agriculture and biodiversity, as summarized in Chapter 2.

The key farming debates: the ideology of "intensification"

At the end of World War II, agriculture throughout the world began on an alternative pathway. Propelled in no small part by the need for the chemical industry to find new markets, having lost their main market in military necessity (pesticides and explosives), agriculture went on a chemical spree.[31] Artificial fertilizers and pesticides combined with "improved" varieties of many crops to change agriculture from what had of necessity been an ecologically based activity to one that was effectively an application of chemical engineering. The US and the rest of the Global North were consumed by this new industrial agriculture. The new agrochemical corporations began looking to the Global South to expand markets for their new agricultural products, even before Norman Borlaug created the new varieties that were able to take full advantage of them. The overall program was complicated, but for our purposes the important component was an ideological one. That ideology was productivism. All research efforts, all developmental programs, all propaganda were aimed at increasing production. Secondary consequences, such as pollution of waterways and wells, or decimation of soil resources, were accepted as necessary collateral damage from this "productivist" mentality. Bolstering this ideology was a meta-ideology claiming that the growing human population was in need of more food, thus justifying the incessant search for higher yields on the same land base. Both of these assumptions – (1) we must increase production to (2) feed a growing population – are deeply embedded in the recent debates about biodiversity and agriculture, especially the FT model and the land-sparing/land-sharing debate, as described in the previous section. They thus merit further discussion.

The sort of high energy-demanding, chemically-intensive agriculture associated with modernity is assumed to generate an apparently ever-increasing productivity (amount produced per unit of land). A simple accounting from this assumption is precisely what results in the land-sparing model, the FT model and the optimistic assessments that rural–urban migration, as it decreases the number of "peasant" producers (automatically presumed to be inefficient), will result in equivalent – or even higher – production on less land, generating more forest recovery. However, what is the evidence that this fundamental assumption is correct?

Anecdotes can easily support the assumption, especially when highly subsidized farmers from the US and other industrialized regions are compared with small-scale farmers of the Global South, and the measure of productivity is either yield of the main commercial crop or net profit. However, if the measure of productivity is simply total output per area, relevant data do not seem to support the basic assumption. For example, analyzing data relating farm size to productivity (output per unit area), Cornia[32] found that in all cases the trend was of decreasing total output per area as farm size increased. Indeed, the "productivity–size inverse relationship" has been a well-known fact among agricultural economists since at least 1962.[33] It seems that small-scale owner-operated farms tend to be more efficient in that the farmer knows the land and its ecology well, and plants crops with that knowledge, usually using a multi-cropping strategy to take advantage of local peculiarities. Large, highly capitalized farms seek economies of scale in which local ecological peculiarities are purposefully ignored. Ironically, the recent enthusiasm for so-called "precision farming"[34] acknowledges precisely this underlying ecological structure, but proposes to accommodate it with a high-tech strategy of sensors and delivery systems. As one of our students, reviewing the literature on precision farming, quipped, small-scale farmers already do precision farming. Thus, neither the logic nor the data support the hypothesis that large-scale agriculture is necessarily more productive, on a per unit area basis, than small-scale agriculture.

Looking at the actual evidence more closely, the idea that only industrial agriculture is capable of feeding the global human population that we expect this century is not supported either,[35] but the argument is a complex one. The Michigan Study examined a global data set of 293 examples and failed to find a difference between organic and conventional yields.[36] A more recent study, while finding the same results for the same time frame as the Michigan Study, reported a change in more recent data, with the supposed advantage for industrial agriculture increasing in studies reported subsequent to the data available to the Michigan Study.[37] It is difficult to explain this dependence of the result on "date of study," without con-cluding that the studies themselves are biased. It thus remains, technically, an open question. While most of the previous studies failed to detect major differences in the productivity of organic versus industrial agriculture, the question of bias within experimental and observational studies must be taken seriously. At the present time, the best we can say is that there is a lack of evidence to support the claim that industrial agriculture is inherently more productive than alternative methods, including (but not exclusively) organic. Other meta-analyses (including the Michigan Study) that have incorporated these kinds of comparisons have shown no statistical differences between organic and conventional yields for studies in developed countries,[38] and a substantial yield increase in organic systems in developing counties when compared to conventional systems (not necessarily highly intensified systems).[39]

Yet even if available data did support the claim that conventional agriculture yields more than organic, the question itself is usually not adequately framed. As Chappell has noted repeatedly,[40] when speaking of technical issues that are evidently tied to human values and social structures, it makes no sense to make comparisons without taking those human values and social structures into consideration. For example, it is the bane of technologists everywhere that practitioners tend to be conservative about adopting new technologies, which is tied to the set of assumptions about how society is organized. Comparing, for example, the utility of the automobile to horses and buggies at a time when roads were generally not paved would certainly not have played out well for the automobile industry, and John Henry would have clearly won if coal had suddenly not been available for the famous steam engine. Comparing organic to industrial when the bulk of research and development over the past half century has been devoted to improving the industrial system, when direct and indirect subsidies flow to that system, and when the socioeconomic superstructure is finely attuned to that system, is bound to give biased results.

There is also a problem in the construction of the "counterfactual," the propo-sition or data set to which any proposed alternative is to be compared. Taking the overall health of the human population and ecosystem into account, if we acknowl-edge that "feeding the people" implies providing the people with long-term health, the negative collateral consequences of the industrial system must also be taken into account, as we noted above with respect to high-fat/high-sugar/high-salt diets and environmental risks such as higher cancer rates. Many other examples could be cited. If the counterfactual includes all the secondary consequences of the industrial system, from human health to environmental health, unfavourable comparison of the indus-trial system to the organic or semi-organic system, while difficult to quantify, seems evident. Ironically, attempts to be rigorous and "scientific" may obscure the obvious, if improper counterfactuals are constructed in service of a flawed notion of rigor. Will automobiles outperform horse-and-buggy transportation? Absolutely not, since roads are not paved. That would have been the conclusion in 1900 if the counterfactual of paved roads had not been considered.

In summary, contrary to the conventional wisdom that intensification of agriculture, which almost always implies a move to the more industrial-style agriculture, is needed to produce enough food to feed the world, the empirical evidence suggests that peasant and small-scale family farm operations adopting agroecological methods can be as (or more) productive than industrial agriculture. Given that most of the world's poor live in rural areas or are city poor who have been recently displaced from rural areas, an agricultural matrix composed of small-scale sustainable farms is precisely what we might expect if policies and politics permit.

However, there exists yet a more complicated question involving the larger socioeconomic framework of agriculture. The search for more productivity – part and parcel of the research agenda of most agricultural researchers – is not necessarily a rational project. In many cases, and here coffee is an excellent recent example, the major agricultural problem is overproduction and the consequent low prices paid to farmers. The recent (and probably temporary – writing in 2013) increase in food prices notwithstanding, it is often the case that farmers receive inadequate compensation for their efforts largely because markets have become saturated. If unregulated markets must be the rule, an assumption that itself might be questioned, overproduction and low prices will continue to plague farmers – not continuously, but on a boom-and-bust cycle. Indeed, the IAASTD, an intergovernmental assessment process that involved three years of research and 400 experts from many disciplines and all over the world, concluded that conventional/industrial agriculture is not a rational option either for alleviating poverty and ending hunger and malnutrition, or for sustainable development, further noting that greater equality is needed in pursuit of alleviating hunger and malnutrition.[41] This equality is more likely to be achieved through land reform that redistributes land currently in the hands of big agrobusiness and planted in commercial monocultures (in many cases for the production of biofuel, animal feed or other export commodities, rather than of food for people) and puts it in the hands of small- and medium-scale family farmers, who are more likely to construct a landscape mosaic that promotes biodiversity and produces more food for the local population.

The matrix quality model

As noted above, the assumptions that agricultural intensification will produce more food and that more food needs to be produced, are part and parcel of several popular conservation schemes. It is then taken as a zero sum game, in principle; the amount of land you need to produce food must be subtracted from the total amount of land to give the amount of land you can let transition to forest, or spare for conservation. And if the population keeps growing, that just makes things worse. Within that framework, it is not surprising that the programmatic conclusion is to increase food production by intensifying agriculture at the same time as reducing the growth of the population. This ideologically based framework is reflected in Norman Borlaug's words when accepting his Nobel Prize:

> There can be no permanent progress in the battle against hunger until the agencies that fight for increased food production and those that fight for population control unite in a common effort.[42]

This argument is also extended to the conservation of forests and biodiversity. The technical error surrounding the biodiversity aspect of the argument lies in the failure to recognize the landscape as the proper unit of analysis, as we argued in Chapter 3. As a minimal framework, we suggest that the landscape be the largest category (some researchers use the word "countryside,"[43] which we feel is equivalent). The landscape is composed of two generalized categories: conservation areas (or natural habitats) and matrix. Conservation areas may be further divided into many categories (community forests, biological reserves, biodiversity refuges, etc.), and the matrix likewise divided into various categories (annual cropping systems, pasture, agro-forestry systems, etc.). However, right from the beginning, the simple dichotomy between preserved areas and matrix needs to be recognized.

It seems to be the case that most scholars are in agreement on this basic definitional framing. The main disagreement arises from a difference in how landscape is viewed from a dynamic perspective. The static framework at the extreme sees the preserved areas and matrix as separate domains, which is quite convenient for empirical work. Viewed in this way, it is a simple task to ask questions about each category separately and take some sort of average, or focus the questions as cost/benefit in which each section has its own costs and benefits (as presented in Chapter 2). This framework is potentially useful, but must be applied with caution and with the full realization that the interconnections between conservation areas and matrix are presumed to be close to zero. If, however, there is an important connection between conservation areas and matrix, the empirical questions of how much productivity and how much biodiversity become less meaningful.

The importance of extinction in the matrix model

It is far from controversial to suggest that local extinctions occur,[44] as we noted in great detail in Chapter 3. Indeed, extinction is a key component of the theory of island biogeography, as one of the major processes driving patterns of biodiversity on islands. Although the fact of local extinctions is well established, their pattern is clearly not random, and certainly deserves more study.[45] Nevertheless, there is little doubt that amid many complications, populations living in isolated fragments of natural vegetation can expect to experience extinctions, if enough time passes. If conservation is to be a long-term goal, this elementary and undeniable fact must be incorporated into planning.

A further complication may result from spatial self-organization. Consider, for example, plant communities in which the constituent species tend to expand in space through seed dispersal, but are attacked by natural enemies in a density-dependent fashion according to the Janzen–Connell effect.[46] It can be shown that such an arrangement will result in the clumping of organisms even in a uniform environment,[47] as discussed extensively in Chapter 4. Because of the dynamic interplay of seed dispersal and density-dependent control, any particular spatial cluster is expected to go locally extinct over the long run. In such a situation, fragmenting the continuous habitat does not change much about the local extinction rates, which are a consequence of density-dependent operation of natural enemy dynamics. However, normal migration (e.g., seed dispersal) will be reduced.

Unfortunately, long-term studies that uncover such patterns of extinctions in continuous habitat are not common in the literature. In a particularly interesting study in the northern Great Lakes region of North America,[48] it was found that there had been dramatic changes in plant species composition in plots embedded in natural forest communities over a 20-year period. Environmental drivers in this case included forces such as deer hunting and invasive species, but one of the researchers' key findings was that, even in this unfragmented forest, species loss at a local level was dramatic. Another example comes from a 20-year study of the amphibians occupying small ponds in a forested matrix.[49] In this case, approximately 30 local extinction events were reported. However, the researchers were able to demonstrate that "reinvasions" – which is to say migration events – completely balanced those local extinctions. Many other studies could be cited. Thus, both ecological theory and empirical studies allow us to conclude that local extinctions are quite normal and occur even in areas of continuous natural habitats. And, as we have emphasized throughout this book, migrations throughout the matrix, whether composed of natural or anthropogenic habitats, can balance those extinctions and maintain a metapopulation structure that will prevent regional extinction.

What is in the matrix?

For these reasons, we and others have argued that an approach that focuses on the matrix is critically important.[50] If the fragments of preserved areas become completely isolated (as they are in a very low-quality matrix), local extinctions will tend to dominate the landscape, eventually leading to regional extinctions. To insure a metapopulation-like structure (which would include various structures, such as source–sink populations), the nature of the matrix should be "friendly" to biodiversity. However, this friendly nature need not mean that populations of concern are able to live in perpetuity in the matrix, but rather that they are able to temporarily use matrix elements as they migrate (or disperse) among fragments of preserved areas. This inevitably brings up the question of what the nature of that high-quality matrix is.

We proposed in Chapter 3 that one way of categorizing matrix components is in a dichotomous fashion, where some patches are "ephemeral sources" while others are "propagating sinks."[51] For example, if the matrix includes farmers engaged in slash-and-burn agriculture, the fallow parts of the matrix can be important "sources" for various organisms. A species of butterfly, for example, may find refuge in these fallow areas, which may contain the food plants necessary for its larval forms, and proceed for many years to send offspring out into the more general landscape. However, that fallow land is eventually turned into an agricultural field, at which point that species of butterfly probably will go locally extinct, but will have been sending propagules into the landscape for several years (as long as the fallow land lasts – it is, by definition, an ephemeral source since it must eventually disappear). An example of a propagating sink might be a local shade coffee farm. Clearly, it will be a sink for many organisms in the sense that they might not be able to form a successful population within it, but some may find refuge in it for a long enough period of time to reproduce and send out propagules, thus contributing to the overall potential for the matrix to be "permeable."

The specific categories of ephemeral sources and propagating sinks are abstractions that may be useful when thinking about matrix structure. However, in real-world applications, such abstractions can sometimes be more constraining than enlightening. Matrices in the tropics can be extremely complicated, both by design and in practice. Pigeonholing management options into these, or any other generalized categories, must be done with caution, reflecting John Stuart Mill's admonition that "the price of analogy is eternal vigilance."

Using a different metaphorical system, several authors have suggested that the construction of agroecosystems should follow as much as possible the natural ecological forces operative in local ecosystems. This is an attitude that seems to be common among traditional agriculturalists.[52] It probably made it into the imagination of European-inspired agroecologists via the writings of Sir Albert Howard. Howard was dispatched to India to teach farmers the wisdom of the emerging sciences associated with agriculture, including the recently established principles of nutrient uptake and utilization by plants. The Victorian idea was basically the same as modern notions of increased efficiency gained through application of "modern" science. However, Howard noted that Indian farmers were producing just fine using technologies that today we would likely call "ecological."[53] His charge – to enlighten those "poor backward farmers" by introducing them to the newly emerging industrial model of agriculture – in the end seemed pointless to him and he took to learning from the Indian farmers rather than teaching them. Their message – much as that of traditional agriculturalists before them and traditional farmers today – was that agriculture is basically an ecological activity, not a chemical one. Ecological information from natural (which is to say pre-agricultural) systems should inform the development of agriculture. Although Howard never used the phrase "natural systems agriculture," his insights certainly reflect ideas of what we call today "natural systems agriculture."

Many years later, working on the prairies of Kansas, Wes Jackson applied the same principles to challenge the entire enterprise of big monocultural agriculture.[54] Noting that maize, wheat and soybean production had effectively taken a "perennial polyculture" (the natural prairie ecosystem) and converted it into an annual monoculture, his insight was to use nature as a model and embark on a program of developing a new agricultural system based on the local natural ecosystem, perennial polyculture. In developing this visionary program, he popularized the idea of "natural systems agriculture."

If we add to natural systems agriculture the important social variables, the science of "agroecology" emerges.[55] As has been noted in Chapter 3 (and in many other places[56]), a matrix in which independent small-scale agriculturalists using agroecological techniques dominate is likely to be the most biodiversity-friendly matrix we can expect, both from the point of view of containing biodiversity itself and, more importantly, from the point of view of providing migratory pathways among fragments of natural areas.

Although we regard this idea as very general, applying throughout tropical and many temperate landscapes, our theme in this book is coffee – the agroecosystem that, perhaps better than most others, makes these principles easily observable. The natural system that coffee replaced was almost always a mid-elevation tropical rainforest, and most coffee farmers worldwide are small-scale producers who grow coffee under shade trees, effectively mimicking the natural system that has been replaced. Given this fact, we can (as we outlined in previous chapters) easily see the quality of the matrix, if dominated by coffee farms, as being determined by the style of coffee growing, along the gradient from rustic to sun, from high forest-like shade cover to no shade trees at all (as in Figure 2.12).

Connecting the matrix to broader socioeconomic structures

In their book *The Coffee Paradox*,[57] Daviron and Ponte provide a complex sociopolitical analysis of the international coffee business. Their paradox is that while coffee prices go up for the consumer, coffee farmers remain mired in economic uncertainty, frequently suffering from unexpectedly low prices paid for their produce. At a very simplified level, the paradox is precisely the same as that faced by agricultural producers everywhere, especially in today's hyperconnected and hyperindustrialized world. The farmer produces coffee beans and the consumer drinks a cup of coffee. In between, the coffee passes from farm to warehouse, from warehouse to roaster and from roaster to retailer, and, as in many other forms of agriculture, the most politically powerful players in this system use their power to their advantage. In the case of coffee, the problem is with those who buy the beans and sell the coffee. The $2.50 we pay for a cappuccino contains the amount of coffee for which the farmer received less than $0.10. Things have gotten progressively worse since the end of the Cold War. Between 1962 and 1989, the International Coffee Agreement, strongly supported by the US and its allies in the Cold War, required all producing countries to engage in effective supply management. The logic of this arrangement was that if small-scale farmers in the Global South were exposed to open markets, their inevitably lowered standards of living would make them potential targets for "Communist propaganda" – a logic that merits considerable reflection, but beyond the scope of this book. Because of such Cold War pressures, the so-called coffee cartel was not merely permitted to exist by world capitalism, it was actually encouraged.

With the fall of the Soviet Union in 1989, coffee-producing nations were no longer in danger of being seduced by socialism, and therefore no longer permitted to maintain their cartel, and prices dropped precipitously, as expected. The entrance of other countries, especially Vietnam and Indonesia, into the producer category exacerbated the overproduction problem and, with a brief interlude in the mid-1990s when a freeze caused lower production in Brazil, prices paid to farmers continued their decline, reaching historical lows in 2001. With increased consumption in China and Russia and reduced harvest in the mid-2000s, prices went up again and continued to increase until 2011, after which they fell precipitously again due to overproduction. A news article posted by Bloomberg News on November 14, 2013 reported:

> Record crops from Brazil to Vietnam are compounding a global surplus that drove prices down 26 percent this year, heading for a third annual decline and the longest slump in two decades. [Coffee farmer Nils Solorzano Villareal], who added trees and fertilizer to expand his farm by a third, said he is spending $140 to produce each 132-pound bag of arabica coffee beans that sells for about $132.[58]

Since the elimination of the coffee agreement, coffee prices have been characterized by wild fluctuations that make it hard for farmers to plan ahead. Small-scale farmers are more vulnerable to these price fluctuations because they do not have excess capital that provides them with a buffer. Because of the post-Cold War liberalization of the international coffee market, the world's estimated 25 million families who rely on coffee production remain perpetually insecure.

While the overall world economic system is very complicated, to be sure, and we do not wish to antagonize any of our colleagues whose business it is to study that system in detail, we note that however you cast it, the "middlemen" are taking a vast amount of the value added from the production of the bean to the cup of coffee. And at this point in time, the intermediaries are few and large and becoming fewer and larger. Of the entire coffee retail market, which is huge, there are effectively only four players: Kraft (Maxwell House), Nestlé (Nescafe and Taster's Choice), Sara Lee (many brands), and Procter & Gamble (Folgers). Put in simplified terms, the consumer pays too much, the farmer receives too little, and four giant corporations get most of it. And this is not a problem in coffee alone. Virtually all agricultural commodities under the industrial model face the same structural constraints. There is a major filter between producer and consumer, and that filter consumes the bulk of the value added during the entire process. Given that structure, it is not surprising that movements have arisen to try and level that playing field. One of the most popular and rapidly growing of such movements, at least in the Global North, is the Community Supported Agriculture (CSA) movement.[59]

The basic idea of CSA is to make a direct connection between consumer and producer. Consumers usually purchase a share in a particular farm and the farm provides them with produce later on. So, for example, we pay our farmer $200 in August, for a winter CSA that we belong to. Every week during the winter, he provides us with abundant fresh greens (spinach, arugula, etc.). We provide him with the money up front (eliminating the need for a bank loan at the beginning of the season), and he provides us with the produce during the season. It is, from an energy and ecological perspective, a remarkably efficient system.

One of the key elements in the CSA philosophy is the direct personal connection between producer and consumer. The consumer knows the farmer, and visa versa. Farmers who offer bad produce will not get customers. It is a system that usually works quite well and members of CSAs are generally satisfied, judging from the continued growth of the movement.[60]

The relevance of the CSA movement for this book relates to the paradox that Daviron and Ponte referred to in their book. The only place to grow coffee is in the tropics and many consumers live in temperate climates. There is no way that coffee could be grown in our home state of Michigan (or anywhere else in the continental US), so it would seem quite difficult – indeed impossible – to develop a CSA arrangement. However, several initiatives are trying to do just that. To take an example close to home, students from the University of Michigan, in analyzing the problem of connecting consumers to producers, contacted a local roaster in Michigan and made a direct connection with a coffee-producing cooperative in Chiapas, Mexico. They sent delegations of university students to meet with the cooperative members and brought members of the cooperative to the university to meet with Ann Arbor community members, about as close as one can get to consumer–farmer encounters when a tropical crop exported to a temperate country is involved. One of many such initiatives, this one is inevitably an on-again off-again operation since it is run by undergraduate students, but the basic idea is an intriguing one: develop the spirit of direct producer–consumer contact.

Another example following the same basic logic is the Community Agroecological Network (CAN) based in Santa Cruz, California.[61] It involves a variety of coffee-producing cooperatives in several Latin American countries. Formed in 2001, "CAN was founded to link farming communities throughout the region and consumers in the United States." CAN relies mainly on direct mail sales, thus cutting out most intermediate actors in the process. Other initiatives continue to emerge associated with coffee-producing areas in a variety of tropical countries, some with an explicit goal of directly connecting producer and consumer in the spirit of the CSA movement, and others more in line with the standard fair trade movement.[62]

While direct contact between producer and consumer and direct mail sales offer great promise for the small-scale coffee producer, an older and perhaps more established initiative remains one of the chief vehicles for promoting the kind of coffee agroecosystem that we argue would be good for the formation of a high-quality matrix. This is the idea of certification, long associated with organic coffee and the fair trade movement. Indirect certification occurs all the time when consumers know their producers and understand their methods, thus personally certifying that the farmer produces within some restricted scheme. In agriculture, one of the oldest formal certification schemes is organic certification. Born of consumer desire to both eat healthy food and protect the environment from poisons, various schemes to certify that a farm produces organically (using no industrially produced synthetic chemicals i.e., no chemical pesticides or fertilizers) have emerged. The US Department of Agriculture, for example, has one, as do most European nations.

A more recent trend has focused on the economic side of agriculture. This has been most active in the coffee industry, with "fair trade" coffee now available in many markets throughout the developed world.[63] The basic idea of fair trade is that the consumer should perhaps pay a little more so as to provide a fair price to the producers of coffee. Prices are set at a competitive level at the farm so as to insure that the farmer can make at least a minimal profit and distribution is handled by a central agency, one of which is Fair Trade USA (formerly a member of a larger organization, Fair Trade International). From the consumer's perspective, purchasing fair trade products helps to insure a fair shake for the producer. From the producer's point of view, being certified brings the benefit of market access to the world of conscientious consumers.

Yet another certification scheme is bird-friendly or biodiversity-friendly certification, sometimes only referred to as certified shade coffee. A variety of criteria were devised that researchers from the Smithsonian Migratory Bird Center concluded would be necessary to promote high-quality habitat for birds and other elements of local biodiversity. These criteria required organic production to start with, but then added onto that the need for "high-quality" shade that provides good habitat for birds (e.g., a minimum number of species of shade trees, a minimum depth of the canopy of shade trees, a minimum density of shade trees). Although other schemes have emerged, the Smithsonian bird-friendly certification program is the gold standard for biodiversity-friendly certification.[64]

There has been some criticism of having three separate certification systems for coffee, suggesting the need for a single system that incorporates all three criteria, so-called triple certification. Thus far, such a movement has not crystalized very effectively.

There are criticisms of certification schemes in general. First and foremost is the inherent contradiction in the need for consumer confidence that a claim of certified production, of whatever form, is indeed legitimate, coupled with the cost of certification. For consumer confidence, it is generally agreed that a third party must be the certifier, yet that third party becomes yet another cost of production for the producer. Sometimes the certification process can be quite expensive, which represents an economic barrier to adoption for many producers. Yet without such third-party intervention, the confidence of the consumer cannot be guaranteed. An obvious solution to this issue is the CSA ideology of direct interaction between producer and consumer, which, as we noted above, is not exactly easy in the case of tropical crops and temperate consumers.

An alternative framework: the New Rurality

Frequently cited as a triumph of the Industrial Revolution, the Global North experienced a draining of the countryside as farming became less profitable and jobs in manufacturing and urban services became more attractive. Some view this as a natural and inevitable process, with social value more regularly concentrated in growing urban centers, even if this came at the expense of the social and cultural value embedded in the rural system. Innovation, creativity and social experimentation were the hallmarks of the city while resistance to change, dullness and backwardness became associated with rurality.

Without too much thought, analysts of both right and left political persuasion assumed that the same dynamic would soon drive the urbanization revolution of the Global South. The peasant farmers of what was then called the Third World would succumb to the lure of the city just as French farmers could not be kept "down on the farm, after they've seen Paris." However, while urbanism has grown explosively in the Global South, it has not been driven by the same dynamic. Rather than the "pull" of the city being the main driving force, the "push" out of the rural areas has been prominent, mainly due to three forces.

First, international agricultural commodity policy has created a system in which local farmers find it difficult if not impossible to compete in local markets. The dumping of corn in Mexico since the implementation of NAFTA in 1994 is only one example.[65] The dumping of potatoes in Jamaica,[66] black beans from Michigan dumped in Mexico[67] and many other examples highlight how farms in the Global North take advantage of direct and indirect political support (popularly referred to as subsidies) to undercut the productive ability of local farmers in the Global South.

Second, the original Green Revolution, while providing only modest gains in productivity, did have a major effect in transforming farming in several regions of the Global South. Farming in the Punjab region of India, for example, became extremely concentrated due to the economies of scale involved in the use of pesticides, chemical fertilizers, irrigation and mechanization, the hallmarks of the Green Revolution. That concentration created difficult economic times for farming at a smaller scale. The pattern repeated itself in many places in the Global South.

Third, partly a consequence of assuming that the same dynamic as seems to have occurred in the Global North would engulf the Global South, planning and investment focused on the cities, not the countryside. The investments and government support that could have gone to strengthen rural communities (paved roads, local markets, schools, hospitals) were absent. The consequence was a stagnation of "development" activities aimed at the rural sectors. Regardless of how committed a farming family was to their local community and their farm, when their kids needed to go to school, if the only school was in the city, they were effectively forced to move. Likewise, if the only true healthcare was available only in the city, when they got sick they were forced to move.

Given these three forces, progress toward the "goal" of moving people out of the rural areas and into the cities has been surprisingly slow. The famous refusal of the "peasantry" to disappear in the Global South has been as much news as the latest giddy prediction of its eventual demise in the face of the undeterrable rise of urban sophistication.[68] Nevertheless, one cannot deny that the forces pushing small-scale farmers out of the picture remain strong.

In a break with the deterministic assumption that the rural population is destined to disappear, Hecht proposes a new conceptual framework, specifically tuned to the contemporary situation (and most evidently applicable to Latin America).[69] This new conceptual framework is called the New Rurality and categorizes not so much rural landscapes themselves, but rather the way in which scholars focus on them. She suggests there are four broad and overlapping categories: environmental, socio-environmental, agroindustrial and peasant. Such a categorization would not have made a great deal of sense either before the Cold War or during the heydays of neoliberalism after the Cold War. However, as Hecht argues, it is a framework that strongly aids our understanding of rural dynamics in the contemporary world, as it has been unfolding since the end of the Cold War. Analysts concerned with rural landscapes tend to fall into one of these categories and their analysis is consequently driven by the vision they bring to the table.

Environmentalists seek to preserve native habitats, socioenvironmentalists seek to incorporate indigenous and local communities in their conservation plan and agroindustrialists see tremendous opportunity in the expansion of industrial agriculture, which sometimes includes and sometimes excludes the peasant element. In contradistinction to these three points of view, the framework that recognizes that the rural world is still populated with peasants (small-scale family farms) sees them as acting in a variety of complex ways, sometimes with strong economic and socio-cultural links to cities. These complicated actions and linkages ultimately will determine the fate of rural landscapes, according to this point of view.

Aligning ourselves within Hecht's description of those who see rural areas as still populated with peasants and small-scale family farms, and focusing on the past few decades of development in the science of ecology, we argue that data and theory suggest that conservation should be viewed from a larger landscape perspective and that moving agriculture toward a sustainability priority, rather than a productivist priority, has more potential to positively affect biodiversity conservation.

The notion of New Rurality intersects our notion of matrix quality in important ways. Of particular interest is the way in which human inhabitants of the matrix, who constitute that New Rurality, are currently acting. Earlier views of rural inhabitants as somehow on their way to the city suggested that perhaps benign neglect would be the best course of action, at least with respect to biodiversity conservation. Thus, the FT model, as applied to the tropics, generally applauded the eventual "emptying" of the rural areas since "natural" vegetation could then recolonize them, thus bringing salvation to the biodiversity that had been driven to the edge of extinction. However, if it is true, as much recent social science suggests, that rural residents in tropical regions are not going to go away, a new focus on what in fact they *are* doing is appropriate. And one of the things they are doing is organizing into groups that appear to be far more effective than they have been in the past.

One example is the Movimiento dos Trabalhadores Rurais Sem Terra (the MST – the Movement of Landless Rural Workers) of Brazil.[70] Historically, Brazil came under a military dictatorship after a 1964 coup partially sponsored by the Central Intelligence Agency of the US. The dictatorship enthusiastically adopted the industrialized agriculture strategy being pushed by the US under the famous Green Revolution. Consequently, massive rural unemployment emerged as peasant agriculture was attacked wholesale. Not unexpectedly, social movements resulted. However, it was not easy to organize politically under the dictatorship, so the myriad rural social organizations remained relatively isolated, and for the most part clandestine. Even before the military coup (and partly because of it), the liberation theology movement, promoted by the progressive wing of the Catholic Church, had been encouraging the formation of such groups and taking on the larger landholders in southern Brazil. The tactics of the liberation theology era included large-scale land occupations as a vehicle for political pressure for land reform. Much of this activity was violently suppressed after the coup.

The military dictatorship came under massive popular pressure in the early 1980s, led by students and workers, and was overthrown in 1984, resulting in the return to a nominally democratic government. Also in that year, the already existing rural social organizations came together under an umbrella organization that became the MST of today. Employing similar political action tactics to those of the earlier liberation theology movement, the MST sponsors what the landowners call "land invasions" but what the participants themselves refer to as "land rescues" (Figure 8.4). The political pressure created by these land occupations creates a legal space to

FIGURE 8.4 Representatives of the Instituto de Pesquisas Ecologicas (IPE) greeted by an MST leader in a new encampment in southern Brazil.

FIGURE 8.5 Food warehouse run by the MST in southern Brazil.

push for a negotiation between Brazil's national land reform agency and the original landowner. Thousands of hectares have been redistributed to rural residents through the actions of the MST.

Originally created as a direct action political movement to pressure for agrarian reform, the MST has branched out into other activities. Food wholesale and retail operations are now operated directly by the MST and are reported to service rural residents and to be currently making headway in penetrating urban markets (Figure 8.5). Although the MST is not particularly active in the coffee-producing regions of Brazil, a most important activity for this book is its growing commitment to biodiversity conservation.

A much larger organization is the umbrella organization La Via Campesina (The Peasant Way) with a membership of over 150 local organizations in 70 countries (including the MST), claiming to represent about 200 million farmers the world over. La Via Campesina originated in Latin America, but quickly acknowledged that peasantries around the world share the same problems, leading to a globalization of the movement from below, including the elaboration of a new conceptual framework, the notion of "food sovereignty." According to Martínez-Torres and Rosset:

> food sovereignty argues that food and farming are about much more than trade and that production for local and national markets is more important than production for export from the perspectives of broad-based and inclusive local and national economic development, for addressing poverty and hunger, preserving rural life, economies and environments, and for managing natural resources in a sustainable fashion.[71]

Through this concept, La Via Campesina envisions "agrarian trajectories that would reintegrate food production and nature as an alternative culture of modernity."[72] It is this sort of global, grassroots social movement that has been challenging the neoliberal model of development and has put forward a strong vision of the New Rurality, which would not only contribute to the conservation of biodiversity but also provide a dignified and sustainable livelihood for rural communities.

The convergence of food production with nature conservation

The matrix quality model asserts that it is the kind of agriculture, not the simple fact of its existence, that matters.[73] Whether looked at from the point of view of the quantitative models presented earlier, or from the more qualitative notion that some habitats promote more migration than others, the agricultural matrix is perhaps the most important habitat on which conservation efforts must focus. However, this brings us face to face with the fact that agriculture, in spite of its acknowledged multiple functions, needs to produce food, a necessity that inevitably brings with it a question of productivity. In this question, we face what seems at first to be a dilemma when seeking a high-quality matrix. The sort of high energy-demanding, chemically intensive agriculture associated with modernity generates a very low-quality matrix, whereas alternative types of agriculture (organic, agroecological, natural systems agriculture, etc.) would seem to be precisely those that would produce a high-quality matrix. Yet, such agricultural types are normally assumed to be less productive. As we noted earlier, simple accounting from this assumption is precisely what generates the land-sparing model, the FT model and the optimistic assessment that rural–urban migration, as it decreases the number of "peasant" producers, will result in higher production on less land, generating more recovery of natural ecosystems. However, as we argued earlier in this chapter, the evidence for this fundamental assumption is lacking.

Our framework, as outlined in this chapter, is what we refer to as the matrix quality approach, intended to be an alternative to some other approaches such as the FT model or the land-sparing model. Our analysis does not aim to prove that the predictions of either the FT model or the land-sparing model cannot be true, but rather seeks to frame the problem in such a way to first see that its predictions are, from a theoretical viewpoint, weak, and, from a practical viewpoint, do not inevitably play out as expected in the real world (see especially Chapter 3 in addition to the present chapter). On the other hand, the realities of the current tropical world, which is mainly in a state of extreme fragmentation, coupled with the growing consensus among ecologists that metapopulations, metacommunities and landscape processes are important determinants of biodiversity, suggest that the matrix quality approach is a better and more accurate descriptive tool than the alternatives. Given this model, the practical consequences suggest that the promotion of small-scale agriculture, and in particular those forms that are sustainable (i.e., organic, agroecological, etc.), as an integral part of tropical landscapes, is more likely to preserve biodiversity in the long term. Furthermore, it is the small-scale agriculturalists who are more likely to adopt sustainable agricultural technologies because they use few or no external inputs, use locally and naturally available materials and generate agroecosystems that are more diverse and resistant to stress than capital-intensive technologies.[74] Within this context, diverse shade coffee farms, managed with agroecological methods that "maximize the array of built-in preventive strengths"[75] of the agroecosystem, represent precisely the types of farms that contribute to a high-quality matrix.

In the end, it appears that the real needs of people for a diet that is sufficient in quantity and quality is the same as the need of the landscape for a high-quality matrix within which fragments of high-diversity native vegetation can persist along with biodiversity-friendly agroecosystems to form an integrated landscape. Indeed, recent international documents that evaluate the role of agriculture in alleviating hunger and promoting sustainable development (including the IAASTD report of the United Nations) concur with the conclusion that small-scale sustainable farming systems are the best option for achieving both of these goals.

Notes

1 Chappell and LaValle (2011); Chappell et al. (2013).
2 Simberloff (1986).
3 Hanski (2011).
4 Hughes and Flintan (2001).
5 McShane and Newby (2004).
6 Pfaff (2001).
7 Mather (1992); Mather and Needle (1998); Mather (2008).
8 Thomlinson et al. (1996); Grau et al. (2003); Aide and Grau (2004).
9 According to one study (Zimmerman et al. 2007), forests recuperating from coffee agriculture seem to have entered a completely different soil nitrogen regime compared to the original forest structure. This completely new and novel regime may indeed forestall return to the original forest for decades, if not centuries.
10 Perfecto and Vandermeer (2010).
11 Angelsen and Kaimowitz (2001b).
12 Angelsen and Kaimowitz (2001b: 89).
13 Angelsen and Kaimowitz (2001b: 89).
14 Angelsen and Kaimowitz (2001a).
15 Roebeling and Ruben (2001).
16 Holden (2001).
17 Kleinn et al. (2002); Rodríguez (2004); Klooster (2003); García-Barrios et al. (2009).
18 See summary of literature in Perfecto and Vandermeer (2010).
19 Green et al. (2005); Phalan et al. (2011a).
20 Angelsen and Kaimowitz (2001a).
21 Tscharntke et al. (2007, 2012a); Fischer et al. (2008, 2011).
22 FAO (2002).
23 FAO (2013).
24 This is a frequently cited figure with little backing it up. See, for example, Bruinsma (2009).
25 Lundqvist et al. (2008); Gustavsson et al. (2011); Foley (2011); Foley et al. (2011).
26 Nellemann (2009).
27 Angelsen and Kaimowitz (2001a).
28 Croom et al. (2000); Daviron and Ponte (2005); Dimitri et al. (2005).

29 Pendergast (1999).
30 Tscharntke et al. (2007, 2012a); Fisher et al. (2008, 2011).
31 Russell (2001).
32 Cornia (1985).
33 Sen (1962).
34 McBratney et al. (2005).
35 Pretty (2008).
36 Badgley et al. (2007).
37 Seufert et al. (2012).
38 Stanhill (1990); Badgley et al. (2007).
39 Pretty et al. (2006, 2011); Badgley et al. (2007).
40 Chappell (2007).
41 IAASTD (2009).
42 Borlaug (1970).
43 Ranganathan et al. (2008).
44 This issue is not really debatable in the technical literature, with local extinction being regarded as natural and inevitable. See, for example, Fischer and Stöcklin (1997); Foufopoulos and Ives (1999); Kéry (2004); Matthies et al. (2004); Wilsey et al. (2005); Williams et al. (2005).
45 Bolger et al. (1991); Brooks et al. (1999); Helm et al. (2006).
46 Janzen (1970); Connell (1971); Hyatt et al. (2003).
47 Vandermeer et al. (2008).
48 Rooney et al. (2004).
49 Werner et al. (2007).
50 Daily et al. (2001); Fischer et al. (2005); Donald and Evans (2006); Harvey et al. (2008); Franklin and Lindenmayer (2009); Chazdon et al. (2009); Perfecto et al. (2009); Perfecto and Vendermeer (2010).
51 Vandermeer et al. (2010b).
52 Altieri (2004).
53 Howard (1943).
54 Jackson (1980).
55 Carroll et al. (1990); Gliessman (1990); Altieri (1995).
56 Lindenmayer and Franklin (2002); Perfecto and Vandermeer (2002); Armbrecht and Perfecto (2003); Vandermeer and Perfecto (2007); Perfecto and Vendermeer (2010).
57 Daviron and Ponte (2005).
58 Bloomberg News (2013).
59 Cone and Myhre (2000); Lyson (2004); Schnell (2007).
60 McFadden (2008).
61 Details and updates on CAN can be found at www.canunite.org/about-us.
62 Bray et al. (2002); Raynolds et al. (2004); Bacon (2005, 2013).
63 Renard (2003).
64 Giovannucci and Ponte (2005); Philpott et al. (2007).
65 Wise (2009).
66 Weis (2004).
67 Lucier et al. (2011).
68 van der Ploeg (2009).
69 Hecht (2010).
70 Wright and Wolford (2003).
71 Martínez-Torres and Rosset (2010: 160).
72 McMichael (2006: 416).
73 Perfecto et al. (2009).
74 UNCTAD-UNEP (2008); IAASTD (2009).
75 Lewis et al. (1997: 12243).

9

SYNDROMES OF COFFEE PRODUCTION

Embracing sustainability

Syndromes of production as ecological regimes

The transformation of agriculture has been a continuous process ever since its invention, multiple times, thousands of years ago. As with other technological issues, such transformations are sometimes slow in coming, sometimes fast, sometimes simple and obvious, sometimes complicated and obscure. A detailed accounting, either historically situated or restricted to contemporary operation, is challenging to say the least. Yet there are clearly modes and tempos that are discernable – *longue durées*, so to speak – within which a kind of sociotechnological homeostasis can be recognized,[1] a framework that we employed at the beginning of this book (Chapter 1). Such a framework, obvious at the level of whole-farm operations, exists at a variety of levels in ecosystems, some with perhaps trivial consequences, such as a sharp increase of the population density of an ant species, but others with dramatic consequences, such as the epidemic of coffee rust in 2013 in northern Latin America. Likewise, some are purely technical transformations of underlying mechanisms (e.g., the possible change in mode of repressing colony expansion of *Azteca* in Chapter 4), while others result from the interplay of ecological, economic and political changes.

These quasi-equilibrium situations in the context of agroecosystems are what some have referred to as "syndromes of production."[2] These syndromes of production are frequently punctuated by dramatic changes, so-called "ruptures in the discourse," such that the system begins operating in a completely different fashion approaching an alternative quasi-equilibrium, or alternative syndrome. This well-known historical framework has inadvertently taken its place in contemporary ecology,[3] in which there is much interest in the issue of regime change, the well-documented tendency of ecosystems that may appear to be quite stable and homeostatic to suddenly change dramatically into a completely different state that then itself begins to look quite stable and homeostatic.[4] Certainly, the socioecological system we call agriculture is not likely to be an exception.

The various historical syndromes of production, from the irrigated fields of the Sumerians to the *chinampas* of the Aztecs, are, in a sense, precursors of the "techno-logical packages" offered in the technologically sophisticated modern world. Each technological package can be thought of as a syndrome, and historically situated syndromes are, in effect, packages of technology/ecology/culture that are distinct from one another in a variety of parameters. Paddy rice production in Bali, for exam-ple, has characteristics that are dramatically different from potato farming in the Andes. Yet in both cases the total package, from socioeconomic organization to pest and soil management, constitutes a distinct syndrome. Viewed historically, any given spot on the surface of the globe undergoes dramatic regime change from time to time, transforming from one quasi-equilibrium to another, from one syndrome to another.

At any given time and place, it is frequently the case that alternative syndromes coexist, either side by side in a neutral to symbiotic manner (e.g., fish farming inte-grated with paddy rice production) or in open conflict (e.g., settled farmers versus migrating herders under certain circumstances). Even when distinct syndromes can be recognized, there are inevitably some commonalities between them, even examples of particular production units that may be intermediate between the two. Furthermore, alternative modes of analysis using eclectic variables frequently obscure underlying patterns, and where alternative syndromes actually exist, their reality may be obscured by such methodological vagaries.

Arguably there are historical examples and evident future expectations of changes in syndromes – first that independent evolution of particular systems will continue, and second that distinct syndromes may contribute to the eventual emergence of a hybrid. But third, and most important, jumps from one syndrome to another may occur, and indeed may occur rather rapidly and unexpectedly. In the jargon of contemporary ecology, we might expect "regime shifts" happening when "tipping points" are reached.

The current world agricultural situation is remarkably diverse, ranging from Amazonian Huaorani harvesting fruits from trees planted by their ancestors[5] to the Chiquita company getting the US government to negotiate on its behalf at the WTO.[6] Nevertheless, one can recognize two general syndromes that represent extremes in what is a complicated playing field. On the one hand, there has been a tendency for agriculture to become more technified, more mechanized, more chemicalized, more corporate-like, beginning in the nineteenth century and accel-erating dramatically since World War II.[7] This tendency, born with the Industrial Revolution and modern capitalism, has been referred to as "conventional," although a better descriptor (and the one we prefer) would be "industrial." On the other hand, there has been a tendency to try and focus agriculture on ecological principles, a tendency that has a long historical trajectory. Indeed, as industrial agriculture was beginning its ascendency, an alternative was already brewing with the thoughts of Sir Albert Howard, as briefly introduced in the previous chapter.

Flush with colonial arrogance, Queen Victoria sent Howard to India in an attempt to help the "backward" Indians learn the modern ways of agriculture. Howard, it turns out, was not so naive. Rather than preaching a paternalistic message about the new scientific agriculture, he focused on observation, and learning from local farmers, who seemed to know what they were doing. According to his own testimony, "By 1910 I had learnt how to grow healthy crops, practically free from disease, without the slightest help from . . . all the . . . expensive paraphernalia of the modern experiment station."[8]

We might propose that this event represents the bifurcation point that led to the two main post-World War II syndromes of agricultural production and their relatively independent evolutionary trajectories. On the one hand, there are traditional ways of doing things combined with careful observation and farmer-to-farmer interactions, to generate a line of evolution that resulted in an agriculture that uses ecological principles as both stimulus and restriction, to increase, stabilize and sustain production. On the other hand, we see an ideology of modernization, fueled by a socioeconomic structure that demands returns on investments rather than goals directly related to production. The result is a dichotomy in which most of the farmers in the world do it one way (the "ecological" way) while most of the farmland in the world is devoted to doing it the other way (the "industrial" way). To be sure, one can recognize a continuum between the local agroecological farm supplying the local farmers' market with fresh fruits and vegetables and the mega corporate landscapes of maize, raked clean with pesticides, fed with synthetic fertilizer and filling the regional export silos. Indeed, many examples could be cited in which components of both syndromes coexist locally, even on the same farm. Yet a random sample of points on the globe would most surely show that a vast number would fall near one of the extremes, presenting us with a bimodal situation. This bimodal distribution represents what are, we contend, two syndromes of production,[9] similar to previous syndromes but operative at a global scale in the contemporary world.

In the current global political climate, it is difficult to engage in discussion of these syndromes without injecting normative content. We admit to our normative preference for the syndrome that gives preference to small-scale agriculture with minimal use of external inputs, supplying local markets as much as possible. However, our intention here is not to argue the merits of that normative position. Rather, we argue for the systematic study of the socioecological forces that may be important in determining the extant dynamics of change in agriculture. To that end, we present, in this chapter, a qualitative framework upon which we feel a systematic study of agroecosystem dynamics, from its technological side (e.g., pest control) to its socioeconomic side (e.g., organic certification) might be further developed. Our framework is the dynamics of syndromes of production, focusing especially on the way in which syndromes might be transformed.

In the past decade, there has been an accumulated synthesis of research into this matter, providing a general qualitative theoretical framework for our point of view.[10] Ecologists have long been familiar with the idea of alternative equilibria (in contemporary lexicon, alternative "regimes"). For example, the classical equations of competition, simultaneously invented by Alfred J. Lotka and Vito Volterra in 1926, include the so-called "indeterminate case," in which one or the other species completely takes over the environment, depending on the initiation point of the competitive process. Such alternative regimes are now well known in the ecological literature.[11] In the context of agricultural ecosystems, the syndromes of production are similar to ecological regimes, although their articulation with socioeconomic forces suggests that they be called socioecological regimes. We take the position that these agricultural syndromes can be understood as identical to ecological regimes if the critical socioeconomic and political forces are taken into account.

Dynamic background for syndromes

The theory

As we have repeatedly argued (Chapters 1, 3, 7 and 8), we take it as axiomatic that the dynamics of agroecosystem transformation includes both socioeconomic and ecological forces. Consequently, the most elementary framework needs to be a bit less elementary than its cousins in either economics or ecology – it must include both types of forces.[12]

From the economic side, although a few farmers (e.g., some isolated groups of indigenous people in the Global South) are not directly connected to markets, our analysis applies to farmers who are in some way connected to markets, either local or global. On the economic side, we assume the standard supply-and-demand framework that the amount produced of a product indirectly determines the price paid for that product and that producers of that product will produce more of it in response to consumers being willing to pay more.[13] We recognize that farmers have many motivations for expanding or contracting the production of particular crops, but for the sake of simplicity in developing a heuristic model, we assume that the price of the product affects how much the farmer decides to plant of that product. Since we are dealing with agriculture, this standard framework plays out in non-continuous time (farmers plant and harvest at discrete points in time and harvest necessarily occurs some time after planting – i.e., planting and harvesting cannot occur simultaneously).

On the ecological side, recent literature acknowledges a variety of ideas, most of which are sufficiently similar to one another that we may contend they are effectively the same. For example, it is common to speak of "ecosystem engineers," organisms that construct part of the environment.[14] Or, recent literature finally recognizes the persistent insistence of Richard Lewontin that organism and niche are dialectically related to one another,[15] and the phenomenon of "niche construction" becomes an ecological and evolutionary force.[16] Or, more traditionally, competition theory recognizes that some organisms "effect" changes in the environment to which they and other organisms must "respond." What all of these ideas mean in terms of agricultural production is that the amount produced is a function of the ecological conditions in which it is produced, but the amount produced, in turn, affects those ecological conditions.

The combination of these two generalizations, economic and ecological, is presented in Figure 9.1 in pictorial form. The schematic in the figure says that the

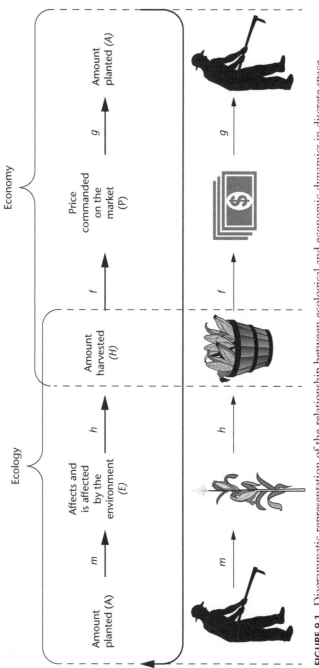

FIGURE 9.1 Diagrammatic representation of the relationship between ecological and economic dynamics in discrete space.

amount of land to be planted in some particular system (*A*) is determined (through the function *g*) by the price of that commodity on the market (*P*). In turn, the price of the commodity is determined (in part) by the supply provided generally, which is determined (through the function *f*) by the amount that is harvested (*H*), which in turn is determined (through the function *h*) by the environment (*E*), which in turn is determined (through the function *m*) by the amount of land planted (*A*). By this schematic, we can basically say that the amount of land planted today is a function of the amount of land planted last year, although that function is a complicated combination of the four functions *m*, *h*, *f* and *g*.[17]

It might be argued that the dynamic picture in Figure 9.1 is the simplest way that one might envision an agricultural socioeconomic system. It is without doubt far too simple to describe in any precise way what the system is about, but its general form provides us with an underlying "frictionless" framework upon which to build more realistic and precise frameworks. However, in addition to providing that frictionless framework, it provides us with a certain amount of "understanding" of what might be expected in the socioeconomic world of agriculture. To use Richard Levins' phrase, it helps to "educate our intuition" about the system as a whole. Not that this sort of simplifying assumption is without precedent. The basic supply-and-demand framework of microeconomics contains this fundamental idea and has certainly been a useful starting point for the theory of the firm. In the same vein, the dynamic picture we portray of ecology (Figure 9.1) is certainly neither a realistic nor a precise representation of ecological engineering or niche construction or any other way in which the environment is modified by and in turn modifies organisms (in this case crops and livestock). However, it too provides us with an underlying conceptual framework that relates ecological forces to agricultural production. We add that the presentation of the model in its most general form is in the tradition of stipulating a model that has universal applicability given the underlying mathematical constraints.[18] Further stipulation of particular conditions is quite unnecessary, since any real-world conditions that satisfy the particular assumptions of the model (and those are given above, with additional constraints below) will give the same results.

In the present work, our goal is to retain the level of simplicity and generalization commonly assumed with formulations such as illustrated in Figure 9.1, but to incorporate the fundamental realism that agroecosystem dynamics indeed *always* include both ecological and socioeconomic forces. Thus, we not only accept the utility of the two approaches shown in Figure 9.1, we also insist that the minimal model must include both of them. Our minimal model then, is the composition of the four functions illustrated in Figure 9.1, and we symbolize it as *Q*, so we have the general model $A_{t+1} = Q(A_t)$.

As long as at least one of the functions (*m*, *h*, *f* or *g*) is non-linear, the generalized function *Q* will also be non-linear. Any of the four key functions could be non-linear, depending on a variety of conditions. Consider, for example, the function *g*, which stipulates how a farmer will respond to market conditions in planning his or her yearly production. It stands to reason that the price of the commodity on today's market will somehow affect the decision that will be made as to how much to plant.

If tomatoes are selling well in the market, that is a clue that it might make sense to plant a lot of tomatoes for next year. Indeed, in a series of interviews with tomato farmers in Costa Rica,[19] when asked how much of their land they would plant with tomatoes next season based on the price of tomatoes today, their response was, as expected, more land with tomatoes proportional to the price of tomatoes on the market today. However, when asked about planting in the face of extreme prices – that is, prices that were way above what they normally would see – many of the farmers (the majority) said they would either not plant tomatoes at all or pull back considerably on how many they would plant. Upon further questioning, it was clear why. With such high prices, we would expect to see a great deal of market volatility and possibly every tomato farmer in the region would plant tomatoes, thus causing the bottom to drop out of the market. The responses of these farmers are reproduced in Figure 9.2.

Figure 9.2 implies that for purely economic reasoning, producers do not necessarily follow a monotonic prediction from market inputs – certainly not a surprise. However, the dynamic consequences of such an observation can be extraordinary. Furthermore, if two or more of the functions that make up the composite function Q are non-linear, the prospect for alternative non-zero solutions arises, which is to say alternative syndromes. To see this basic point, we take a small detour to examine just how Q behaves.

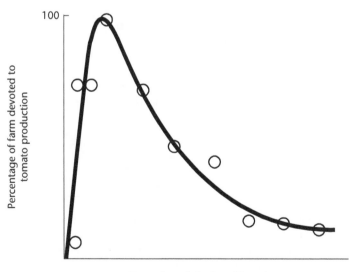

FIGURE 9.2 Results of interviews with Costa Rican tomato farmers regarding what they would do under ten scenarios of current market prices of tomatoes. Smooth curve was drawn by eye.

Source: Modified from Vandermeer (1997)

Educating the intuition about Q

Consider a subset of the general model, the effect of market price on amount planted (as in Figure 9.2), and combine that with the effect of harvest (which for purposes of this section we presume to be equal to the amount planted) on the price (assuming the environment plays no role at all). If we make reasonable assumptions about the

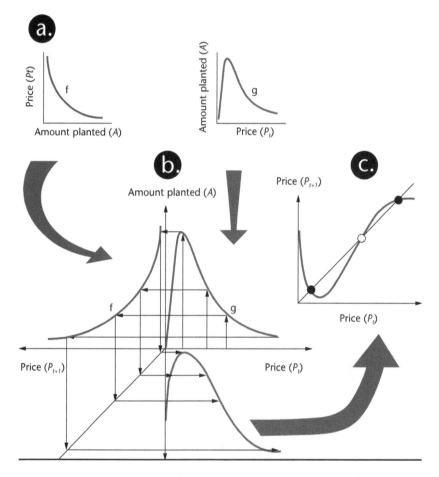

FIGURE 9.3 The composition of functions f and g to make the composite function Q that stipulates how price in one year determines price in the next year. (a) The two functions f and g. (b) The composition process. From the upper right quadrant, the results of the operation of the function g (arrows pointing upward) are translated to the "amount planted" axis and then to the function f (arrows pointing to the left). Finally, the results of the operation of the function f are translated to the 45-degree line (arrows pointing down) so as to be related to the original axis (arrows pointing to the right). (c) The resulting function relating price this year to price next year (a repeat of the lower right quadrant of (b), rotated). Black circles indicate stable equilibrium points, and the open circle indicates an unstable equilibrium point.

relationship between amount planted and price, effectively the supply and resulting price, we can envision two graphs, representing the functions f and g (Figure 9.3(a)), and their composition into a compound function that lets us examine the effect of price this year on price next year.

Note that the general form of Q (the composed function, Figure 9.3(c)) is not strictly as shown in the figure. The particular example is constructed for heuristic purposes only. Indeed, the function Q could take on an enormous number of different forms. The example here is chosen to make a point about regimes and their shifts. In Figure 9.3(c), the Q function has three equilibrium points, two stable and one unstable. Each of the two stable points formally represents a regime and the unstable point marks the dividing line between the two regimes. A system that begins just a small deviation to the right of the unstable point will increase to the upper regime, while a system that begins just a small deviation to the left of the unstable point will decrease to the lower regime. So we see how very simple and reasonable hypotheses about how prices are set in an agroecosystem can naturally lead to two regimes. Further analysis of this modeling style suggests that the upper regime is characterized as "a consequence of a combination of high prices, market volatility, ecological decline, high cultivation intensity, and low overall supply," while the lower regime is characterized as having "low prices, stable market, [a state of] ecologically benign, low cultivation intensity and high overall supply."[20]

A further analysis of the function Q reveals a higher-level general structure, as illustrated in Figure 9.4.

The pattern shown in Figure 9.4 is now a classical way of looking at regime change, normally thought of as one general qualitative arrangement of the way in which ecosystems respond to external forcing. Repeating Figure 9.4, but with different variables, we illustrate this whole range of dynamic outcomes in Figure 9.5.

The case of coffee syndromes

The coffee agroecosystem provides an illustrative example of syndromes of production. As fully described in Chapter 2, coffee is commonly produced using a range of production techniques and ranges from rustic coffee, in which the understory of a natural forest is replaced with coffee bushes, through various forms of shade cover, to the extreme of so-called sun coffee, which is to say coffee monoculture with no shade trees or other plant species.[21] However, examining the way in which coffee is actually produced in the world suggests bimodality. In Brazil, still the world's largest producer of coffee, sun coffee has been the rule for many years and for all practical purposes all coffee produced in Brazil is sun coffee. This implies that the most common type of coffee production system in the world is sun coffee. Add to Brazil the new expansion of coffee in Vietnam and Indonesia, most of which is sun coffee, and the picture is one in which the vast majority of coffee in the world is produced under a sun system. By contrast, the traditional way of coffee production persists in many areas of Latin America, southern India and East Africa. In Central America and Mexico, for example, coffee farms tend to be shaded and in some areas

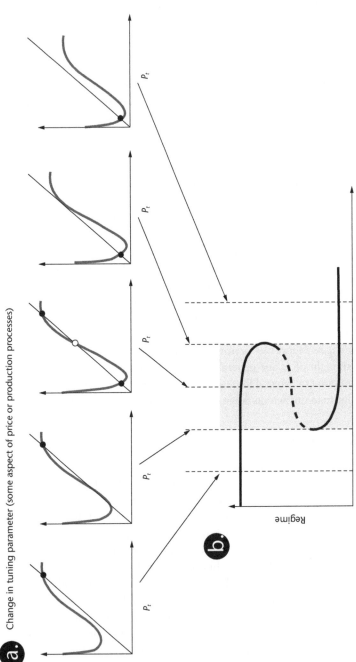

a. Change in tuning parameter (some aspect of price or production processes)

P_t P_t P_t P_t P_t

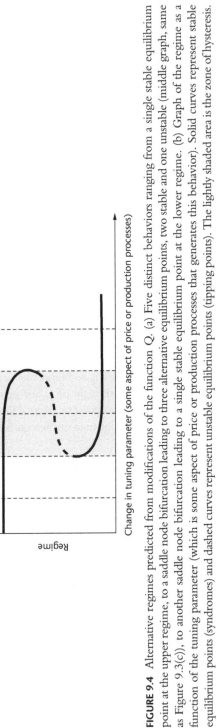

b.

Regime

Change in tuning parameter (some aspect of price or production processes)

FIGURE 9.4 Alternative regimes predicted from modifications of the function Q. (a) Five distinct behaviors ranging from a single stable equilibrium point at the upper regime, to a saddle node bifurcation leading to three alternative equilibrium points, two stable and one unstable (middle graph, same as Figure 9.3(c)), to another saddle node bifurcation leading to a single stable equilibrium point at the lower regime. (b) Graph of the regime as a function of the tuning parameter (which is some aspect of price or production processes that generates this behavior). Solid curves represent stable equilibrium points (syndromes) and dashed curves represent unstable equilibrium points (tipping points). The lightly shaded area is the zone of hysteresis.

Source: Redrawn from Vandermeer and Perfecto (2012)

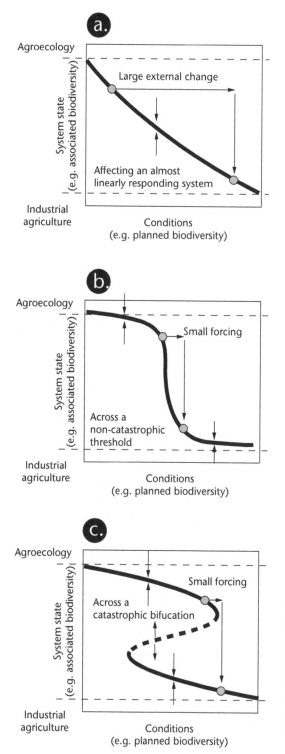

FIGURE 9.5 Repeat of Figure 9.4 with different exemplary conditions and system definition, emphasizing the whole agroecosystem nature of the "system state," a repeat of the generalized argument reflected in Figure 4.21.

it is difficult to find unshaded coffee at all (e.g., south-western Mexico), while in other areas coffee farms are more or less distinctly divided into shaded versus unshaded (e.g., Puerto Rico or Colombia). To be sure, it is possible to find areas in which one can find a whole range of production techniques (e.g., El Salvador, which has farms that are heavily shaded, lightly shaded, very lightly shaded and full sun), but for the most part, when shade coffee has been the tradition, one finds either shade or sun, rarely what would be classified as an intermediate. In all systems, many varieties of coffee are used. Although the more modern varieties tend to be used more in the sun system, contemporary farms of all sorts use many varieties, including sometimes the incorporation of *Coffea canephora* (syn. *C. robusta*, or Robusta coffee) within a dominance of *C. arabica* (Arabica coffee). Despite all these complications, the world distribution of coffee agroecosystems seems to be a bimodal distribution of sun versus shade coffee, with the current situation overwhelmingly biased in favor of sun coffee.

An examination of historical transformations in the coffee agroecosystem provides clues to the dynamic processes that have led to this situation of bimodality.

On a visit to coffee farms in the state of Minas Gerais, Brazil, we met a farmer who had been convinced some years ago to add shade to his sun coffee farm. Based on some questionable technical advice, he began by planting some fast-growing shade tree species, not necessarily trees that in other regions were associated with coffee. The coffee plants, perhaps because of their long-standing exposure to full sunlight, or due to competition for water or nutrients, dropped yields dramatically over the next two years. This, along with some other problems he encountered, convinced this producer that he needed to go back to the security of what he knew well before, a full sun coffee system. Similarly, on the evidence of extensive conversations with small-scale coffee producers in Costa Rica, it seems that other attempts at converting sun coffee into a shade system frequently meet with similar problems. Or, consider the single case of a small-scale producer in Puerto Rico who offered the unsolicited opinion that he would love to have his coffee farm filled with shade trees, knowing they function in the suppression of weeds and fertilization of soil as well as making the farm far more aesthetically pleasing. "Why not plant shade trees, then?", we asked him. It turned out that he was leasing the land and his lease was only guaranteed for five years, with no guarantee that planted trees would not be uprooted were the lease not renewed. Thus, it seems that if one is currently involved with producing coffee under a full sun system, moving to a shade system is not easy, and attempts to do so might very well be met with barriers related to a lack of knowledge about the management of the shade or the appropriate tree species to intercrop with coffee, insecure land tenure and a variety of other technical and socioeconomic issues. If these barriers are not quickly overcome, the farmer will feel a need to maintain or return to the sun system, even when either trying or hoping to change it, as happened with the farmers we met in Brazil, Costa Rica and Puerto Rico. In other words, the lower plateau (taking the sun coffee system to be represented by the lower plateau in Figures 9.5(b) and 9.5(c)) is something of a stable situation and deviations from it are countered by regression to it.

Consider the other extreme. Experience in attempting to transform a shade farm into a sun farm can be equally daunting for a producer. For example, in one case in southern Mexico, a farmer interested in increasing the productivity on his farm began removing the shade, but quickly discovered that production declined dramatically due to excessive weed growth. Having invested a great deal of money for labor to cut the shade, he had little left to pay labor costs for cutting the weeds, so he had to move to a herbicide, which was a dramatic risk since the farm was certified organic, and had he been caught using this non-organic technique, the premium he was counting on for the organic product would have been lost. As risky as it was, he felt he needed to control the weeds but he could not afford the labor cost, so he decided to go with the herbicide, perhaps thinking it would only be a one-time temporary activity. However, he then discovered that he could not afford the desired product, and was forced to purchase a lower-priced product, which turned out to be not nearly as effective, and at last sight he was facing massive drops in yields. Clearly, such a series of events would convince many producers to return to the shade system. Another example comes from Costa Rica in the early 1990s, when the government was providing support for farmers to intensify their plantations by eliminating all shade trees and planting new varieties of coffee. In spite of the government support for the intensification, after a couple of years of trying the new sun system, many small-scale farmers decided to re-introduce shade trees on their farms. When questioned, they mentioned lower yields as the main reason. It turned out that the sun system was high yielding as long as the whole technical package was implemented. That included expensive herbicides and fertilizers that the small- and medium-scale farmers could not afford. In addition to these economic factors, it turned out that ecological factors were at play as well. Without shade trees, the coffee plants were more vulnerable to the strong winds that come through the region during the dry season, precisely at the time when the coffee plants are flowering. Many other aspects of the way in which a shade system functions – from provisioning nitrogen from leguminous shade trees[22] to autonomous pest control from the embedded biodiversity in the system[23] – will be slowly lost as the move away from shade is pursued. These considerations may suggest to the farmer that going back to the shade system is the best strategy. In other words, the upper plateau illustrated in Figures 9.5(b) and 9.5(c) might also be something of a stable situation and deviations from it are countered by regression to it.

Given this basically qualitative analysis of coffee production, it is clear that socioecological conditions are always changing, yet there is some stability in the style of production. To be sure, that stability is frequently ruptured, but it seems very much like the *longue durée* that we postulated in Chapter 1. If we treat the system as others have been treated with regard to their characteristics and potential for change, we can postulate the three classical potential patterns of Figure 9.5, as redrawn in Figure 9.6. Since there does seem to be some stability in both extremes, as we argued above, it seems that the condition of catastrophic transition (9.6(c)) may characterize this socioecological system.[24]

This example of coffee production we believe is repeated in other agroecosystems, as indicated in Figure 9.5. Perhaps there are multiple syndromes, or perhaps it would

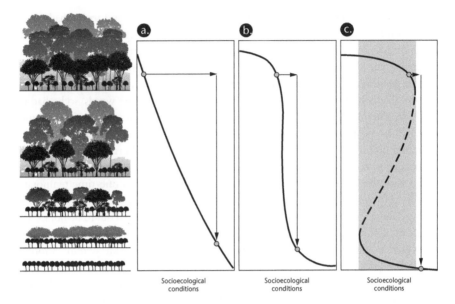

FIGURE 9.6 The basic hypotheses about regime change applied to the coffee agroecosystem (a recasting of Figure 9.5 so as to apply specifically to the coffee agroecosystem). The shaded area in (c) indicates the zone of hysteresis.

benefit understanding if we were to conceive of them as multiple, but as a starting point constructing a dual classification is useful, not only for coffee agroecosystems, but for agroecosystems in general. And such a classification already exists in the mind of most analysts, of many different persuasions. The normative judgment as to which syndrome is best is wildly divergent, but there seems be agreement that two syndromes indeed coexist, especially during the post-World War II years. The details were presented in Chapter 8. To recount briefly, on the one hand, there is small-scale agriculture, usually connected to some sort of family or extended family structure. According to some analysts, it is inefficient and anachronistic; according to others, it is productive and ecologically sustainable. It is, by actual calculations, more productive on a per unit area basis,[25] and frequently more energy efficient[26] than its alternative, but, by modern economic standards, frequently far less labor efficient and less profitable.[27] On the other hand, there is industrial agriculture, which has gone through a variety of stages since Victorian times,[28] but which took its present form subsequent to the ideological reforms in the wake of World War II.[29] According to some analysts, it is efficient and modern;[30] according to others, it is wasteful and ecologically unsustainable.[31] Our analysis does not attempt to demonstrate that the two syndromes exist; we take that as evident from simple observation and universally accepted opinion. Our argument is that these two syndromes represent dynamically stable alternatives, perhaps not over a very long period, but over the *longue durée* in which deviations toward one or the other alternative are resisted.

What forces enable both the apparent stability of the *longue durée* and its eventual rupture? This is a question that historians have long sought to answer in the context of human socioeconomic and political systems more generally. Our emphasis is on the agroecosystems that humans create, and our framework is in the dialectical relationships between forces that are fundamentally economic and those that are fundamentally ecological.

Self-generating dynamics of agricultural syndromes

We propose that the conditions under which agricultural syndromes of production change can be thought of as a complex socioecological system which generates its own dynamics. Our focus in this book is the coffee agroecosystem, and our comments are tied to it. Nevertheless, we feel the principles apply to a broad range of agricultural systems. The above simplified modeling approach, incorporating both ecological and socioeconomic forces, is not meant to capture the situation in a precise fashion, but rather to suggest broad features of dynamic expectations that might emerge from the non-linear nature of the socioecological formulation. Indeed, the "conditions" we envision ultimately include management strategies, land tenure, economic structures, political structures, climate, soil fertility, pests and their natural enemies, the amount of mycorrhyzal fungi and a host of other complex and inter-penetrating factors. Our simple model is intended to suggest the minimal level of complexity that should be expected from the complex conditions involved in changing syndromes.

Although the sense of our framework is intended to apply to contemporary or recent agricultural syndromes, there is nothing in its elaboration that immediately excludes its application to more remote situations. For example, the transition from hunting and gathering to agriculture likely involved the interplay of ecological and socioeconomic factors. Better radiocarbon dating and paleoclimatic interpretations suggest that the sudden change of the Natufians from hunter-gatherers to agriculturalists was the result of such interplay.[32] The open oak woodlands and wild cereals that were exploited by the Natufians in the Levant and northern Mesopotamia changed dramatically and abruptly with the onset of the Younger Dryas 12,900 years ago.[33] The cooler and drier climatic conditions likely affected the production of wild cereals to the point that gathering alone was not enough to provide sustenance to the Natufians, forcing a relocation of settlements to areas where cultivation of the cereals was possible.[34] Permanent location and agriculture also require dramatic socioeconomic changes, including changes in the division of labor and work schedule seasonality. Furthermore, to engage in agricultural production, the Natufians needed a large enough population to be able to plant and maintain the fields (i.e., the necessary population). However, the more stable provision of food that resulted from cultivation resulted in increased fertility (and therefore a larger sustainable population)[35] and consequently population growth.[36] This interplay of ecological and socioeconomic factors created the conditions for a dramatic change in the syndrome of production, which was key in the subsequent social evolution.[37]

In more modern times, we have examples in which a decrease in the local population caused a clear shift in the syndrome. For example, the system of oak and truffles[38] is a syndrome that was maintained in France for generations through grazing and fire management. Then, beginning at the turn of the twentieth century and accelerated due to the military needs of World War I, outmigration of the local population caused a crisis in available labor, which meant that the population density necessary for maintaining the savannah dipped below a critical level, and the oak savannah gave way to secondary succession and the rapid loss of this highly productive system. Furthermore, its recuperation seems to have conformed to the classical notion of hysteresis. A similar situation has been reported in more recent times in the Las Lagunas Valley of Mexico,[39] where an agroecosystem based on terracing and other sophisticated traditional methods was effectively destroyed (a major and rapid syndrome shift) as a result of sudden outmigration due to employment opportunities elsewhere. The lack of labor to maintain the terraces combined with the steep slopes and fragile soils of the region to cause a collapse of a highly sophisticated traditional system of production.

In the growing literature on regime change in ecology,[40] a common theme is the search for indicators of impending regime change. A variety of suggestions, ranging from changes in spatial patterns to "critical slowing-down," have appeared. Although this book is not intended to contribute to that discourse, it is worth noting that, when applied to agroecosystems, such a search has a precedent in the search for sustainability indices.[41] Empirically, there are a variety of measurement strategies that purport to be indicators of sustainability. If, as we suggest, the moniker sustainability represents a syndrome of agroecosystems, these previous studies may represent the early "warning" signs that a change to such a syndrome is imminent. Naturally, the disappearance of those indicators may indicate that a reverse regime change is on the way.

The history of agriculture is filled with examples of sudden or relatively rapid changes in syndromes of production. In many cases, these systems appeared to be stable until something happened to move them toward a tipping point that led to an alternate state. Sometimes, the push comes from changes in the ecological conditions; other times, it comes from socioeconomic changes; yet other times, it comes from the interplay between the two. However, regardless of what factors generate the tipping point and subsequently push the system toward it, understanding the process of change in the syndromes of production requires the incorporation of both ecological and socioeconomic factors of production and suggests a way of envisioning possible future changes.

Biodiversity and function, conservation and matrix quality: the ecology and political ecology of coffee syndromes

Returning to the opening lines of this book, the syndromes represented by the two farms described in Chapter 1 fall well within the two extremes of the coffee intensification gradient, and if indeed the world currently is composed of two basic regimes, they are certainly emblematic of those regimes. In those two particular cases, the style of management seems well suited to maintain them in a quasi-permanent state – a *longue durée*, so to speak – with what seems to be a catastrophic transition between the two (Figure 9.6(c)). And within those syndromes, and perhaps within the conditions that force them to change, we see the themes of this book repeatedly.

The first evident issue is the basic question of biodiversity. The coffee system has been perhaps the best studied agroecosystem with regard to this issue, and is generally acknowledged as an important repository of biodiversity. Undoubtedly, other agroecosystems will emerge with a similar pattern, but thus far research on this question lags. An additional problematic issue is that much of the biodiversity research done in agroecosystems is static. For logistical reasons, most studies that compare biodiversity among habitats, especially between natural habitats and different agricultural habitats take one or a few samples over a short period of time. These kinds of studies provide a snapshot view of the local biodiversity but tell us nothing about the long-term dynamics, and thus carry the risk of providing a mistaken estimate of biodiversity at the landscape level in the long run. Ecologists have long recognized that these studies ignore those organisms that are present in a site but are on their way to local extinction, the "extinction debt."[42] It is well known from simple theory that the species richness of a community can vary over time by several orders of magnitude,[43] suggesting that an estimate of biodiversity at one point in time is not especially useful.

It is thus widely acknowledged that the biodiversity sampled at any given point in time is not necessarily the same as the biodiversity that will exist later, even if no further changes in the background habitat or climate occur. Yet much of the available data constitute a single snapshot of this dynamic process, and thus one could argue that there is no reason to suspect that it will adequately reflect the biodiversity in the future, even in the short-term future (five to ten years). This problem is legitimately perplexing, calling for more careful thought about how biodiversity assessments should be approached. However, the science is frequently overtaken by political considerations and today's snapshot is used as fundamental input data to plan for the future. This problem has become especially severe with the "easy data" of current biodiversity science. It is almost as if research is motivated by the data available rather than the scientific questions outstanding, a problem beyond the scope of this book, but one that we acknowledge is especially severe.

Equally perplexing is the acknowledged unpredictability of long-term ecological dynamics. For example, when the fire ant (*S. invicta*) was introduced to Texas, an alarming study demonstrated that there was a sudden and dramatic drop in species diversity (a 50 percent decrease in ant species richness within a decade).[44] Yet, when the same area was studied 12 years later, 86 percent of the original biodiversity had recuperated.[45] The original observation of a 50 percent decline had been a temporary aberration, and the longer-term ecological dynamics had resulted in a recuperation of a large part of the original biodiversity. It is not difficult to imagine the reverse situation in which a sudden introduction seems to have no effect initially, yet massive extinction debt occurs, causing a time-lagged major biodiversity loss.

Similar to the unconscious bias associated with the static study of biodiversity, the question of the distribution of biodiversity and its complications over space has seen a great deal of theoretical study in academic ecology, but notably less so when it comes to agroecosystems. Historically, the elaboration of the basics of metapopulation dynamics and the theory of island biogeography are rightly cited as revolutionary ideas that unite spatial patterns with population dynamics.[46] The explicit acknowledgment of space as an important element of biodiversity dynamics is increasingly widespread and has seen important developments in studies of the coffee agroecosystem.[47] The autonomous creation of spatial patterns frequently provides the basis for biodiversity permanence through creating habitat structures and through isolating subpopulations, thus providing for reduced regional extinction probability. Space is thus tied up with questions of biodiversity generally and especially in agroecosystems, as we extensively discuss in Chapter 4.

Another aspect of biodiversity that warrants more detailed analysis is the trophic and non-trophic effects in food webs, as we discuss in Chapter 5. Especially important, yet understudied, is the notion of trait-mediated cascades. There is now a recognized body of research that not only contends that trait-mediated indirect effects are important, but that they frequently overwhelm both the direct effects and the density-mediated indirect effects.[48]

One step further in complication is the intersection of space and trophic structures, an important yet understudied subject. One of ecology's most famous and pioneering studies was that of Huffaker,[49] who, with this one experiment, anticipated a great deal that would happen in the field for the next 40 years. Basically, what he showed was that when a particular predator–prey system was held in isolation, both species went extinct – the prey because it was completely eaten by the predator, and the predator because a lack of prey meant it died of starvation. However, when subhabitats were connected so that the two species could migrate from patch to patch, the populations as a whole persisted for the length of the experiment. Huffaker essentially anticipated much of the recent debate over the construction of agroecosystems. Indeed, he was one of the pioneers in warning the world about the ecological disasters that could befall us with the growing industrial agricultural system based on monocultures. As a small contribution to the literature that emerged from Huffaker's landmark experiments, we discuss in Chapter 6 the intricacies of ecological forces operative at different spatial scales in the ant communities of the coffee system.

All of these observations feed into the all-encompassing subject of ecosystem services – one that, despite its obvious importance, has only recently been explored systematically.[50] The material of the first six chapters effectively comes together in Chapter 7, a partial analysis of ecosystem services. Here we have limited ourselves to reflections on only those ecosystem services that we or our students have directly worked on: pollination, pest control and climate mitigation.

As we note in Chapter 7, the overall structure of landscapes can be conceptualized along a gradient. At one end of the gradient, imagine a 100 percent forested (or other pre-agricultural vegetation) landscape and, at the other end, 100 percent agriculture. As we move along that gradient, we move from the complete forest to forest with small farms widely scattered, such that the farms appear to be habitat islands in the matrix of natural vegetation. As we approach the other end of the gradient, we see an agricultural landscape with small patches of forest widely scattered, such that the forest patches appear to be habitat islands in the matrix of agriculture. One is tempted to ask where we should be along this gradient if we wish to maximize both food production and biodiversity conservation.[51] However, rarely are we allowed to act as such "master planners;" rather, the local and regional political, ecological and economic dynamics will dictate how that landscape evolves. Thus, in the real world, we are almost always faced with an extant landscape, not a blank slate on which either ecological planners or command economies can apply their trades. And it is that landscape that we must understand with regard to ecosystem services.

Most often, the landscape begins with small patches of agriculture in a matrix of pre-agricultural vegetation. Those fragments of agriculture in a matrix of natural habitat are thought to be the beginnings of further conversion of natural habitat to agriculture. How to stabilize the system – which is to say, how to maintain the system as is without further conversion of natural habitat to agriculture – is sometimes the subject of serious study,[52] but sometimes fodder for what are effectively political arguments for converting agriculture into an industrial form.[53] Attention to this matter is a two-way street. First, when farms are isolated within a matrix of natural vegetation, the salient questions are associated with what ecosystem services are brought to the farm by the natural vegetation, frequently coupled with concern over what "damage" the farm does to the natural vegetation. Second, at the other extreme, when natural vegetation fragments are trivial and the landscape is dominated by agriculture, the nature of that agriculture may still provide the "service" of allowing interfragment migration of the organisms of conservation concern, if indeed the fragments of natural vegetation are substantial enough to contain them. Finally, in the complete absence of natural vegetation fragments, the nature of the agriculture becomes important for the direct conservation of some species, as discussed in Chapter 2. The forces that determine if, when and how these landscape transformations happen – whether at the level of advancing the agricultural frontier or moving a particular farm to a less or more intensive operation – are complicated intersections of the socioeconomic and ecological, a fact that calls for some sort of theoretical framing.

It is clear that such a theoretical framing must tie together the more traditional ecological topics with the socioeconomic facts of agricultural landscapes. Although the underlying dynamics are complex, the nature of contemporary tropical landscapes reflects a certain structure that makes some analytical treatments quite obvious to even casual observers. As landscapes continue to become fragmented – a process that, unfortunately, seems to be unabated – it is generally agreed that some pieces of natural habitat should be preserved. A long-term view of conservation suggests that we should ideally seek to structure landscapes such that as many as possible and as large as possible fragments remain, as landscapes are transformed by agriculture. Indeed, the contemporary world, especially in the tropics, where the most biodiversity is located, is already quite fragmented. That fragmentation, as explained in

Chapter 3, has set in place a process that will ultimately lead to massive extinctions if a functioning metapopulation structure is not put into place. This imperative means that the matrix in which the fragments are located must be of the highest quality possible, a conclusion that derives from elementary ecological theory. This result is sometimes regarded as counterintuitive. If we are concerned with preserving bio-diversity, it seems that we should concentrate on those areas where biodiversity is richest. However, we argue (Chapter 3) that it is precisely the areas between the biodiversity-rich fragments that must be the target of intervention. Although the details are bound to be complicated, there is an underlying principle that is obvious – the agricultural matrix must be sustained at as high a quality as possible to allow the migrations that maintain metapopulations.

Almost by definition, human activity in the agricultural matrix determines its quality, which draws us inexorably into the apparent dialectic of human versus nature. And subdividing the land into human-influenced versus "natural" is a point of view that puts biodiversity in jeopardy. If human activity is to create large industrial monocultures of soybean (or oil palm, or pineapples, or bananas, or maize), the quality of the matrix will be low. Yet the arguments of many conservationists are aimed in just that direction. If the important parameter determining the quality of the matrix is the ability of organisms to migrate through it (as argued in Chapter 3), it would be wise to search for structures that promote that migratory potential. However, it is not an easy task to determine precisely what general vegetative patterns and structure will be conducive to migration for any particular species. Use of genetic information may be our best way to ultimately determine what long-term migratory patterns are associated with what vegetation structure and pattern,[54] and such studies should be encouraged in our view. Short of such information, we have argued that the promotion of small-scale agroecological farming is probably the best way to insure a high-quality matrix.[55]

There is considerable debate about the actual nature of the agriculture performed by those small-scale agriculturalists, but the bulk of evidence suggests that the majority are involved in something approaching agroecological production, or at least an agriculture that is not as intensive as the industrial mode. The dynamic tendency seems locked in a struggle between the forces of pseudo-modernization and ecology. On the one hand, there is a tendency to consolidate farms, creating large estates, which then tend to move to a highly intensive and industrial form of production. This tendency has been in operation for 500 years, but has been especially intense since World War II. On the other hand, there is a tendency to break up the largest farming operations through various political movements, generally referred to as agrarian reform. These are, we have argued in both this chapter and in Chapter 8, distinct regimes. Furthermore, we propose that they may frequently be organized along the general unintentional plan of a catastrophic transition (Figures 9.5(c) and 9.6(c)).

A recent trend has been noted within the small-scale farming sector of a move toward more ecologically based agriculture, promoted by a combination of concerns ranging from the health of producers (farmers and farmworkers) to the health of consumers, to the health of the environment, to the sustainability of the farm. Large-scale operations tend to focus on short-term profits while smaller-scale operations tend to view the farm as more of an investment in the future. Consequently, the literature on environmental and human health and its relationship to the style of farming has been spreading rapidly among small-scale farmers worldwide.

Of great importance in this context is the tendency to promote farmer-to-farmer education, a method that is advocated by small-scale farmers' organizations like La Via Campesina.[56] Furthermore, the promulgation of the food sovereignty model has brought with it a focus on agroecological techniques, which inevitably act to create a matrix that is of higher quality than that produced by the industrial system. Indeed, we have argued elsewhere[57] that support for the small-scale farming sector could be one of the most important activities for those concerned with biodiversity conservation. It is, of course, ironic that the so-called land-sparing argument, from this point of view, if widely adopted, would likely have a generalized negative effect on biodiversity conservation.

In a world in which people go hungry amidst an abundance of food and in which the great majority of the poor live in rural areas or are forced, by economic necessity, to abandon their rural livelihood only to face similar conditions in the city, models of agricultural intensification that continue this trend are bound to fail. Furthermore, there is little evidence to support the idea that agricultural intensification necessarily leads to aforestation or less deforestation, or that sparing some land for conservation by destroying biodiversity on neighboring land is a good strategy. In fact, most of the evidence suggests the contrary, and a detailed look at the ecological complexity involved within the agroecosystem suggests that removing that complexity through intensification could also remove some of the ecosystem services that may have gone unnoticed. The traditional farmer's view that a "balance" of nature should guide the planning of the farm may in the end be closer to the truth than the modern agronomist's view that a magic bullet will solve all problems, especially when the environment is drastically simplified. A New Rurality based on a matrix quality approach supporting the growing rural agricultural movements is more likely to lead to situations in which biodiversity is conserved, locally and regionally, at the same time as more food is available to those who need it the most, while effectively supporting the farmers who produce it.

In this book, we have used the coffee agroecosystem as a model system. Although coffee is not a food, it does provide the livelihood of millions of small-scale farmers in the world and is an excellent model system that allows us to investigate the dialectical interplay between the ecological and the social, economic and political.

Notes

1 Braudel (1984); Wallerstein (2001).
2 See Chapter 1; Andow and Hidaki (1989); Vandermeer and Perfecto (2012).
3 Beddoe et al. (2009).
4 Scheffer (2009).

5   As reported by Rival (2002), this native American group from the upper Amazon plant palm trees in their local clearing, but the clearing only lasts for a couple of years and the trees do not bear fruit until four or five years old. Thus, their "agriculture" might be called "informal" or perhaps proto-agriculture.

6   Myers (2004); Witter (2004).

7   Robinson and Sutherland (2002).

8   Howard (1943).

9   Andow and Hidaki (1989); Vandermeer (1990, 1997).

10 Scheffer et al. (2012).

11 See May (1977) for a summary of the older literature.

12 Vandermeer and Perfecto (2012); Vandermeer (1990, 1997).

13 See any elementary economics text, such as Nicholson (2001) or Mankiw (2009).

14 Cuddington et al. (2007).

15 Lewontin (1991).

16 Odling-Smee et al. (2003); Vandermeer (2008, 2010).

17 Formally the functions are "composed" such that $A_{t+1} = m\{h[f(g(A_t))]\} = G(A_t)$.

18 The formal mathematical representation of these ideas can be found in Vandermeer (1990, 1997); Vandermeer and Perfecto (2012).

19 Interviews conducted in the general farmers' market near Turrialba, Costa Rica in 1987; Vandermeer (1997).

20 Vandermeer (1997: 70).

21 Moguel and Toledo (1999).

22 Glover and Beer (1986).

23 Vandermeer et al. (2010a).

24 Vandermeer and Perfecto (2012).

25 Carter (1984); Cornia (1985); Binswanger et al. (1995); Heltberg (1998).

26 Pimentel et al. (1983, 2005); Lin et al. (2011).

27 Sen (1962); Hanson et al. (1997).

28 Holt-Giménez and Shattuck (2011).

29 Russell (2001).

30 Avery (2000).

31 IAASTD (2009).

32 Weiss and Bradley (2001).

33 Moore and Hillman (1992).

34 Hole (1998); Bar-Yosef (2000).

35 The use of necessary and sustainable population is treated generally in Vandermeer (2008) and specifically with regard to the human population in Vandermeer (2010).

36 Bentley (1996).

37 Weiss and Bradley (2001).

38 Sourzat (2009).

39 García-Barrios and García-Barrios (1990).

40 Scheffer (2009); Scheffer et al. (2012).

41 Astier et al. (2009).

42 Kuussaari et al. (2009).

43 Huisman and Weissing (1999); Benincà et al. (2008); Vandermeer and Pascual (2006).

44 Porter et al. (1988).

45 Morrison (2002).

46 This is a point made by Hubbell about MacArthur and Wilson's theory of island biogeography in his book (Hubbell 2001). It could have been said about metapopulation theory also.

47 Vandermeer et al. (2008); Jackson et al. (2009); Jha and Vandermeer (2009); Jackson et al. (2012a).

48 Werner and Peacor (2003).

49 Huffaker (1958) studied two species of mites. He set up an ingenious experiment in which oranges (the natural habitat of the prey species of mite) were arranged in a spatial grid.

The predator–prey system was followed for three distinct cycles, wherein without the spatial structure, both species went extinct.

50 Alexander et al. (1998); Costanza et al. (1997); Daily (1997); Millennium Ecosystem Assessment (2005); Costanza (2008).
51 Green et al. (2005).
52 Tscharntke et al. (2007).
53 Green et al. (2005); Phalan et al. (2011a).
54 Jha and Dick (2008, 2010); Otero-Jiménez (2013).
55 Perfecto et al. (2009); Perfecto and Vandermeer (2010).
56 Rosset and Martínez-Torres (2010).
57 Vandermeer and Perfecto (2005, 2007, 2012); Perfecto and Vandermeer (2008a, 2010); Chappell et al. (2013).

REFERENCES

Abrams, P. A. (1984) "Foraging time optimization and interactions in food webs", *The American Naturalist*, vol. 124, no. 1, 80–96

Abrams, P. A. (1991) "Strengths of indirect effects generated by optimal foraging", *Oikos*, vol. 62, no. 2, 167–176

Abrams, P. A. (1995) "Implications of dynamically variable traits for identifying, classifying, and measuring direct and indirect effects in ecological communities", *The American Naturalist*, vol. 146, no. 1, 112–134

Adams, R. M., Houston, L. I., McCarl, B. A., Tiscareno, L. M., Matus, G. J. and Weiher, R. E. (2003) "The benefits to Mexican agriculture of an El Niño Southern Oscillation (ENSO) early warning system", *Agricultural and Forest Meteorology*, vol. 115, nos 3–4, 183–194

Adler, P. B., RisLambers, J. H. and Levine, J. M. (2007) "A niche for neutrality", *Ecology Letters*, vol. 10, no. 2, 95–104

Aguilar-Ortiz, F. (1982) "Estudio ecológico de las aves del cafetal", in Avila-Jiménez, E. (ed.) *Estudios Ecológicos en el Agroecosistema Cafetalero*, Instituto National de Investigaciones Sobre Recursos Bióticos, Xalapa, Mexico, 103–128

Aide, T. M. and Grau, H. R. (2004) "Globalization, migration, and Latin American ecosystems", *Science*, vol. 305, no. 5692, 1915–1916

Alegre, C. (1959) "Climates et caféirs d'Arabie", *Aronomie Tropicale*, vol. 14, 25–48

Alexander, A. M., List, J. A., Margolis, M. and d'Arge, R. C. (1998) "A method for valuing global ecosystem services", *Ecological Economics*, vol. 27, no. 2, 161–170

Allen, D. N. (2012) "Feedback between ecological interactions and spatial pattern in a transitional Michigan forest", PhD thesis, University of Michigan, Ann Arbor, MI

Alonso, D., Bartemeus, F. and Catalan, J. (2002) "Mutual interference between predators can give rise to Turing spatial pattern", *Ecology*, vol. 83, no. 1, 28–34

Altieri, M. A. (1995) *Agroecology: The Science of Sustainable Agriculture*, second edition, Westview Press, Boulder, CO

Altieri, M. A. (1999) "Agroecology: the science of natural resource management for poor farmers in marginal environments", *Agriculture, Ecosystems and Environment*, vol. 93, no. 1, 1–3

Altieri, M. A. (2004) "Linking ecologists and traditional farmers in the search for sustainable agriculture", *Frontiers in Ecology and the Environment*, vol. 2, no. 1, 35–42

Amarasekare, P. (1998) "Allee effects in metapopulation dynamics", *The American Naturalist*, vol. 152, no. 2, 298–302

Ambinakudige, S. and Sathish, B. N. (2009) "Comparing tree diversity and composition in coffee farms and sacred forests in the Western Ghats of India", *Biodiversity Conservation*, vol. 18, no. 4, 987–1000

American Ornithologists Union (1998) *Check-list of North American Birds*, American Ornithologists Union, Washington, DC

Andow, D. A. and Hidaka, K. (1989) "Experimental natural history of sustainable agriculture: syndromes of production", *Agriculture, Ecosystems and Environment*, vol. 27, no. 1, 447–462

Angelsen, A. and Kaimowitz, D. (eds) (2001a) *Agricultural Technologies and Tropical Deforestation*, CABI Publishing, Oxon, UK

Angelsen, A., and Kaimowitz, D. (2001b) "When does technological change in agriculture promote deforestation?", in Lee, D. R. and Barrett, C. B. (eds) *Tradeoffs or Synergies? Agricultural Intensification, Economic Development and the Environment*, CABI Publishing, Oxon, UK, 89–114

Arellano, L., Favila, M. E. and Huerta, C. (2005) "Diversity of dung and carrion beetles in a disturbed Mexican tropical montane cloud forest and on shade coffee plantations", *Biodiversity and Conservation*, vol. 14, no. 3, 601–615

Arim, M. and Marquet, P. A. (2004) "Intraguild predation: a widespread interaction related to species biology", *Ecology Letters*, vol. 7, no. 7, 557–564

Armbrecht, I. and Perfecto, I. (2003) "Litter-twig dwelling ant species richness and predation potential within a forest fragment and neighboring coffee plantations of contrasting habitat quality in Mexico", *Agriculture, Ecosystems and Environment*, vol. 97, no. 1, 107–115

Armbrecht, I. and Gallego, M. C. (2007) "Testing ant predation on the coffee berry borer in shaded and sun coffee plantations in Colombia", *Entomologia Experimentalis et Applicata*, vol. 124, no. 3, 261–267

Armbrecht, I., Perfecto, I. and Vandermeer, J. (2004) "Enigmatic biodiversity correlations: ant diversity responds to diverse resources", *Science*, vol. 304, no. 5668, 284–286

Armbrecht, I., Rivera, L. and Perfecto, I. (2005) "Reduced diversity and complexity in the leaf-litter ant assemblage of Colombian coffee plantations", *Conservation Biology*, vol. 19, no. 3, 897–907

Armbrecht, I., Perfecto, I. and Silverman, E. (2006) "Limitation of nesting resources for ants in Colombian forests and coffee plantations", *Ecological Entomology*, vol. 31, no. 5, 403–410

Armstrong, R. A. and McGehee, R. (1980) "Competitive exclusion", *The American Naturalist*, vol. 115, no. 2, 151–170

Assad, E. D., Pinto, H. S., Zullo Junior, J. and Ávila, A. M. H. (2004) 'Impacto das mudanças climáticas no zoneamento agroclimático do café no Brasil', *Pesquisa Agropecuária Brasileira*, vol. 39, no. 11, 1057–1064

Astier, M., Masera, O. and Galván, Y. (2009) *Evaluación de sustentabilidad: un enfoque dinámico y multi-dimensional*, CIGA-UNAM; CIEco-UNM; ECOSUR; GIRA A. C. SEAE, Mexico

Avelino, J., Willocquet, L. and Savary, S. (2004) "Effects of crop management patterns on coffee rust epidemics", *Plant Pathology*, vol. 53, no. 5, 541–547

Avery, D. T. (2000) *Saving the Planet with Pesticides and Plastic: The Environmental Triumph of High-Yield Farming*, Hudson Institute, Washington, DC

Bäckstrand, K. and Lövbrand, E. (2006) "Planting trees to mitigate climate change: contested discourses of ecological modernization, green governmentality and civic environmentalism", *Global Environmental Politics*, vol. 6, no. 1, 50–75

Bacon, C. (2005) "Confronting the coffee crisis: can fair trade, organic, and specialty coffees reduce small-scale farmer vulnerability in northern Nicaragua?", *World Development*, vol. 33, no. 3, 497–511

Bacon, C. M. (2013) "Quality revolutions, solidarity networks, and sustainability innovations: following Fair Trade coffee from Nicaragua to California", *Journal of Political Ecology*, vol. 20, 98–115

Bacon, C. M., Mendez, V. E., Gómez, M. E. F., Stuart, D. and Flores, S. R. D. (2008) "Are sustainable coffee certifications enough to secure farmer livelihoods? The millenium development goals and Nicaragua's Fair Trade cooperatives", *Globalizations*, vol. 5, no. 2, 259–274

Badano, E. I. and Vergara, C. H. (2011) "Potential negative effects of exotic honey bees on the diversity of native pollinators and yield of highland coffee plantations", *Agricultural and Forest Entomology*, vol. 13, no. 4, 365–372

Badgley, C., Moghtader, J., Quintero, E., Zakem, E., Chappell, M. J., Aviles-Vazquez, K., . . . and Perfecto, I. (2007) "Organic agriculture and the global food supply", *Renewable Agriculture and Food Systems*, vol. 22, no. 2, 86–108

Bak, P. (1999) *How Nature Works: The Science of Self-Organized Criticality*, Copernicus Press, New York

Bakermans, M. H., Vitz, A. C., Rodewald, A. D. and Rengifo, C. G. (2009) "Migratory songbird use of shade coffee in the Venezuelan Andes with implications for conservation of cerulean warbler", *Biological Conservation*, vol. 142, no. 11, 2476–2483

Bakermans, M. H., Rodewald, A. D., Vitz, A. C. and Rengifo, C. G. (2012) "Migratory bird use of shade coffee: the role of structural and floristic features", *Agroforestry Systems*, vol. 85, no. 1, 85–94

Bandeira, F. P., Martorell, C., Meave, J. A. and Caballero, J. (2005) "The role of rustic coffee plantations in the conservation of wild tree diversity in the Chinantec region of Mexico", *Biodiversity and Conservation*, vol. 14, no. 5, 1225–1240

Bar-Yosef, O. (2000) "The impact of radiocarbon dating on Old World archeology: past achievements and future expectations", *Radiocarbon*, vol. 42, no. 1, 23–29

Bascompte, J. and Solé, R. V. (1995) "Rethinking complexity: modelling spatiotemporal dynamics in ecology", *Trends in Ecology and Evolution*, vol. 10, no. 9, 361–366

Batáry, P., Báldi, A., Kleijn, D. and Tscharntke, T. (2011) "Landscape-moderated biodiversity effects of agri-environmental management: a meta-analysis", *Proceedings of the Royal Society B: Biological Sciences*, vol. 278, no. 1713, 1894–1902

Bawa, K. S. (1990) "Plant–pollinator interactions in tropical rain forests", *Annual Review of Ecology and Systematics*, vol. 21, 399–422

Beddoe, R., Costanza, R., Farley, J., Garza, E., Kent, J., Kubiszewski, I., . . . and Woodward, J. (2009) "Overcoming systemic roadblocks to sustainability: the evolutionary redesign of worldviews, institutions, and technologies", *Proceedings of the National Academy of Sciences*, vol. 106, no. 8, 2483–2489

Beer, J. (1988) "Litter production and nutrient cycling in coffee (*Coffea arabica*) or cacao (*Theobroma cacao*) plantations with shade trees", *Agroforestry Systems*, vol. 7, no. 2, 103–114

Beer, J., Muschler, R., Kass, D. and Somarriba, E. (1998) "Shade management in coffee and cacao plantations", *Agroforestry Systems*, vol. 38, no. 2, 134–164

Benincà, E., Huisman, J., Heerkloss, R., Jöhnk, K. D., Branco, P., Van Nes, E. H., . . . and Ellner, S. P. (2008) "Chaos in a long-term experiment with a plankton community", *Nature*, vol. 451, no. 7180, 822–825

Benincà, E., Jöhnk, K. D., Heerkloss, R., and Huisman, J. (2009) "Coupled predator–prey oscillations in a chaotic food web", *Ecology Letters*, vol. 12, no. 12, 1367–1378

Bentley G. R. (1996) "How did prehistoric women bear 'Man the Hunter'? Reconstructing fertility from the archaeological record", in Wright, R. P. (ed.) *Gender and Archaeology*, University of Pennsylvania Press, Philadelphia, PA, 23–51

Berry, P. M., Rounsevell, M. D. A., Harrison, P. A. and Audsley, E. (2006) "Assessing the vulnerability of agricultural land use and species to climate change and the role of policy in facilitating adaptation", *Environmental Science and Policy*, vol. 9, no. 2, 189–204

Bess, H. A. (1958) "The green scale, *Coccus viridis* (Green) (*Homoptera: Coccidae*), and ants", *Proceedings of the Hawaii Entomological Society*, vol. 16, 349–355

Binswanger, H. P., Deininger, K. and Feder, G. (1995) "Power, distortions, revolt and reform in agricultural land relations", in Behrman, J. and Srinivasan, T. N. (eds) *Handbook of Development Economics*, Vol. 3A, Elsevier, Amsterdam, The Netherlands, 2659–2772

Bloomberg News (2013) "Coffee surplus driving down prices at Starbucks, elsewhere, but hurting bean farmers", nj.com, November 14. Available online at www.nj.com/business/index.ssf/2013/11/coffee_surplus_driving_down_pr.html (accessed August 6, 2014)

Blüthgen, N. and Stork, N. E. (2007) "Ant mosaics in a tropical rainforest in Australia and elsewhere: a critical review", *Austral Ecology*, vol. 32, no. 1, 93–104

Blüthgen, N. and Feldhaar, H. (2010) "Food and shelter: how resources influence ant ecology", in Lach, L., Parr, C. L. and Abbott, K. L. (eds) *Ant Ecology*, Oxford University Press, Oxford, 115–136

Boerlijst, M. C. and Hogeweg, P. (1995) "Attractors and spatial patterns in hypercycles with negative interactions", *Journal of Theoretical Biology*, vol. 176, no. 2, 199–210

Bolger, D. T., Alberts, A. C. and Soulé, M. E. (1991) "Occurrence patterns of bird species in habitat fragments: sampling, extinction, and nested species subsets", *The American Naturalist*, vol. 137, no. 2, 155–166

Bolker, B., Holyoak, M., Kfiivan, V., Rowe, L. and Schmitz, O. (2003) "Connecting theoretical and empirical studies of trait-mediated interactions", *Ecology*, vol. 84, no. 5, 1101–1114

Bonds, M. H., Dobson, A. P., Keenan, D. C. (2012) "Disease ecology, biodiversity, and the latitudinal gradient in income", *PLOS Biol* vol. 10, no. 12, e1001456, doi:10.1371/journal.pbio.1001456

Borkhataria, R. R., Collazo, J. A. and Groom, M. J. (2006) "Additive effects of vertebrate predators on insects in a Puerto Rican coffee plantation", *Ecological Applications*, vol. 16, no. 2, 696–703

Borlaug, N. (1970). *Acceptance Speech*, on the occasion of the award of the Nobel Peace Peace Prize in Oslo, December 10. Available online at www.nobelprize.org/nobel_prizes/peace/laureates/1970/borlaug-acceptance.html (accessed August 6, 2014)

Borrero, J. I. (1986) "La substitución de cafetales de sombrío por caturrales y su efecto negativo sobre la fauna de vertebrados", *Caldasia*, vol. 15, nos 71–75, 725–732

Botero, J. E. and Baker, P. S. (2002) "Coffee and biodiversity: a producer-country perspective", in Baker, P. S. (ed.) *Coffee Futures*, CABI-FNC-USDA-ICO, Chinchiná, Colombia, 94–103

Brash, A. R. (1987) "The history of avian extinctions and forest conversion on Puerto Rico", *Biological Conservation*, vol. 39, no. 2, 97–111

Braudel, F. (1958) "Histoire et sciences sociales: La longue durée", *Annales, Histoire, Sciences Sociales*, 13e Année, vol. 13, no. 4, 725–753

Braudel, F. (1984) *Civilization and Capitalism, 15–18th Century*, HarperCollins, New York

Bray, D. B., Sanchez, J. L. P. and Murphy, E. C. (2002) "Social dimensions of organic coffee production in Mexico: lessons for eco-labeling initiatives", *Society and Natural Resources*, vol. 15, no. 5, 429–446

Briand, F. and Cohen, J. E. (1987) "Environmental correlates of food chain length", *Science*, vol. 238, no. 4829, 956–960

Brooks, T. M., Pimm, S. L. and Oyugi, J. O. (1999) "Time lag between deforestation and bird extinction in tropical forest fragments", *Conservation Biology*, vol. 13, no. 5, 1140–1150

Brosi, B. J., Daily, G. C., Shih, T. M., Oviedo, F. and Durán, G. (2008) "The effects of forest fragmentation on bee communities in tropical countryside", *Journal of Applied Ecology*, vol. 45, no. 3, 773–783

Bruinsma, J. (2009) "How to feed the world in 2050", *Proceedings of a Technical Meeting of Experts*, Rome, Italy, June 24–26. Available online at www.fao.org/docrep/012/ak542e/ak542e00.htm (accessed April 18, 2013)

Bustillo, A. E. (2002) "El manejo de cafetales y su relación con el control de la broca del café en Colombia", *Boletín Técnico*, vol. 24, 1–40

Bustillo, A. E., Cardenas, R. and Posada, F. J. (2002) "Natural enemies and competitors of *Hypothenemus hampei* (Ferrari) (*Coleoptera: Scolytidae*) in Colombia", *Neotropical Entomology*, vol. 31, no. 4, 635–639

Calo, M. and Wise, T. A. (2005) *Revaluing Peasant Coffee Production: Organic and Fair Trade Markets in Mexico*, Globalization and Sustainable Development Program, Global Development and Environment Institute, Tufts University, Medford, MA

Calvo, L. and Blake, J. (1998) "Bird diversity and abundance on two different shade coffee plantations in Guatemala", *Bird Conservation International*, vol. 8, no. 3, 297–308

Camargo, A. P. (1985) "O clima e a cafeicultura no Brasil", *Informe Agropecuário*, vol. 11, 13–26

Canaday, C. (1996) "Loss of insectivorous birds along a gradient of human impact in Amazonia", *Biological Conservation*, vol. 77, no. 1, 63–77

Cane, J. H., Minckley, R. L., Kervin, L. J., Roulston, T. A. H. and Williams, N. M. (2006) "Complex responses within a desert bee guild (*Hymenoptera: Apiformes*) to urban habitat fragmentation", *Ecological Applications*, vol. 16, no. 2, 632–644

Cannell, M. G. R. (1976) "Crop physiological aspects of coffee bean yield: a review", *Kenya Coffee*, vol. 41, 245–253

Cannell, M. G. R. (1983) "Coffee", *Biologist*, vol. 30, 257–263

Cannell, M. G. R. (1985) "Physiology of the coffee crop", in Clifford, M. N. and Willson, K. C. (eds) *Coffee: Botany, Biochemistry and Production of Bean and Beverage*, Springer-Verlag, New York, 108–134

Carbone, C., Rowcliffe, J. M., Cowlishaw, G. and Isaac, N. J. (2007) "The scaling of abundance in consumers and their resources: implications for the energy equivalence rule", *The American Naturalist*, vol. 170, no. 3, 479–484

Carpenter, S. R. and Kitchell, J. F. (eds) (1996) *The Trophic Cascade in Lakes*, Cambridge University Press, Cambridge

Carpenter, S. R., Ludwig, D. and Brock, W. A. (1999) "Management of eutrophication for lakes subject to potentially irreversible change", *Ecological Applications*, vol. 9, no. 3, 751–771

Carr, M. K. V. (2001) "The water relations and irrigation requirements of coffee", *Experimental Agriculture*, vol. 37, no. 1, 1–36

Carreño Rocabado, I. G. (2006) *Evaluación de los cafetales bajo sombra y fragmentos de bosque adyacentes como hábitats para conservar la diversidad de los helechos en el Estado de Veracruz, México*, PhD thesis, UNAM, Veracruz, Mexico

Carroll, C. R., Vandermeer, J. H. and Rosset, P. M. (1990) *Agroecology*, McGraw-Hill Inc, New York

Carter, M. R. (1984) "Identification of the inverse relationship between farm size and productivity: an empirical analysis of peasant agricultural production", *Oxford Economic Papers*, vol. 36, 131–145

Castro Soto, G. (1998) *El impacto de la crisis chiapaneca en la economía*, Centro de Investigaciones Económicas y Políticas de Acción Comunitaria, Chiapas, Mexico

Cerda, X., Arnan, X. and Retana, J. (2013) "Is competition a significant hallmark of ant (*Hymenoptera: Formicidae*) ecology?", *Myrmecological News*, vol. 18, 131–147

Chapman, A. D. (2009) *Numbers of Living Species in Australia and the World*, 2nd ed., Australian Biological Resources Study, Canberra. Available online at www.environment.gov.au/biodiversity/abrs/publications/other/species-numbers/2009/pubs/nlsaw-2nd-complete.pdf (accessed June 21, 2014)

Chappell, M. J. (2007) "Shattering myths: can sustainable agriculture feed the world?", *Institute for Food and Development Policy Backgrounder*, vol. 13, 1–4

Chappell, M. J. and LaValle, L. A. (2011) "Food security and biodiversity: can we have both? An agroecological analysis", *Agriculture and Human Values*, vol. 28, no. 1, 3–26

Chappell, M. J., Wittman, H., Bacon, C. M., Ferguson, B. G., Barrios, L. G., Barrios, R. G., . . . and Perfecto, I. (2013) "Food sovereignty: an alternative paradigm for poverty reduction and biodiversity conservation in Latin America", *F1000Research*, vol. 2, 235. Available online at http://f1000research.com/articles/2-235/v1 (accessed August 6, 2014)

Chazdon, R. L., Harvey, C. A., Komar, O., Griffith, D. M., Ferguson, B. G., Martínez-Ramos, M., . . . and Philpott, S. M. (2009) "Beyond reserves: a research agenda for conserving biodiversity in human-modified tropical landscapes", *Biotropica*, vol. 41, no. 2, 142–153

Chomsky, N. (2011) "Language and the cognitive science revolution(s)", lecture given at Carleton University, April 8, 2011. Available online at www.chomsky.info/talks/20110408.htm (accessed June 21, 2014)

Cohen, J. E. (1978) *Food Webs and Niche Space*, Princeton University Press, Princeton, NJ

Cohen, J. E., Pimm, S. L., Yodzis, P. and Saldaña, J. (1993) "Body sizes of animal predators and animal prey in food webs", *Journal of Animal Ecology*, vol. 62, 67–78

Comins, H. N., Hassell, M. P. and May, R. M. (1992) "The spatial dynamics of host–parasitoid systems", *Journal of Animal Ecology*, vol. 61, no. 3, 735–748

Condit, T. (1995) "Research in large, long-term tropical forest plots", *Trends in Ecology and Evolution*, vol. 10, no. 1, 18–22

Cone, C. A. and Myhre, A. (2000) "Community-supported agriculture: a sustainable alternative to industrial agriculture?", *Human Organization*, vol. 59, no. 2, 187–197

Connell, J. H. (1971) "On the role of natural enemies in preventing competitive exclusion in some marine animals and in rain forest trees", in den Boer, P. J. and Gradwell, G. R. (eds) *Dynamic Populations*, Center for Agricultural Publishing and Documentation, Wageningen, Netherlands, 298–312

Connell, J. H. (1978) "Diversity in tropical rain forests and coral reefs", *Science*, vol. 199, no. 4335, 1302–1310

Conroy, M. E., Rosset, P. M. and Murray, D. L. (1996) *A Cautionary Tale: Failed U.S. Development Policy in Central America*, Lynne Rienner Publishers, Boulder, CO

Cornia, G. A. (1985) "Farm size, land yields and the agricultural production function: an analysis for fifteen developing countries", *World Development*, vol. 13, no. 4, 513–534

Costanza, R. (2008) "Ecosystem services: multiple classification systems are needed", *Biological Conservation*, vol. 141, no. 2, 350–352

Costanza, R., d'Arge, R., De Groot, R., Farber, S., Grasso, M., Hannon, B., . . . and Van den Belt, M. (1997) "The value of the world's ecosystem services and natural capital", *Nature*, vol. 387, no. 6630, 253–260

Cousins, S. H. (1980) "A trophic continuum derived from plant structure, animal size and a detritus cascade", *Journal of Theoretical Biology*, vol. 82, no. 4, 607–618

Cousins, S. H. (1987) "The decline of the trophic level concept", *Trends in Ecology and Evolution*, vol. 2, no. 10, 312–316

Cressey, D. (2013) "Coffee rust regains foothold", *Nature*, vol. 493, no. 7434, 587

Croom, S., Romano, P. and Giannakis, M. (2000) "Supply chain management: an analytical framework for critical literature review", *European Journal of Purchasing and Supply Management*, vol. 6, no. 1, 67–83

Cruz-Agnón, A. and Greenberg, R. (2005) "Are epiphytes important for birds in coffee plantations? An experimental assessment", *Journal of Applied Ecology*, vol. 42, no. 1, 150–159

Cruz-Agnón, A., Sillett, T. S. and Greenberg, R. (2008) "An experimental study of habitat selection by birds in a coffee plantation", *Ecology*, vol. 89, no. 4, 921–927

Cruz-Agnón, A., Baena, M. L. and Greenberg, R. (2009) "The contribution of epiphytes to the abundance and species richness of canopy insects in a Mexican coffee plantation", *Journal of Tropical Ecology*, vol. 25, no. 5, 453–463

Cruz, E. L., Lorenzo, C., Soto, L., Naranjo, E. and Ramírez, N. (2004) "Diversidad de mamíferos en cafetales y selvas medianas de las cañadas de La Selva Lacandona, Chiapas, México", *Acta Zoológica Mexicana*, vol. 20, no. 1, 63–81

Cuddington, K., Byers, J. E., Wilson, W. G. and Hastings, A. (2007) *Ecosystem Engineers: Plants to Protists*, Academic Press (Elsevier), Amsterdam

Daily, G. C. (ed.) (1997) *Nature's Services: Societal Dependence on Natural Ecosystems*, Island Press, Washington, DC

Daily, G. C., Ehrlich, P. R. and Sanchez-Azofeifa, G. A. (2001) "Countryside biogeography: use of human-dominated habitats by the avifauna of southern Costa Rica", *Ecological Applications*, vol. 11, no. 1, 1–13

Damon, A. (2000) "A review of the biology and control of the coffee berry borer, *Hypothenemus hampei (Coleoptera: Scolytidae)*", *Bulletin of Entomological Research – London*, vol. 90, no. 6, 453–466

Daugherty, P. M., Harmon, J. P. and Briggs, C. J. (2007) "Trophic supplements to intraguild predation", *Oikos*, vol. 116, no. 4, 662–677

Davidson, D. W. (1985) "An experimental study of diffuse competition in harvester ants", *The American Naturalist*, vol. 125, no. 4, 500–506

Daviron, B. and Ponte, S. (2005) *The Coffee Paradox: Global Markets, Commodity Trade and the Elusive Promise of Development*, Zed Books, London

De Beenhouwer, M., Aerts, R. and Honnay, O. (2013) "A global meta-analysis of the biodiversity and ecosystem service benefits of coffee and cacao agroforestry", *Agriculture, Ecosystems and Environment*, vol. 175, no. 1, 1–7

de la Mora, A. and Philpott, S. M. (2010) "Wood-nesting ants and their parasites in forests and coffee agroecosystems", *Environmental Entomology*, vol. 39, no. 5, 1473–1481

de la Mora, A., Livingston, G. and Philpott, S. M. (2008) "Arboreal ant abundance and leaf miner damage in coffee agroecosystems in Mexico", *Biotropica*, vol. 40, no. 6, 742–746

de la Mora, A., Murnen, C. J. and Philpott, S. M. (2013) "Local and landscape drivers of biodiversity of four groups of ants in coffee landscapes", *Biodiversity and Conservation*, vol. 22, vol. 4, 871–888

De Marco Jr, P. and Coelho, F. M. (2004) "Services performed by the ecosystem: forest remnants influence agricultural cultures' pollination and production", *Biodiversity and Conservation*, vol. 13, no. 7, 1245–1255

de Roos, A. M., McCauley, E. and Wilson, W. G. (1998) "Pattern formation and the spatial scale of interaction between predators and their prey", *Theoretical Population Biology*, vol. 53, no. 2, 108–130

DeClerck, F. A., Chazdon, R., Holl, K. D., Milder, J. C., Finegan, B., Martinez-Salinas, A., Imbach, P., Canet, L. and Ramos, Z. (2010) "Biodiversity conservation in human-modified landscapes of Mesoamerica: past, present and future", *Biological Conservation*, vol. 143, no. 10, 2301–2313

Dejean, A., Djieto-Lordon, C. and Durand, J. L. (1997) "Ant mosaic in oil palm plantations of the southwest province of Cameroon: impact on leaf miner beetle (*Coleoptera: Chrysomelidae*)", *Journal of Economic Entomology*, vol. 90, no. 5, 1092–1096

Denevan, W. M. (1992) "The pristine myth: the landscape of the Americas in 1492", *Annals of the Association of American Geographers*, vol. 82, no. 3, 369–385

Denevan, W. M. (2001) *Cultivated Landscapes of Native Amazonia*, Oxford University Press, Oxford

Denslow, J. (1987) "Tropical rainforest gaps and tree species diversity", *Annual Review of Ecology and Systematics*, vol. 18, 431–451

Dial, R. and Roughgarden, J. (1988) "Theory of marine communities: the intermediate disturbance hypothesis", *Ecology*, vol. 79, no. 4, 1412–1424

Dieckmann, U., Law, R. and Metz, J. A. (eds) (2000) *The Geometry of Ecological Interactions: Simplifying Spatial Complexity*, Cambridge University Press, Cambridge

Dietsch, T. V. (2003) "Conservation and ecology of birds in coffee agroecosystems of Chiapas, Mexico", PhD thesis, University of Michigan, Ann Arbor, MI

Dietsch, T.V., Perfecto, I. and Greenberg, R. (2007) "Avian foraging behavior in two coffee agroecosystems of Chiapas, Mexico", *Biotropica*, vol. 39, no. 2, 232–240

Dimitri, C., Effland, A. B. and Conklin, N. C. (2005) *The 20th Century Transformation of US Agriculture and Farm Policy*, US Department of Agriculture, Economic Research Service, Washington, DC

Donald, P. F. (2004) "Biodiversity impacts of some agricultural commodity production systems", *Conservation Biology*, vol. 18, no. 1, 17–38

Donald, P. F. and Evans, A. D. (2006) "Habitat connectivity and matrix restoration: the wider implications of agri-environment schemes", *Journal of Applied Ecology*, vol. 43, no. 2, 209–218

Donaldson, J., Nänni, I., Zachariades, C. and Kemper, J. (2002) "Effects of habitat fragmentation on pollinator diversity and plant reproductive success in renosterveld shrublands of South Africa", *Conservation Biology*, vol. 16, no. 5, 1267–1276

Done, T. J. (1992) "Phase shifts in coral reef communities and their ecological significance", *Hydrobiologia*, vol. 247, 121–132

Dornhaus, A. and Powell, S. (2010) "Foraging and defence strategies", in Lach, L., Parr, C. L. and Abbott, K. L. (eds) *Ant Ecology*, Oxford University Press, Oxford, 210–230

Duffy, J. E., Cardinale, B. J., France, K. E., McIntyre, P. B., Thébault, E. and Loreau, M. (2007) "The functional role of biodiversity in ecosystems: incorporating trophic complexity", *Ecology Letters*, vol. 10, no. 6, 522–538

Dunbar, R. B., Wellington, G. M., Colgan, M. W. and Glynn, P. W. (1994) "Eastern Pacific sea surface temperature since 1600 AD: The $\delta18O$ record of climate variability in Galápagos corals", *Paleoceanography*, vol. 9, no. 2, 291–315, doi: 10.1029/93PA03501

Dunne, J. A., Williams, R. J., and Martinez, N. D. (2002) "Network structure and biodiversity loss in food webs: robustness increases with connectance", *Ecology Letters*, vol. 5, no. 4, 558–567

Durham, S. (2004) "Stopping the coffee berry borer from boring into profits", *Agricultural Research Magazine*, vol. 52, no. 11, 10–11

Durrett, R. and Levin, S. (1994) "The importance of being discrete (and spatial)", *Theoretical Population Biology*, vol. 46, no. 3, 363–394

Durrett, R. and Levin, S. (1998) "Spatial aspects of interspecific competition", *Theoretical Population Biology*, vol. 53, no. 1, 30–43

Edwards, K. F. and Schreiber, S. J. (2010) "Preemption of space can lead to intransitive coexistence of competitors", *Oikos*, vol. 119, no. 7, 1201–1209

Elton, C. S. (1927) *Animal Ecology*, University of Chicago Press, Chicago, IL

Emmerson, M. and Yearsley, J. M. (2004) "Weak interactions, omnivory and emergent food-web properties", *Proceedings of the Royal Society of London. Series B: Biological Sciences*, vol. 271, no. 1537, 397–405

Emmerson, M. C. and Raffaelli, D. (2004) "Predator–prey body size, interaction strength and the stability of a real food web", *Journal of Animal Ecology*, vol. 73, no. 3, 399–409

Engels, F. (1883) *Dialectics of Nature*. Available online at http://www.marxists.org/archive/marx/works/1883/don/ (accessed August 5, 2014)

Erwin, T. L. (1983) "Tropical forest canopies: the last biotic frontier", *Bulletin of the Ecological Society of America*, vol. 29, no. 1, 14–18

Estrada, A. and Coates-Estrada, R. (2005) "Diversity of neotropical migratory landbird species assemblages in forest fragments and man-made vegetation in Los Tuxtlas, Mexico", *Biodiversity and Conservation*, vol. 14, no. 7, 1719–1734

Estrada, A., Coates-Estrada, R. and Meritt Jr, D. A. (1997) "Anthropogenic landscape changes and avian diversity at Los Tuxtlas, Mexico", *Biodiversity and Conservation*, vol. 6, no. 1, 19–43

Estrada, A., Coates-Estrada, R., Dadda, A. A. and Cammarano, P. (1998) "Dung and carrion beetles in tropical rain forest fragments and agricultural habitats at Los Tuxtlas, Mexico", *Journal of Tropical Ecology*, vol. 14, no. 5, 577–593

Estrada, C. G., Damon, A., Sánchez Hernández, C., Soto-Pinto, L. and Ibarra-Nuñez, G. (2006) "Bat diversity in montane rainforest and shaded coffee under different management regimes in southeastern Chiapas, Mexico", *Biological Conservation*, vol. 132, no. 3, 351–361

Fagan, W. F. (1997) "Omnivory as a stabilizing feature of natural communities", *The American Naturalist*, vol. 150, no. 5, 554–567

Fahrig, L. (2003) "Effects of habitat fragmentation on biodiversity", *Annual Review of Ecology, Evolution, and Systematics*, vol. 34, 487–515

FAO (Food and Agriculture Organization) (2002) *Reducing Poverty and Hunger: The Critical Role of Financing for Food, Agriculture, and Rural Development*, FAO, International Fund for Agricultural Development, World Food Program, Rome

FAO (Food and Agriculture Organization) (2013) *The State of Food Insecurity in the World. The Multiple Dimensions of Food Security*, FAO, Rome

Federico, P., Hallam, T. G., McCracken, G. F., Purucker, S. T., Grant, W. E., Correa-Sandoval, A. N., . . . and Kunz, T. H. (2008) "Brazilian free-tailed bats as insect pest regulators in transgenic and conventional cotton crops", *Ecological Applications*, vol. 18, no. 4, 826–837

Feener, D. H. (2000) "Is the assembly of ant communities mediated by parasitoids?", *Oikos*, vol. 90, no. 1, 79–88

Feener, D. H. and Brown, B. V. (1992) "Reduced foraging of *Solenopsis geminata* (*Hymenoptera: Formicidae*) in the presence of parasitic *Pseudacteon* spp. (*Diptera: Phoridae*)", *Annals of the Entomological Society of America*, vol. 85, no. 1, 80–84

Feener, D. H., Orr, M. R., Wackford, K. M., Longo, J. M., Benson, W. W., and Gilbert, L. E. (2008) "Geographic variation in resource dominance-discovery in Brazilian ant communities", *Ecology*, vol. 89, no. 7, 1824–1836

Finke, D. L. and Denno, R. F. (2002) "Intraguild predation diminished in complex-structured vegetation: implications for prey suppression", *Ecology*, vol. 83, no. 3, 643–652

Fischer, J., Fazey, I., Briese, R. and Lindenmayer, D. B. (2005) "Making the matrix matter: challenges in Australian grazing landscapes", *Biodiversity and Conservation*, vol. 14, no. 3, 561–578

Fischer, J., Brosi, B., Daily, G. C., Ehrlich, P. R., Goldman, R., Goldstein, J., . . . and Tallis, H. (2008) "Should agricultural policies encourage land sparing or wildlife-friendly farming?", *Frontiers in Ecology and the Environment*, vol. 6, no. 7, 380–385

Fischer, J., Batary, P., Bawa, K. S., Brussaard, L., Chappell, M. J., Clough, Y., . . . and von Wehrden, H. (2011) "Conservation: limits of land sparing", *Science*, vol. 334, no. 6056, 593

Fischer, M. and Stöcklin, J. (1997) "Local extinctions of plants in remnants of extensively used calcareous grasslands 1950–1985", *Conservation Biology*, vol. 11, no. 3, 727–737

Fisher, R. A. (1956) *Statistical Methods and Scientific Inference*, Hafner Publishing, Oxford

Foley J. A. (2011) "Can we feed the world and sustain the planet?", *Scientific American*, vol. 305, 60–65

Foley, J. A., Ramankutty, N., Brauman, K. A., Cassidy, E. S., Gerber, J. S., Johnston, M., . . . and Zaks, D. P. (2011) "Solutions for a cultivated planet", *Nature*, vol. 478, no. 7369, 337–342

Foufopoulos, J. and Ives, A. R. (1999) "Reptile extinctions on land-bridge islands: life-history attributes and vulnerability to extinction", *The American Naturalist*, vol. 153, no. 1, 1–25

Fournier, L. A. (1988) "El cultivo del cafeto (*Coffea arabica* L.) al sol o a la sombra: un enfoque agronómico y ecofisiológico", *Agronomía Costarricense*, vol. 12, no. 1, 131–146

Fowler, A. M. and Hennessy, K. J. (1995) "Potential impacts of global warming on the frequency and magnitude of heavy precipitation", *Natural Hazards*, vol. 11, no. 3, 283–303

Fragoso, D. B., Guedes, R. N. C., Picanço, M. C. and Zambolim, L. (2002) "Insecticide use and organophosphate resistance in the coffee leaf miner *Leucoptera coffeella* (*Lepidoptera: Lyonetiidae*)", *Bulletin of Entomological Research*, vol. 92, no. 3, 203–212

Franco, R. A., Cárdenas, M. R., Montoya, E. C. and Zenner De Polania, I. G. (2003) "Ants associated with sucking insects in the aerial part of the coffee tree", *Revista Colombiana de Entomología*, vol. 29, no. 2, 95–105

Franklin, J. F. and Lindenmayer, D. B. (2009) "Importance of matrix habitats in maintaining biological diversity", *Proceedings of the National Academy of Sciences*, vol. 106, no. 2, 349–350

Fulton, R. H. (1984) "Chemical control of coffee leaf rust in Central America", in *Coffee Rust in the Americas*, American Phytopathological Society, St. Paul, MN, 75–83

Gallina, S., Mandujano, S. and Gonzalez-Romero, A. (1996) "Conservation of mammalian biodiversity in coffee plantations of Central Veracruz, Mexico", *Agroforestry Systems*, vol. 33, no. 1, 13–27

Gallina, S., Gonzalez-Romero, A. and Mandon R. H. (2008) "Mamíferos pequeños y medianos", in Manson, R. H., Hernández-Ortiz, V., Gallina, S. and Mehltreter, K. (eds) *Agroecosistemas Cafetaleros de Veracruz: Biodiversidad, Manejo y Conservación*, Instituto Nacional de Ecología, Mexico, 161–180

García-Barrios, L., Galván-Miyoshi, Y. M., Valdivieso-Pérez, I. A., Masera, O. R., Bocco, G. and Vandermeer, J. (2009) "Neotropical forest conservation, agricultural intensification, and rural out-migration: the Mexican experience", *BioScience*, vol. 59, no. 10, 863–873

García-Barrios, R. and García-Barrios, L. (1990) "Environmental and technological degradation in peasant agriculture: a consequence of development in Mexico", *World Development*, vol. 18, no. 11, 1569–1585

Garcia-Estrada, C., Damon, A., Sánchez Hernández, C., Soto-Pinto, L. and Ibarra-Nuñez, I. (2012) "Diets of frugivorous bats in montane rain forest and coffee plantations in southeastern Chiapas, Mexico", *Biotropica*, vol. 44, no. 3, 394–401

Garcia-Franco, J. G. and Toledo Aceves, T. (2008) "Epífitas vasculares: bromelias y orquídeas", in Manson, R. H., Hernández-Ortiz, V., Gallina, S. and Mehltreter, K. (eds) *Agroecosistemas Cafetaleros de Veracruz: Biodiversidad, Manejo y Conservación*, Instituto Nacional de Ecología, Mexico, 69–78

Gilinsky, E. (1984) "The role of fish predation and spatial heterogeneity in determining benthic community structure", *Ecology*, vol. 65, no. 2, 455–468

Gillespie, J. (1974) "The role of environmental grain in the maintenance of genetic variation", *The American Naturalist*, vol. 108, no. 964, 831–836

Giovannucci, D. and Ponte, S. (2005) "Standards as a new form of social contract? Sustainability initiatives in the coffee industry", *Food Policy*, vol. 30, no. 3, 284–301

Gliessman, S. R. (ed.) (1990) *Agroecology: Researching the Ecological Basis for Sustainable Agriculture*, Ecological Studies Series no. 78, Springer-Verlag, New York

Gliessman, S. R. (1998) *Agroecology: Ecological Process in Sustainable Agriculture*, Ann Arbor Press, Ann Arbor, MI

Glover, N. and Beer, J. (1986) "Nutrient cycling in two traditional Central American agroforestry systems", *Agroforestry Systems*, vol. 4, no. 2, 77–87

Golubski, A. J. and Abrams, P. A. (2011) "Modifying modifiers: what happens when interspecific interactions interact?", *Journal of Animal Ecology*, vol. 80, no. 5, 1097–1108

Gonthier, D. J., Ennis, K. K., Philpott, S. M., Vandermeer, J. and Perfecto, I. (2013) "Ants defend coffee from berry borer colonization", *BioControl*, vol. 58, no. 4, 1–6

Gonzalez-Romero, A. and Murrieta Balindo, R. (2008) "Anfibios y reptiles", in Manson, R. H., Hernández-Ortiz, V., Gallina, S. and Mehltreter, K. (eds) *Agroecosistemas Cafetaleros de Veracruz: Biodiversidad, Manejo y Conservación*, Instituto Nacional de Ecología, Xalapa, Mexico, 69–78

Gordon, C., Manson, R., Sundberg, J. and Cruz-Angón, A. (2007) "Biodiversity, profitability, and vegetation structure in a Mexican coffee agroecosystem", *Agriculture, Ecosystems, and Environment*, vol. 118, no. 1, 256–266

Gordon, C. E., McGill, B., Ibarra-Núñez, G., Greenberg, R. and Perfecto, I. (2009) "Simplification of a coffee foliage-dwelling beetle community under low-shade management", *Basic and Applied Ecology*, vol. 10, no. 3, 246–254

Gotelli, N. J. and Ellison, A. M. (2002) "Assembly rules for New England ant assemblages", *Oikos*, vol. 99, no. 3, 591–599

Gove, A. D., Hylander, K., Nemomisa, S. and Shimelis, A. (2008) "Ethiopian coffee cultivation: implications for bird conservation and environmental certification", *Conservation Letters*, vol. 1, no. 5, 208–216

Grau, H. R., Aide, T. M., Zimmerman, K. J., Thomlinson, R. J., Helmer, E. and Zou, X. (2003) "The ecological consequences of socioeconomic and land-use changes in post-agriculture Puerto Rico", *BioScience*, vol. 53, no. 12, 1159–1168

Green, R. E., Cornell, S. J., Scharlemann, J. P. and Balmford, A. (2005) "Farming and the fate of wild nature", *Science*, vol. 307, no. 5709, 550–555

Greenberg, R., Bichier, P. and Sterling, J. (1997a) "Bird populations in rustic and planted shade coffee plantations of eastern Chiapas, Mexico", *Biotropica*, vol. 29, no. 4, 501–514

Greenberg, R., Bichier, P., Agnón, A. C. and Reitsma, R. (1997b) "Bird populations in shade and sun coffee plantations in central Guatemala", *Conservation Biology*, vol. 11, no. 2, 448–459

Greenberg, R., Bichier, P., Agnón, A. C., MacVean, C., Perez, R. and Cano, E. (2000) "The impact of avian insectivory on arthropods and leaf damage in some Guatemalan coffee plantations", *Ecology*, vol. 81, no. 6, 1750–1755

Greenberg, R., Perfecto, I. and Philpott, S. M. (2008) "Agroforests as model systems for tropical ecology", *Ecology*, vol. 89, no. 4, 913–914

Gregory, P. J. and Ingram, J. (2000) "Global change and food and forest production: future scientific challenges", *Agriculture Ecosystems and Environment*, vol. 82, nos 1–3, 3–14

Grime, J. P. (1973) "Competitive exclusion in herbaceous vegetation", *Nature*, vol. 242, no. 5396, 344–347

Griscom, L. (1932) "The distribution of bird-life in Guatemala", *Bulletin of the American Museum of Natural History*, vol. 64, 1–439

Gruner, D. S. (2004) "Attenuation of top-down and bottom-up forces in a complex terrestrial community", *Ecology*, vol. 85, no. 11, 3010–3022

Guhl, A. (2008) "Coffee production intensification and landscape change in Colombia 1970–2002", in Jepson, W. and Millington, A. (eds) *Land Change Science in the Tropics*, Springer-Verlag, New York

Gunnarsson, B. (2007) "Bird predation on spiders: ecological mechanisms and evolutionary consequences", *Journal of Arachnology*, vol. 35, no. 3, 509–529

Gurney, W. S. and Veitch, A. R. (2000) "Self-organization, scale and stability in a spatial predator–prey interaction", *Bulletin of Mathematical Biology*, vol. 62, no. 1, 61–86

Gustavsson, J., Cedelberg, C. and Sonesson, U. (2011) *Global Food Losses and Food Waste*, FAO, Rome

Hairston, Jr, N. G. and Hairston, Sr, N. G. (1997) "Does food web complexity eliminate trophic-level dynamics?", *The American Naturalist*, vol. 149, no. 5, 1001–1007

Hairston, N. G., Smith, F. E. and Slobodkin, L. B. (1960) "Community structure, population control, and competition", *The American Naturalist*, vol. 94, no. 879, 421–425

Halffter, G., Pineda, E., Arellano, L. and Escobar, F. (2007) "Instability of copronecrophagous beetle assemblages (*Coleoptera: Scarabaeinae*) in a mountainous tropical landscape of Mexico", *Environmental Entomology*, vol. 36, no. 6, 1397–1407

Hall, S. J. and Raffaelli, D. (1991) "Food-web patterns: lessons from a species-rich web", *The Journal of Animal Ecology*, vol. 60, 823–841

Hamilton, A. J., Basset, Y., Benke, K. K., Grimbacher, P. S., Miller, S. E., Novotn?, V., . . . and Yen, J. D. (2010) "Quantifying uncertainty in estimation of tropical arthropod species richness", *The American Naturalist*, vol. 176, no. 1, 90–95

Hansen, J. (2005) "A slippery slope: how much global warming constitutes 'dangerous anthropogenic interference'?", *Climate Change*, vol. 68, no. 3, 269–279

Hanski, I. A. (2011) "Eco-evolutionary spatial dynamics in the Glanville fritillary butterfly", *Proceedings of the National Academy of Sciences*, vol. 108, no. 35, 14397–14404

Hanski, I., Moilanen, A. and Gyllenberg, M. (1996) "Minimum viable metapopulation size", *The American Naturalist*, vol. 146, no. 4, 527–541

Hanson, J. C., Lichtenberg, E. and Peters, S. E. (1997) "Organic versus conventional grain production in the mid-Atlantic: an economic and farming system overview", *American Journal of Alternative Agriculture*, vol. 12, no. 1, 2–9

Harrison, S. and Taylor, A. D. (1997) "Empirical evidence for metapopulation dynamics", in Haski, I. and Gilpin, A. D. (eds) *Metapopulation Biology: Ecology, Genetics, and Evolution*, Academic Press, San Diego, CA, 27–42

Harvey, C. A., Medina, A., Sánchez, D. M., Vílchez, S., Hernández, B., Saenz, J. C., . . . and Sinclair, F. L. (2006) "Patterns of animal diversity in different forms of tree cover in agricultural landscapes", *Ecological Applications*, vol. 16, no. 5, 1986–1999

Harvey, C. A., Komar, O., Chazdon, R., Ferguson, B. G., Finegan, B., Griffith, D. M., . . . and Wishnie, M. (2008) "Integrating agricultural landscapes with biodiversity conservation in the Mesoamerican hotspot", *Conservation Biology*, vol. 22, no. 1, 8–15

Hassell, M. P., Comins, H. N. and May, R. M. (1991) "Spatial structure and chaos in insect population dynamics", *Nature*, vol. 353, no. 6341, 255–258

Hassell, M. P., Miramontes, O., Rohani, P. and May, R. M. (1995) "Appropriate formulations for dispersal in spatially structured models: comments on Bascompte and Solé", *Journal of Animal Ecology*, vol. 64, no. 5, 662–664

Hastings, A. and Powell, T. (1991) "Chaos in a three-species food chain", *Ecology*, vol. 72, no. 3, 896–903

Hastings, H. M. and Conrad, M. (1979) "Length and evolutionary stability of food chains", *Nature*, vol. 282, 838–839

Hecht, S. (2010) "The new rurality: globalization, peasants and the paradoxes of landscapes", *Land Use Policy*, vol. 27, no. 2, 161–169

Helm, A., Hanski, I. and Pärtel, M. (2006) "Slow response of plant species richness to habitat loss and fragmentation", *Ecology Letters*, vol. 9, no. 1, 72–77

Heltberg, R. (1998) "Rural market inperfections and the farm size–productivity relationship: evidence from Pakistan", *World Development*, vol. 26, no. 10, 1807–1826

Henderson, R. W. and Powell, R. (2001) "Responses by the West Indian herpetofauna to human-influenced resources", *Caribbean Journal of Science*, vol. 37, nos 1–2, 41–54

Heredia Abarca, G. and Arias Mota, R. M. (2008) "Hongos saprobios y endomicorrizógenos en el suelo", in Manson, R. H., Hernández-Ortiz, V., Gallina, S. and Mehltreter, K. (eds) *Agroecosistemas Cafetaleros de Veracruz: Biodiversidad, Manejo y Conservación*, Instituto Nacional de Ecología, Mexico, 193–203

Hernández, S. M., Mattsson, B. J., Peters, V. E., Cooper, R. J. and Carroll, C. R. (2013) "Coffee agroforests remain beneficial for Neotropical bird community conservation across seasons", *PLOS ONE*, vol. 8, no. 9, e65101

Hernández-Ortiz, V. and Dzul-Cauich, J. F. (2008) "Moscas (*Insecta: Diptera*)", in Manson, R. H., Hernández-Ortiz, V., Gallina, S. and Mehltreter, K. (eds) *Agroecosistemas Cafetaleros de Veracruz: Biodiversidad, Manejo y Conservación*, Instituto Nacional de Ecología, Mexico, 95–105

Hietz, P. (2005) "Conservation of vascular epiphyte diversity in Mexican coffee plantations", *Conservation Biology*, vol. 19, no. 2, 391–399

Hillebrand, H. (2004) "On the generality of the latitudinal diversity gradient", *The American Naturalist*, vol. 163, no. 2, 192–211

Hodge, M. A. (1999) "The implications of intraguild predation for the role of spiders in biological control", *Journal of Arachnology*, vol. 27, 351–362

Hodges, A. (1992) *Alan Turing: The Enigma*, Random House, London

Hofbauer, J. and Sigmund, K. (1998) *Evolutionary Games and Population Dynamics*, Cambridge University Press, Cambridge

Holden, S. (2001) "A century of technological change and deforestation in the Miombo woodlands of northern Zambia", in Angelsen, A. and Kaimowitz, D. (eds) *Agricultural Technologies and Tropical Deforestation*, CABI Publishing, Oxon, UK, 251–270

Hole, F. (1998) "The spread of agriculture to the eastern arc of the Fertile Crescent: food for the herders", in Damania, A. B., Valkoun, J., Willcox, G. and Qualset, C. O. (eds) *The Origins of Agriculture and Crop Domestication*, ICARDA, Aleppo, Syria, 83–92

Holling, C. S. (1973) "Resilience and stability of ecological systems", *Annual Review of Ecology and Systematics*, vol. 4, 1–23

Holmes, R. T. (1990) "Ecological and evolutionary impacts of bird predation on forest insects: an overview", *Studies in Avian Biology*, vol. 13, 6–13

Holmes, R. T., Schultz, J. C. and Nothnagle, P. (1979) "Bird predation on forest insects: an exclosure experiment", *Science*, vol. 206, no. 4417, 462–463

Holt, R. D. (1985) "Population dynamics in two-patch environments: some anomalous consequences of an optimal habitat distribution", *Theoretical Population Biology*, vol. 28, no. 2, 181–208

Holt, R. D. and Polis, G. A. (1997) "A theoretical framework for intraguild predation", *The American Naturalist*, vol. 149. no. 4, 745–764

Holt-Giménez, E. (2002) "Measuring farmers' agroecological resistance after Hurricane Mitch in Nicaragua: a case study in participatory, sustainable land management impact monitoring", *Agriculture, Ecosystems and Environment*, vol. 93, no. 1, 87–105

Holt-Giménez, E. and Shattuck, A. (2011) "Food crises, food regimes and food movements: rumblings of reform or tides of transformation?" *Journal of Peasant Studies*, vol. 38, 109–144

Holyoak, M. and Sachdev, S. (1998) "Omnivory and the stability of simple food webs", *Oecologia*, vol. 117, no. 3, 413–419

Hooks, C. R., Pandey, R. R. and Johnson, M. W. (2003) "Impact of avian and arthropod predation on lepidopteran caterpillar densities and plant productivity in an ephemeral agroecosystem", *Ecological Entomology*, vol. 28, no. 5, 522–532

Hooper, D. U., Chapin III, F. S., Ewel, J. J., Hector, A., Inchausti, P., Lavorel, S., . . . and Wardle, D. A. (2005) "Effects of biodiversity on ecosystem functioning: a consensus of current knowledge", *Ecological Monographs*, vol. 75, no. 1, 3–35

Horgan, F. G. (2009) "Invasion and retreat: shifting assemblages of dung beetles amidst changing agricultural landscapes in central Peru", *Biodiversity and Conservation*, vol. 18, no. 13, 3519–3541

Horn, H. (1975) "Forest succession", *Scientific American*, vol. 232, no. 1975, 90–98

Horner-Devine, M. C., Daily, G. C., Ehrlich, P. R. and Boggs, C. L. (2003) "Countryside biogeography of tropical butterflies", *Conservation Biology*, vol. 17, no. 1, 168–177

Houghton, J. T., Ding, Y., Griggs, D. J., Noguer, M., Van der Linden, P. J., Dai, X., . . . and Johnson, C. A. (2001) *IPCC, 2001: Climate Change 2001: The Scientific Basis. Contribution of Working Group I to the Third Assessment Report of the Intergovernmental Panel on Climate Change*, Cambridge University Press, Cambridge, UK and New York, USA,

Howard, A. (1943) *An Agricultural Testament*, Oxford University Press, Oxford

Hoyos, C. D., Agudelo, P. A., Webster, P. J. and Curry, J. A. (2006) "Deconvolution of the factors contributing to the increase in global hurricane intensity", *Science*, vol. 312, no. 5770, 94–97

Hsieh, H. and Perfecto, I. (2012) "Ecological impacts of phorid parasitoids on ant communities", *Psyche: A Journal of Entomology*. Available online at http://dx.doi.org/10.1155/2012/380474 (accessed June 24, 2014)

Hsieh, H. Y., Liere, H., Soto, E. J. and Perfecto, I. (2012) "Cascading trait-mediated interactions induced by ant pheromones", *Ecology and Evolution*, vol. 2, no. 9, 2181–2191

Hubbell, S. P. (2001) *The Unified Neutral Theory of Biodiversity and Biogeography*, Princeton University Press, Princeton, NJ

Huffaker, C. B. (1958) "Experimental studies on predation: dispersion factors and predator–prey oscillations", *Hilgardia: A Journal of Agricultural Science*, vol. 27, 795–834

Hughes, R. and Flintan, F. (2001) *Integrating Conservation and Development Experience: A Review and Bibliography of the ICDP Literature*, International Institute for Environment and Development, London

Huisman, J. and Weissing, F. J. (1999) "Biodiversity of plankton by species oscillations and chaos", *Nature*, vol. 402, no. 6760, 407–410

Huston, M. A. (1997) "Hidden treatments in ecological experiments: re-evaluating the ecosystem function of biodiversity", *Oecologia*, vol. 110, no. 4, 449–460

Hutchinson, G. E. (1957) "Concluding remarks", *Cold Spring Harbor Symposia on Quantitative Biology*, vol. 22, no. 2, 415–427

Hutchinson, G. E. (1961) "The paradox of the plankton", *The American Naturalist*, vol. 95, no. 882, 137–145

Huxley, P. A. and Ismail, S. A. H. (1969) "Floral atrophy and fruit set in Arabica coffee in Kenya", *Turrialba*, vol. 19, no. 3, 345–352

Hyatt, L. A., Rosenberg, M. S., Howard, T. G., Bole, G., Fang, W., Anastasia, J., Brown. K., Grella, R., Hinman, K., Kurdziel, J. P. and Gurevitch, J. (2003) "The distance dependence prediction of the Janzen–Connell hypothesis: a meta-analysis", *Oikos*, vol. 503, no. 3, 590–602

IAASTD (2009) *International Assessment of Agricultural Knowledge, Science and Technology for Development: The Synthesis Report*, McIntyre, B. D., Herren, H. R., Wakhungu, J. and Watson, R. T. (eds) Island Press, Washington, DC

Ibarra-Nuñez, G. (1990) "Los artrópodos asociados a cafetos en un cafetal mixto del Soconusco, Chiapas, México", *Folia Entomológica Mexicana*, vol. 79, 207–231

Ibarra-Nuñez, G. (2001) "Prey analysis in the diet of some ponerine ants (*Hymenoptera: Formicidae*) and web-building spiders (*Araneae*) in coffee plantations in Chiapas, Mexico", *Sociobiology*, vol. 37, no. 3B, 723–755

ICT (2006) *Anuário Estadísticas de Demanda 2006* (in Spanish), Intituto Costarricense de Turismo, Departamento de Estadísticas, San José. Available online at www.visitcostarica.com/ict/backoffice/treeDoc/files/Anuario%20de%20Turismo%202006%20(VERSION%20FINAL).pdf (accessed July 29, 2008), Table 44 and 45

Inside Costa Rica (2013), "Coffee rust could become unmanageable, agronomists warn". Available online at http://insidecostarica.com/2013/01/31/coffee-rust-could-become-unmanageable-agronomists-warn/ (accessed February 4, 2013)

International Coffee Organization (2013) *Report on the Outbreak of Coffee Leaf Rust in Central America and Action Plan to Combat the Pest*. Available online at www.dev.ico.org/documents/cy2012-13/ed-2157e-report-clr.pdf? (accessed November 25, 2013)

Jackson, D., Vandermeer, J. and Perfecto, I. (2009) "Spatial and temporal dynamics of a fungal pathogen promote pattern formation in a tropical agroecosystem", *Open Ecology Journal*, vol. 2, 62–73

Jackson, D., Skillman, J. and Vandermeer, J. (2012a) "Indirect biological control of the coffee leaf rust, *Hemileia vastatrix*, by the entomogenous fungus *Lecanicillium lecanii* in a complex coffee agroecosystem", *Biological Control*, vol. 61, no. 1, 89–97

Jackson, D., Zemenick, K. and Huerta, G. (2012b) "Occurrence in the soil and dispersal of *Lecanicillium lecanii*, a fungal pathogen of the green coffee scale (*Coccus viridis*) and coffee rust (*Hemileia vastatrix*)", *Tropical and Subtropical Agroecosystems*, vol. 15, no. 2, 389–401

Jackson, D., Allen, D., Perfecto, I. and Vandermeer, J. (2014a) "Self-organization of background habitat determines the nature of population spatial structure", *Oikos*, vol. 123, no. 6, 751–761

Jackson, D., Vandermeer, J., Perfecto, I. and Philpott, S. M. (2014b) "Population responses to environmental change in a tropical ant: the interaction of spatial and temporal dynamics", *PLOS ONE*, vol. 9, no. 5, e97809, doi:10.1371/journal.pone.0097809

Jackson, W. (1980) *New Roots for Agriculture*, Friends of the Earth, San Francisco, CA

Janssen, A., Sabelis, M. W., Magalhães, S., Montserrat, M. and Van der Hammen, T. (2007) "Habitat structure affects intraguild predation", *Ecology*, vol. 88, no. 11, 2713–2719

Janzen, D. H. (1970) "Herbivores and the number of tree species in tropical forests", *The American Naturalist*, vol. 104, no. 940, 501–528

Jaramillo, J., Borgemeister, C. and Baker, P. (2006) "Coffee berry borer *Hypothenemus hampei* (*Coleoptera: Curculionidae*): searching for sustainable control strategies", *Bulletin of Entomological Research*, vol. 96, no. 3, 223–234

Jedlicka, J. A., Greenberg, R., and Letourneau, D. K. (2011) "Avian conservation practices strengthen ecosystem services in California vineyards", *PLOS ONE*, vol. 6, no. 11, e27347

Jha, S. and Dick, C. W. (2008) "Shade coffee farms promote genetic diversity of native trees", *Current Biology*, vol. 18, no. 24, R1126–R1128

Jha, S. and Vandermeer, J. H. (2009) "Contrasting bee foraging in response to resource scale and local habitat management", *Oikos*, vol. 118, no. 8, 1174–1180

Jha, S. and Dick, C. W. (2010) "Native bees mediate long-distance pollen dispersal in a shade coffee landscape mosaic", *Proceedings of the National Academy of Sciences*, vol. 107, no. 31, 13760–13764

Jha, S. and Vandermeer, J. (2010) "Impacts of coffee agroforestry management on tropical bee communities", *Biological Conservation*, vol. 143, no. 6, 1423–1431

Jha, S., Allen, D., Liere, H., Perfecto, I. and Vandermeer, J. (2012) "Mutualisms and population regulation: mechanism matters", *PLOS ONE*, vol. 7, no. 8, e43510

Jiménez-Soto, E., Cruz-Rodríguez, J. A., Vandermeer, J. and Perfecto, I. (2013) "*Hypothenemus hampei* (*Coleoptera*: *Curculionidae*) and its interactions with *Azteca instabilis* and *Pheidole synanthropica* (*Hymenoptera*: *Formicidae*) in a shade coffee agroecosystem", *Environmental Entomology*, vol. 42, no. 5, 915–924

Johnson, C. R. and Seinen, I. (2002) "Selection for restraint in competitive ability in spatial competition systems", *Proceedings of the Royal Society of London. Series B: Biological Sciences* vol. 269, no. 1492, 655–663

Johnson, M. D. (2000) "Effects of shade-tree species and crop structure on the winter arthropod and bird communities in a Jamaican shade coffee plantation", *Biotropica*, vol. 32, no. 1, 133–145

Johnson, M. D. and Sherry, T. W. (2001) "Effects of food availability on the distribution of migratory warblers among habitats in Jamaica", *Journal of Animal Ecology*, vol. 70, no. 4, 546–560

Johnson, M. D., Levy, N. J., Kellermann, J. L. and Robinson, D. E. (2009) "Effects of shade and bird exclusion on arthropods and leaf damage on coffee farms in Jamaica's Blue Mountains", *Agroforestry Systems*, vol. 76, no. 1, 139–148

Johnson, M. D., Kellerman, K. L. and Stercho, A. M. (2010) "Pest reduction services by birds in shade and sun coffee in Jamaica", *Animal Conservation*, vol. 13, no. 2, 140–147

Kalka, M. B., Smith, A. R. and Kalko, E. K. V. (2008) "Bats limit arthropods and herbivory in a tropical forest", *Science*, vol. 320, no. 5872, 71

Kamran-Disfani, A. R. and Golubski, A. J. (2013) "Lateral cascade of indirect effects in food webs with different types of adaptive behavior", *Journal of Theoretical Biology*, vol. 339, 58–69

Kareiva, P. and Wennergren, U. (1995) "Connecting landscape patterns to ecosystem and population processes", *Nature*, vol. 373, no. 6512, 299–302

Kareiva, P., Mullen, A. and Southwood, R. (1990) "Population dynamics in spatially complex environments: theory and data [and discussion]", *Philosophical Transactions of the Royal Society of London. Series B: Biological Sciences*, vol. 330, no. 1257, 175–190

Karp, D. S. and Daily, G. C. (2014) "Cascading effects of insectivorous birds and bats in tropical coffee plantations", *Ecology*, vol. 95, no. 4, 1065–1074

Karp, D. S., Rominger, A. J., Zook, J., Ranganathan, J., Ehrlich, P. R. and Daily, G. C. (2012) "Intensive agriculture erodes ,-diversity at large scales", *Ecology Letters*, vol. 15, no. 9, 963–970

Karp, D. S., Mendenhall, C. D., Sandí, R. F., Chaumont, N., Ehrlich, P. R., Hadly, E. A. and Daily, G. C. (2013) "Forest bolsters bird abundance, pest control and coffee yield", *Ecology Letters*, vol. 16, no. 11, 1339–1347

Kearns, C. A., Inouye, D. W. and Waser, N. M. (1998) "Endangered mutualisms: the conservation of plant–pollinator interactions", *Annual Review of Ecology and Systematics*, vol. 29, 83–112

Kéfi, S., Rietkerk, M., Alados, C. I., Pueyo, Y., Papanastasis, V. P., ElAich, A. and de Ruiter, P. C. (2007) "Spatial vegetation patterns and imminent desertification in Mediterranean arid ecosystems", *Nature*, vol. 449, no. 7159, 213–217

Kéfi, S., Rietkerk, M., Roy, M., Franc, A., De Ruiter, P. C. and Pascual, M. (2011) "Robust scaling in ecosystems and the meltdown of patch size distributions before extinction", *Ecology Letters*, vol. 14, no. 1, 29–35

Kellermann, J. L., Johnson, M. D., Stercho, A. M. and Hackett, S. C. (2008) "Ecological and economic services provided by birds on Jamaican Blue Mountain coffee farms", *Conservation Biology*, vol. 22, no. 5, 1177–1185

Kéry, M. (2004) "Extinction rate estimates for plant populations in revisitation studies: importance of detectability", *Conservation Biology*, vol. 18, no. 2, 570–574

Kiepe, P. and Rao, M. R. (1994) "Management of agroforestry for the conservation and utilization of land and water resources: soil management and conservation", *Outlook on Agriculture*, vol. 23, no. 1, 17–25

Kirlinger, G. (1986) "Permanence in Lotka–Volterra equations: linked prey–predator systems", *Mathematical Biosciences*, vol. 82, no. 2, 165–191

Klein, A. M., Steffan-Dewenter, I., Buchori, D. and Tscharntke, T. (2002) "Effects of land-use intensity in tropical agroforestry systems on coffee flower-visiting and trap-nesting bees and wasps", *Conservation Biology*, vol. 16, no. 4, 1003–1014

Klein, A. M., Steffan-Dewenter, I. and Tscharntke, T. (2003a) "Bee pollination and fruit set of *Coffea arabica* and *C. canephora* (*Rubiaceae*)", *American Journal of Botany*, vol. 90, no. 1, 153–157

Klein, A. M., Steffan-Dewenter, I. and Tscharntke, T. (2003b) "Fruit set of highland coffee increases with the diversity of pollinating bees", *Proceedings of the Royal Society of London. Series B: Biological Sciences*, vol. 270, no. 1518, 955–961

Klein, A. M., Steffan-Dewenter, I. and Tscharntke, T. (2003c) "Pollination of *Coffea canephora* in relation to local and regional agroforestry management", *Journal of Applied Ecology*, vol. 40, no. 5, 837–845

Klein, A. M., Vaissiere, B. E., Cane, J. H., Steffan-Dewenter, I., Cunningham, S. A., Kremen, C. and Tscharntke, T. (2007) "Importance of pollinators in changing landscapes for world crops", *Proceedings of the Royal Society B: Biological Sciences*, vol. 274, no. 1608, 303–313

Klein, A. M., Cunningham, S. A., Bos, M. and Steffan-Dewenter, I. (2008) "Advances in pollination ecology from tropical plantation crops", *Ecology*, vol. 89, no. 4, 935–943

Kleinn, C., Corrales, L. and Morales, D. (2002) "Forest area in Costa Rica: a comparative study of tropical forest cover estimates over time", *Environmental Monitoring and Assessment*, vol. 73, no. 1, 17–40

Klooster, D. (2003) "Forest transitions in Mexico: institutions and forests in a globalized countryside", *The Professional Geographer*, vol. 55, no. 2, 227–237

Knowlton, N. (1999) "Thresholds and multiple stable states in coral reef community dynamics", *American Zoologist*, vol. 32, no. 6, 674–682

Komar, O. (2006) "Priority contribution. Ecology and conservation of birds in coffee plantations: a critical review", *Bird Conservation International*, vol. 16, no. 1, 1–23

Komar, O. and Domínguez, J. P. (2001) *Lista de Aves de El Salvador*, Fundación Ecológica de El Salvador–SalvaNATURA, San Salvador

Komar, O. and Domínguez, J. P. (2002) "Efectos del estrato de sombra sobre poblaciones de anfibiosreptiles y aves en plantaciones de café de El Salvador: implicaciones para programas de certificación", Actas del Simposio de Café y Biodiversidad, *Revista Protección Vegetal*, vol. 12, no. 2, 61–62

Kondo, S. and Miura, T. (2010) 'Reaction–diffusion model as a framework for understanding biological pattern formation', *Science*, vol. 329, no. 5999, 1616–1620

Kremen, C., Williams, N. M. and Thorp, R. W. (2002) "Crop pollination from native bees at risk from agricultural intensification", *Proceedings of the National Academy of Sciences*, vol. 99, no. 26, 16812–16816

Kremen, C., Williams, N. M., Bugg, R. L., Fay, J. P. and Thorp, R. W. (2004) "The area requirements of an ecosystem service: crop pollination by native bee communities in California", *Ecology Letters*, vol. 7, no. 11, 1109–1119

Kremen, C., Iles, A. and Bacon, C. (2012) "Diversified farming systems: an agroecological, systems-based alternative to modern industrial agriculture", *Ecology and Society*, vol. 17, no. 4, art. 44. Available online at www.ecologyandsociety.org/vol17/iss4/art44/ (accessed June 21, 2014)

Krishna, S. N., Krishna, S. B. and Vijayalaxmi, K. K. (2005) "Variation in anuran abundance along the streams of the Western Ghats, India", *The Herpetological Journal*, vol. 15, no. 3, 167–172

Krishnan, S., Kushalappa, C. G., Shaanker, R. U. and Ghazoul, J. (2012) "Status of pollinators and their efficiency in coffee fruit set in a fragmented landscape mosaic in South India", *Basic and Applied Ecology*, vol. 13, no. 3, 277–285

Kumar, A. C. (1988) "A review of work on nematology at Central Coffee Research Institute", *Indian Coffee (India)*, vol. 52, no. 6, 5–8

Kuussaari, M., Bommarco, R., Heikkinen, R. K., Helm, A., Krauss, J., Lindborg, R., . . . and Steffan-Dewenter, I. (2009) "Extinction debt: a challenge for biodiversity conservation", *Trends in Ecology and Evolution*, vol. 24, no. 10, 564–571

Lach, L., Parr, C. L. and Abbott, K. L. (eds) (2010) *Ant Ecology*, Oxford University Press, Oxford

Lander, T. A., Bebber, D. P., Choy, C. T., Harris, S. A. and Boshier, D. H. (2011) "The Circe Principle explains how resource-rich land can waylay pollinators in fragmented landscapes", *Current Biology*, vol. 21, no. 15, 1302–1307

Larsen, A. and Philpott, S. M. (2010) "Twig-nesting ants: the hidden predators of the coffee berry borer in Chiapas, Mexico", *Biotropica*, vol. 42, no. 3, 342–347

Laurence, W. F. and Bierregaard R. O. (eds) (1997) *Tropical Forest Remnants: Ecology, Management, and Conservation of Fragmented Communities*, University of Chicago Press, Chicago, IL

Law, R. and Blackford, J. C. (1992) "Self-assembling food webs: a global viewpoint of coexistence of species in Lotka–Volterra communities", *Ecology*, vol. 73, no. 2, 567–578

Le Pelley, R. H. (1973) "Coffee insects", *Annual Review of Entomology*, vol. 18, no. 1, 121–142

LeBrun, E. G., and Feener, D. H. (2002) "Linked indirect effects in ant–phorid interactions: impacts on ant assemblage structure", *Oecologia*, vol. 133, no. 4, 599–607

LeBrun, E. G., and Feener, D. H. (2007) "When trade-offs interact: balance of terror enforces dominance discovery trade-off in a local ant assemblage", *Journal of Animal Ecology*, vol. 76, no. 1, 58–64

Leopold, A. (1949) *A Sand County Almanac*, Oxford University Press, Oxford

Letourneau, D. K., Armbrecht, I., Rivera, B. S., Lerma, J. M., Carmona, E. J., Daza, M. C., . . . and Trujillo, A. R. (2011) "Does plant diversity benefit agroecosystems? A synthetic review", *Ecological Applications*, vol. 21, no. 1, 9–21

Levins, R. (1968) *Evolution in Changing Environments: Some Theoretical Explorations*, Princeton University Press, Princeton, NJ

Levins, R. (1969) "Some demographic and genetic consequences of environmental heterogeneity for biological control", *Bulletin of the ESA*, vol. 15, no. 3, 237–240

Levins, R. (1974) "Discussion paper: the qualitative analysis of partially specified systems", *Annals of the New York Academy of Sciences*, vol. 231, no. 1, 123–138

Levins, R. (1979) "Coexistence in a variable environment", *The American Naturalist*, vol. 114, no. 6, 765–783

Levins, R. and Lewontin, R. (1980) "Dialectics and reductionism in ecology", *Synthese*, vol. 43, no. 1, 47–78

Levins, R. and Lewontin, R. (1985) *The Dialectical Biologist*, Harvard University Press, Cambridge, MA

Lewis, W. J., Van Lenteren, J. C., Phatak, S. C. and Tumlinson, J. H. (1997) "A total system approach to sustainable pest management", *Proceedings of the National Academy of Sciences*, vol. 94, no. 23, 12243–12248

Lewontin, R. (1982) "Agricultural research and the penetration of capital", *Science for the People*, vol. 14, no. 1, 12–17

Lewontin, R. (1991) *Biology as Ideology: The Doctrine of DNA*, HarperCollins, New York

Li, L., Li, S. M., Sun, J. H., Zhou, L. L., Bao, X. G., Zhang, H. G. and Zhang, F. S. (2007) "Diversity enhances agricultural productivity via rhizosphere phosphorus facilitation on phosphorus-deficient soils", *Proceedings of the National Academy of Sciences*, vol. 104, no. 27, 11192–11196

Liere, H. and Perfecto, I. (2008) "Cheating on a mutualism: indirect benefits of ant attendance to a coccidophagous coccinellid", *Environmental Entomology*, vol. 37, no. 1, 143–149

Liere, H. and Larsen, A. (2010) "Cascading trait-mediation: disruption of a trait-mediated mutualism by parasite-induced behavioral modification", *Oikos*, vol. 119, no. 9, 1394–1400

Liere, H., Jackson, D. and Vandermeer, J. (2012) "Ecological complexity in a coffee agroecosystem: spatial heterogeneity, population persistence and biological control", *PLOS ONE*, vol. 7, no. 9, e45508

Lin, B. B. (2007) "Agroforestry management as an adaptive strategy against potential microclimate extremes in coffee agriculture", *Agricultural and Forest Meteorology*, vol. 144, no.1, 85–94

Lin, B. B., Perfecto, I. and Vandermeer, J. (2008) "Synergies between agricultural intensification and climate change could create surprising vulnerabilities for crop production", *BioScience*, vol. 58, no. 9, 847–854

Lin, B. B., Chappell, M. J., Vandermeer, J., Smith, G. R., Quintero, E., Bezner-Kerr, R., . . . and Perfecto, I. (2011) "Effects of industrial agriculture on global warming and the potential of small-scale agroecological farming to mitigate those effects", *CAB Reviews: Perspectives in Agriculture, Veterinary Science, Nutrition and Natural Resources*, vol. 6, no. 20, 1–18

Lindeman, R. L. (1942) "The trophic-dynamic aspect of ecology", *Ecology*, vol. 23, no. 4, 399–417

Lindenmayer, D. B. and Franklin, J. F. (2002) *Conserving Forest Biodiversity: A Comprehensive Multiscaled Approach*, Island Press, Washington, DC

Livingston, G. F. and Philpott, S. M. (2010) "A metacommmunity approach to co-occurrence patterns and the core-satellite hypothesis in a community of tropical arboreal ants", *Ecological Research*, vol. 25, no. 6, 1129–1140

Lomeli-Flores, J. R., Barrera, J. F. and Bernal, J. S. (2009) "Impact of natural enemies on coffee leafminer *Leucoptera coffeella* (*Lepidoptera: Lyonetiidae*) population dynamics in Chiapas, Mexico", *Biological Control*, vol. 51, no. 1, 51–60

Longino, J. T. (2009) "Additions to the taxonomy of New World *Pheidole* (*Hymenoptera: Formicidae*)", *Zootaxa*, vol. 2181, 1–90

Loreau, M., Mouquet, N. and Gonzalez, A. (2003) "Biodiversity as spatial insurance in heterogeneous landscapes", *Proceedings of the National Academy of Sciences*, vol. 100, no. 22, 12765–12770

Lucier, G., Glaser, L. and Jerardo, A. (2011) "Dry bean crop report: record large black bean production", *Bean Commission News*, vol. 17, 2–3

Lundqvist, J., de Fraiture, C. and Molden, D. (2008) *Saving Water: From Field to Fork – Curbing Losses and Wastage in the Food Chain*, SIWI Policy Brief, Stockholm International Water Institute, Stockholm

Lyson, T. A. (2004) *Civic Agriculture: Reconnecting Farm, Food, and Community*, UPNE, Paris

Maas, B., Clough, Y. and Tscharntke, T. (2013) "Bats and birds increase crop yield in tropical agroforestry landscapes", *Ecology Letters*, vol. 16, no. 12, 1480–1487

MacArthur, R. and Levins, R. (1964) "Competition, habitat selection, and character displacement in a patchy environment", *Proceedings of the National Academy of Sciences of the United States of America*, vol. 51, no. 6, 1207–1210

MacArthur, R. H. and Wilson, E. O. (1967) *The Theory of Island Biogeography*, Princeton University Press, Princeton, NJ

McBratney, A., Whelan, B., Ancev, T. and Bouma, J. (2005) "Future directions of precision agriculture", *Precision Agriculture*, vol. 6, no. 1, 7–23

McCann, K. and Hastings, A. (1997) "Re-evaluating the omnivory–stability relationship in food webs", *Proceedings of the Royal Society of London. Series B: Biological Sciences*, vol. 264, no. 1385, 1249–1254

McCann, K. and Yodzis, P. (1998) "On the stabilizing role of stage structure in piscene consumer–resource interactions", *Theoretical Population Biology*, vol. 54, no. 3, 227–242

McCann, K., Hastings, A. and Huxel, G. R. (1998) "Weak trophic interactions and the balance of nature", *Nature*, vol. 395, no. 6704, 794–798

McCann, K. S., Rasmussen, J. B. and Umbanhowar, J. (2005) "The dynamics of spatially coupled food webs", *Ecology Letters*, vol. 8, no. 5, 513–523

McCook, L. J. (1999) "Macroalgae, nutrients and phase shifts on coral reefs: scientific issues and management consequences for the Great Barrier Reef", *Coral Reefs*, vol. 18, no. 4, 357–367

McCook, S. (2006). "Global rust belt: *Hemileia vastatrix* and the ecological integration of world coffee production since 1850", *Journal of Global History*, vol. 1, no. 2, 177–195

McFadden, S. (2008) *The History of Community Supported Agriculture Part II: CSA's World of Possibilities.* Available online at http://newfarm.rodaleinstitute.org/features/0204/csa2/part2.shtml (accessed June 22, 2014)

Macip-Ríos, R., and Muñoz-Alonso, A. (2008) "Diversidad de lagartijas en cafetales y bosque primario en el Soconusco chiapaneco", *Revista Mexicana de Biodiversidad*, vol. 79, no. 1, 185–195

McMichael, P. (2006) "Peasant prospects in the neoliberal age", *New Political Economy*, vol. 11, no. 3, 407–418

McShane, T. O. and Newby, S. A. (2004) "Expecting the unattainable: the assumptions behind ICDPs", in McShane, T. O. and Wells, M. P. (eds) *Getting Biodiversity Projects to Work*, Columbia University Press, New York, 49–74

Magalhaes, A. C. and Angelocci, L. R. (1976) "Sudden alterations in water balance associated with flower bud opening in coffee plants", *Journal of Horticultural Science*, vol. 51, no. 3, 419–421

Majer, J. D., LaSalle, J. and Gauld, I. D. (1993) "Comparison of the arboreal ant mosaic in Ghana, Brazil, Papua New Guinea and Australia: its structure and influence on arthropod diversity", in LaSalle, J. and Gauld, I. D. (eds) *Hymenoptera and Biodiversity*, CABI Publishing, Oxon, UK, 115–141

Mankiw, N. G. (2011) *Principles of Economics*, South-Western Cengage Learning, Mason, OH

Mann, C. C. (2005) *1491: New Revelations of the Americas before Columbus*, Random House Digital, Inc., New York

Manrique, A. J. and Thimann, R. E. (2002) "Coffee (*Coffea arabica*) pollination with africanized honeybees in Venezuela", *Interciencias*, vol. 27, no. 8, 414–416

Manson, R., Hernandez-Ortiz, V., Gallina, S. and Mehltreter, K. (eds) (2008) *Agroecosistemas Cafetaleros de Veracruz: Biodiversidad, Manejo y Conservación*, Instituto Nacional de Ecología, Mexico

Marquis, R. J. and Whelan, C. J. (1994) "Insectivorous birds increase growth of white oak through consumption of leaf-chewing insects", *Ecology*, vol. 75, no. 7, 2007–2014

Marsh, G. P. (1884) *Man and Nature or Physical Geography as Modified by Human Action*, Charles Scribner, New York

Marshall, C. (2008) "Costa Rica bids to go carbon neutral", *BBC News*, August 11. Available online at http://news.bbc.co.uk/2/hi/americas/7508107.stm (accessed June 22, 2014)

Martínez-Torres, M. E. (2006) *Organic Coffee: Sustainable Development by Mayan Farmers*, Ohio University Press, Columbus, OH

Martínez-Torres, M. E. and Rosset, P. M. (2010) "La Via Campesina: the birth and evolution of a transnational social movement", *The Journal of Peasant Studies*, vol. 37, no. 1, 149–175

Mas, A. H. and Dietsch, T. V. (2003) "An index of management intensity for coffee agro-ecosystems to evaluate butterfly species richness", *Ecological Applications*, vol. 13, no. 5, 1491–1501

Mas, A. H. and Dietsch, T. V. (2004) "Linking shade coffee certification to biodiversity conservation: butterflies and birds in Chiapas, Mexico", *Ecological Applications*, vol. 14, no. 3, 642–654

Mather, A. S. (1992) "The forest transition", *Area*, vol. 24, no. 4, 367–379

Mather, A. S. (2008) "Forest transition theory and the reforesting of Scotland", *Scottish Geographical Journal*, vol. 120, nos 1–2, 83–98

Mather, A. S. and Needle C. L. (1998) "The forest transition: a theoretical basis", *Area*, vol. 30, no. 2, 117–124

Mathis, K. A., Philpott, S. M. and Moreira, R. F. (2011) "Parasite lost: chemical and visual cues used by *Pseudacteon* in search of *Azteca instabilis*", *Journal of Insect Behavior*, vol. 24, no. 3, 186–199

Matthies, D., Bräuer, I., Maibom, W. and Tscharntke, T. (2004) "Population size and the risk of local extinction: empirical evidence from rare plants", *Oikos*, vol. 105, no. 3, 481–488

May, R. M. (1977) "Thresholds and breakpoints in ecosystems with a multiplicity of stable states", *Nature*, vol. 269, no. 5628, 471–477

May, R. M. (1990) "Taxonomy as destiny", *Nature*, vol. 347, no. 6289, 129–130

May, R. M. (2010) "Tropical arthropod species, more of less?", *Science*, vol. 329, no. 1257, 41–42

Mayne, W. W. (1935) *Annual Report of the Coffee Scientific Officer, 1933–34*, The Mysore Coffee Bulletin No. 12, Bangalore, India

Megahan, W. (1978) "Erosion processes on steep granitic road fills in central Idaho", *Soil Science Society of America Journal*, vol. 43, no. 2, 350–357

Mehltreter, K. (2008) "Helechos", in Manson, R., Hernández-Ortiz, V., Gallina, S. and Mehltreter, K. (eds) *Agroecosistemas Cafetaleros de Veracruz: Biodiversidad, Manejo y Conservacion*, Instituto Nacional de Ecología, Mexico, 83–93

Memmott, J., Martinez, N. D. and Cohen, J. E. (2000) "Predators, parasitoids and pathogens: species richness, trophic generality and body sizes in a natural food web", *Journal of Animal Ecology*, vol. 69, no. 1, 1–15

Mendenhall, C. D., Sekercioglu, C. H., Brenes, F. O., Ehrlich, P. R. and Daily, G. C. (2011) "Predictive model for sustaining biodiversity in tropical countryside", *Proceedings of the National Academy of Sciences*, vol. 108, no. 39, 16313–16316

Mendez, V. E., Gliessman, S. R. and Gilbert, G. S. (2007) "Tree biodiversity in farmer cooperatives of a shade coffee landscape in western El Salvador", *Agriculture, Ecosystems and Environment*, vol. 119, no. 1, 145–159

Mendoza, V. M., Villanueva, E. E. and Adem, J. (1997) "Vulnerability of basins and watersheds in Mexico to global climate change", *Climate Research*, vol. 9, nos 1–2, 139–145

Mera Velasco, Y. A., Gallego Ropero, M. C. and Armbrecht, I. (2010) "Interactions between ants and insects in foliage of sun and shade coffee plantations, Cauca-Colombia", *Revista Colombiana de Entomología*, vol. 36, no. 1, 116–126

Meszéna, G., Gyllenberg, M., Pásztor, L. and Metz, J. A. (2006) "Competitive exclusion and limiting similarity: a unified theory", *Theoretical Population Biology*, vol. 69, no. 1, 68–87

Meyling, N. and Eilenberg, J. (2007) "Ecology of the entomopathogenic fungi *Beauveria bassiana* and *Metarhizium anisopliae* in temperate agroecosystems: potential for conservation biology control", *Biological Control*, vol. 43, no. 2, 145–155

Michener, C. D. (2000) *The Bees of the World, Vol. 1*, Johns Hopkins University Press, Baltimore, MA

Milder, J. C., Scherr, S. J. and Bracer, C. (2010) "Trends and future potential of payment for ecosystem services to alleviate rural poverty in developing countries", *Ecology and Society*, vol. 15, no. 2, art. 4. Available online at http://www.ecologyandsociety.org/vol15/iss2/art4/ (accessed June 24, 2014)

Millennium Ecosystem Assessment (2005) *Ecosystems and Human Well-Being: Biodiversity Synthesis*, World Resources Institute, Washington, DC

Miller, J. H. and Page, S. E. (2007) *Complex Adaptive Systems: An Introduction to Computational Models*, Princeton University Press, Princeton, NJ

Moguel, P. and Toledo, V. M. (1999) "Biodiversity conservation in traditional coffee systems of Mexico", *Conservation Biology,* vol. 13, no. 1, 11–21

Molina, J. (2000) "Diversidad de escarabajos coprófagos (*Scarabaeidae: Scarabaeinae*) en matrices de la zona cafetera (Quindío-Colombia)", in Federación Nacional de Cafeteros de Colombia (eds) *Memorias Foro Internacional Café y Biodiversidad, Agosto 10–12, Chinchiná, Colombia*

Mols, C. M. and Visser, M. E. (2002) "Great tits can reduce caterpillar damage in apple orchards", *Journal of Applied Ecology*, vol. 39, no. 6, 888–899

Mondragón, D., Santos-Moreno, A. and Damon, A. (2009) "Epiphyte diversity on coffee bushes: a management question? *Journal of Sustainable Agriculture*, vol. 33, no. 7, 703–715

Montoya, J. M. and Solé, R. V. (2003) "Topological properties of food webs: from real data to community assembly models", *Oikos*, vol. 102, no. 3, 614–622

Montoya, L. A. and Sylvain, P. G. (1962) "Aplicación de soluciones de azúcar en aspersions folliares, para prevenir la caída premature de grano verde del café", *Turrialba*, vol. 12, no. 2, 100–101

Mooney, H. A. and Ehrlich, P. R. (1997) "Ecosystem services: a fragmentary history", in Daily, G. C. (ed.) *Nature's Services: Societal Dependence on Natural Ecosystems*, Island Press, Washington, DC, 11–22

Mooney, K. A., Gruner, D. S., Barber, N. A., Van Bael, S. A., Philpott, S. M. and Greenberg, R. (2010) "Interactions among predators and the cascading effects of vertebrate insectivores on arthropod communities and plants", *Proceedings of the National Academy of Sciences*, vol. 107, no. 16, 7335–7340

Moore, A. M. T. and Hillman, G. C. (1992) "The Pleistocene to Holocene transition and human economy in southwest Asia: the impact of the Younger Dryas", *American Antiquity*, vol. 57, no. 3, 482–494

Moorhead, L. C., Philpott, S. M. and Bichier, P. (2010) "Epiphyte biodiversity in the coffee agricultural matrix: canopy stratification and distance from forest fragments", *Conservation Biology*, vol. 24, no. 3, 737–746

Mora, C., Tittensor, D. P., Adl, S., Simpson, A. G. B. and Worm, B. (2011) "How many species are there on Earth and in the ocean?", *PLOS Biology*, vol. 9, no. 8, e1001127

Morales, H. E. and Perfecto, I. (2000) "Traditional knowledge and pest management in the Guatemalan highlands", *Agriculture and Human Values*, vol. 17, no. 1, 49–63

Moron, M. A. and López-Méndez, J. A. (1985) "Analysis of the necrophilic entomofauna in a coffee plantation in the Soconusco, Chiapas, Mexico", *Folia Entomológica Mexicana*, vol. 63, 47–59

Morozov, A. and Poggiale, J. C. (2012) "From spatially explicit ecological models to mean-field dynamics: the state of the art and perspectives", *Ecological Complexity*, vol. 10, 1–11

Morrison, L. W. (1999) "Indirect effects of phorid fly parasitoids on the mechanisms of interspecific competition among ants", *Oecologia*, vol. 121, no. 1, 113–122

Morrison, L. W. (2002) "Long-term impacts of an arthropod-community invasion by the imported fire ant, *Solenopsis invicta*", *Ecology*, vol. 83, no. 8, 2337–2345

Morrison, L. W., Kawazoe E. A., Guerra, R. Gilbert, L. E. (2000) "Ecological interactions of *Pseudacteon* parasitoids and *Solenopsis* ant hosts: environmental correlates of activity and effects on competitive hierarchies", *Ecological Entomology*, vol. 25, no. 4, 433–444

Müller, C. B. and Brodeur, J. (2002) "Intraguild predation in biological control and conservation biology", *Biological Control*, vol. 25, no. 3, 216–223

Munyuli, T. (2011) "Factors governing flower visitation patterns and quality of pollination services delivered by social and solitary bee species to coffee in central Uganda", *African Journal of Ecology*, vol. 49, no. 4, 501–509

Murdoch, W. W., Briggs, C. J. and Nisbet, R. M. (2013) *Consumer–Resource Dynamics*, Princeton University Press, Princeton, NJ

Murrieta Galindo, R., González-Romero, A., López-Barrera, F. and Parra-Olea, G. (2013a) "Coffee agrosystems: an important refuge for amphibians in central Veracruz, Mexico", *Agroforestry Systems*, vol. 87, no. 4, 1–13

Murrieta Galindo, R., López-Barrera, F., González-Romero, A. and Parra-Olea, G. (2013b) "Matrix and habitat quality in a montane cloud-forest landscape: amphibians in coffee plantations in central Veracruz, Mexico", *Wildlife Research*, vol. 40, no. 1, 25–35

Mutersbaugh, T. (2002) "The number is the beast: a political economy of organic-coffee certification and producer unionism", *Environment and Planning A*, vol. 34, no. 7, 1165–1184

Myers, G. (2004) *Banana Wars: The Price of Free Trade, A Caribbean Perspective*, Zed Books, London

Myers, N., Mittermeier, R. A., Mittermeier, C. G., da Fonseca, G. A. B. and Kent, J. (2000) "Biodiversity hotspots for conservation priorities", *Nature*, vol. 403, no. 6772, 853–858

Mylius, S. D., Klumpers, K., de Roos, A. M. and Persson, L. (2001) "Impact of intraguild predation and stage structure on simple communities along a productivity gradient", *The American Naturalist*, vol. 158, no. 3, 259–276

Namba, T., Tanabe, K. and Maeda, N. (2008) "Omnivory and stability of food webs", *Ecological Complexity*, vol. 5, no. 2, 73–85

Nellemann, C. (ed.) (2009) *The Environmental Food Crisis: The Environment's Role in Averting Future Food Crises: A UNEP Rapid Response Assessment*, UNEP/Earthprint, New York

Nestel, D., Dickschen, F. and Altieri, M. A. (1993) "Diversity patterns of soil macro-coleoptera in Mexican shaded and unshaded coffee agroecosystems: an indication of habitat perturbation", *Biodiversity and Conservation*, vol. 2, no. 1, 70–78

Neuhauser, C. and Pacala, S. W. (1999) "An explicitly spatial version of the Lotka–Volterra model with interspecific competition", *Annals of Applied Probability*, vol. 9, no. 4, 1226–1259

Newman, M. (2005) "Power laws, Pareto distributions and Zipf's law", *Contemporary Physics*, vol. 46, no. 5, 323–351

Newman, M. E. and Palmer, R. G. (2003) *Modeling Extinction*, Oxford University Press, Oxford

Newmark, W. D. (1995) "Extinction of mammal populations in western North American national parks", *Conservation Biology*, vol. 9, no. 3, 512–526

Nicholson, A. J. and Bailey, V. A. (1935) "The balance of animal populations", *Proceedings of the Zoological Society of London*, vol. 1935, 551–598

Nicholson, W. (2001) *Microeconomic Theory: Basic Principles and Extensions*, eighth edition, South-Western College Pub, Mason, OH

Nir, M. A. (1988) "The survivors: orchids on a Puerto Rican coffee finca", *American Orchid Society Bulletin*, vol. 57, 989–995

Noy-Meir, I. (1975) "Stability of grazing systems: an application of predator–prey graphs", *Journal of Ecology*, vol. 63, no. 2, 459–481

Numa, C., Verdú, J. R. and Sánchez-Palomino, P. (2005) "Phyllostomid bat diversity in a variegated coffee landscape", *Biological Conservation*, vol. 122, no. 1, 151–158

Nuñes, M. A., Bierhuizen, J. F. and Ploegman, C. (1968) "Studies on the productivity of coffee, I: effect of light, temperature, and CO_2 concentration on photosynthesis of *Coffea arabica*", *Acta Botanica Neerlandica*, vol. 17, no. 2, 93–102

O'Brian, T. G. and Kinnaird, M. F. (2003) "Caffeine and conservation", *Science*, vol. 300, no. 5619, 587

Oaten, A. and Murdoch, W. W. (1975) "Switching, functional response, and stability in predator–prey systems", *The American Naturalist*, vol. 109, no. 967, 299–318

Odling-Smee, F. J., Laland, K. N. and Feldman, M. W. (2003) *Niche Construction: The Neglected Process in Evolution*, Princeton University Press, Princeton, NJ

Oerlemans, O. (2004) *Romanticism and the Materiality of Nature*, University of Toronto Press, Toronto

Oksanen, L. and Oksanen, T. (2000) "The logic and realism of the hypothesis of exploitation ecosystems", *The American Naturalist*, vol. 155, no. 6, 703–723

Olschewski, R., Tscharntke, T., Benítez, P. C., Schwarze, S. and Klein, A. M. (2006) "Economic evaluation of pollination services comparing coffee landscapes in Ecuador and Indonesia", *Ecology and Society*, vol. 11, no. 1, art. 7. Available online at www.ecologyandsociety.org/vol11/iss1/art7/ (accessed August 6, 2014)

Oram, P. A. (1989) "Sensitivity of agricultural production to climatic change: an update", in *Climate and Food Security*, International Symposium on Climate Variability and Food Security in Developing Countries, New Delhi (India), February 5–9 1987, organized by American Association for the Advancement of Science, Indian National Science Academy, and the International Rice Research Institute, 25–44

Orindi, V. A. and Ochieng, A. (2005) "Case study 5: Kenya seed fairs as a drought recovery strategy in Kenya", *IDS Bulletin*, vol. 36, no. 4, 87–102

Osborn, H. F. (1948) *Our Plundered Planet*, Little Brown, Boston, MA

Otero-Jiménez, B. (2013) "The effect of matrix composition in agroecosystems: assessing population structure in *Heteromys* mice", MSc thesis, University of Michigan, Ann Arbor, MI

Ottonetti, A. L., Tucci, L., Frizzi, F., Chelazzi, G. and Santini, G. (2010) "Changes in ground-foraging ant assemblages along a disturbance gradient in a tropical agricultural landscape", *Ethology, Ecology and Evolution*, vol. 22, no. 1, 73–86

Ovadia, O. and Schmitz, O. J. (2002) "Linking individuals with ecosystems: experimentally identifying the relevant organizational scale for predicting trophic abundances", *Proceedings of the National Academy of Sciences*, vol. 99, no. 20, 12927–12931

Pagiola, S. and Ruthenberg, I. M. (2002) "'Selling biodiversity in a coffee cup: shade-grown coffee and conservation in Mesoamerica", in Pagiola, S., Bishop, J. and Landell-Mills, N. (eds) *Selling Forest Environmental Services: Market-Based Mechanisms for Conservation and Development*, Earthscan, London, 103–126

Paine, R. T. (1969) "A note on trophic complexity and community stability", *The American Naturalist*, vol. 103, no. 929, 91–93

Paine, R. T. (1980) "Food webs: linkage, interaction strength and community infrastructure", *Journal of Animal Ecology*, vol. 49, no. 3, 667–685

Pardee, G. L. and Philpott, S. M. (2011) "Cascading indirect effects in a coffee agroecosystem: effects of parasitic phorid flies on ants and the coffee berry borer in a high-shade and low-shade habitat", *Environmental Entomology*, vol. 40, no. 3, 581–588

Parr, C. L and Gibb, H. (2010) "Competition and the role of dominant ants", in Lach, L., Parr, C. L. and Abbot, K. L. (eds) *Ant Ecology*, Oxford University Press, Oxford, 77–96

Parrish, L. J. and Petit, D. R. (1995) "Evaluating the importance of human-modified lands for neotropical bird conservation", *Conservation Biology*, vol. 17, no. 3, 687–694

Parry, M. L., Canziani, O. F., Palutikof, J. P., van der Linden, P. J. and Hanson, C. E. (eds) (2007) *Contribution of Working Group II to the Fourth Assessment Report of the Intergovernmental Panel on Climate Change*, Cambridge University Press, Cambridge

Pascual, M. (1993) "Diffusion-induced chaos in a spatial predator–prey system", *Proceedings of the Royal Society of London. Series B: Biological Sciences*, vol. 251, no. 1330, 1–7

Pascual, M. and Levin, S. A. (1999a) "From individuals to population densities: searching for the intermediate scale of nontrivial determinism", *Ecology*, vol. 80, no. 7, 2225–2236

Pascual, M. and Levin, S. A. (1999b) "Spatial scaling in a benthic population model with density-dependent disturbance", *Theoretical Population Biology*, vol. 56, no. 1, 106–122

Pascual, M. and Dunne, J. A. (eds) (2005) *Ecological Networks: Linking Structure to Dynamics in Food Webs*, Oxford University Press, Oxford

Pascual, M. and Guichard, F. (2005) "Criticality and disturbance in spatial ecological systems", *Trends in Ecology and Evolution*, vol. 20, no. 2, 88–95

Pascual, M., Roy, M. and Franc, A. (2002a) "Simple temporal models for ecological systems with complex spatial patterns", *Ecology Letters*, vol. 5, no. 3, 412–419

Pascual, M., Roy, M., Guichard, F. and Flierl, G. (2002b) "Cluster size distributions: signatures of self-organization in spatial ecologies", *Philosophical Translations of the Royal Society of London B,* vol. 357, no. 1421, 657–666

Pendergast, M. (1999) *Uncommon Grounds: The History of Coffee and How It Transformed Our World*, Basic Books, New York

Perez, U. M. (2005) "Temen por 4 mil jornaleros y por la cosecha de café", *La Jornada*, Mexico City, October 12

Perfecto, I. (1991) "Dynamics of *Solenopsis geminata* in a tropical fallow field after ploughing", *Oikos*, vol. 62, no. 2, 139–144

Perfecto, I. (1992) "Observations of a *Labidus coecus* (Latreille) underground raid in the central highlands of Costa Rica", *Psyche: A Journal of Entomology*, vol. 99, nos 2–3, 214–220

Perfecto, I. (1994) "Foraging behavior as a determinant of asymmetric competitive interaction between two ant species in a tropical agroecosystem", *Oecologia*, vol. 98, no. 2, 184–192

Perfecto, I. and Snelling, R. (1995) "Biodiversity and the transformation of a tropical agroecosystem: ants in coffee plantations", *Ecological Applications*, vol. 5, no. 4, 1084–1097

Perfecto, I. and Vandermeer, J. (1996) "Microclimatic changes and the indirect loss of ant diversity in a tropical agroecosystem", *Oecologia*, vol. 108, no. 3, 577–582

Perfecto, I. and Castiñeiras, A. (1998) "Deployment of the predaceous ants and their conservation in agroecosystems" in Barbosa, P. (ed.) *Conservation Biological Control*, Academic Press, San Diego, CA, 269–289

Perfecto, I. and Vandermeer, J. (2002) "The quality of the agroecological matrix in a tropical montane landscape: ants in coffee plantations in southern Mexico", *Conservation Biology*, vol. 16, no. 1, 174–182

Perfecto, I. and Armbrecht, I. (2003) "The coffee agroecosystem in the Neotropics: combining ecological and economic goals", in Vandermeer, J. (ed.) *Tropical Agroecosystems*, CRC Press, Boca Raton, FL, 159–194

Perfecto, I. and Vandermeer, J. (2006) "The effect of an ant–hemipteran mutualism on the coffee berry borer (*Hypothenemus hampei*) in southern Mexico", *Agriculture, Ecosystems and Environment*, vol. 117, no. 2, 218–221

Perfecto, I. and Vandermeer, J. (2008a) "Biodiversity conservation in tropical agroecosystems", *Annals of the New York Academy of Sciences*, vol. 1134, no. 1, 173–200

Perfecto, I. and Vandermeer, J. (2008b) "Spatial pattern and ecological process in the coffee agroecosystem", *Ecology*, vol. 89, no. 4, 915–920

Perfecto, I. and Vandermeer, J. (2010) "The agroecological matrix as alternative to the land-sparing/agriculture intensification model", *Proceedings of the National Academy of Sciences*, vol. 107, no. 13, 5786–5791

Perfecto, I. and Vandermeer, J. (2011) "Discovery dominance tradeoff: the case of *Pheidole subarmata* and *Solenopsis geminata* (*Hymenoptera: Formicidae*) in neotropical pastures", *Environmental Entomology*, vol. 40, no. 5, 999–1006

Perfecto, I. and Vandermeer, J. (2013) "Ant assemblage on a coffee farm: spatial mosaic versus shifting patchwork", *Environmental Entomology*, vol. 42, no. 1, 38–48

Perfecto, I. and Vandermeer, J. (in press) "Ant community structure and the ecosystem service of pest control in a coffee agroecosystem in southern Mexico", *Entomologia Neotropical*

Perfecto, I., Rice, R. A., Greenberg, R. and Van der Voort, M. E. (1996) "Shade coffee: a disappearing refuge for biodiversity", *BioScience*, vol. 46, no. 8, 598–608

Perfecto, I., Vandermeer, J., Hanson, P. and Cartín, V. (1997) "Arthropod biodiversity loss and the transformation of a tropical agro-ecosystem", *Biodiversity and Conservation*, vol. 6, no. 7, 935–945

Perfecto, I., Mas, A., Dietsch, T. and Vandermeer, J. (2003) "Conservation of biodiversity in coffee agroecosystems: a tri-taxa comparison in southern Mexico", *Biodiversity and Conservation*, vol. 12, no. 6, 1239–1252

Perfecto, I., Vandermeer, J. H., Bautista, G. L., Nunez, G. I., Greenberg, R., Bichier, P. and Langridge, S. (2004) "Greater predation in shaded coffee farms: the role of resident neotropical birds", *Ecology*, vol. 85, no. 10, 2677–2681

Perfecto, I., Vandermeer, J., Mas, A. and Pinto, L. S. (2005) "Biodiversity, yield, and shade coffee certification", *Ecological Economics*, vol. 54, no. 4, 435–446

Perfecto, I., Armbrecht, I., Philpott, S. M., Soto-Pinto, L. and Dietsch, T. V. (2007) "Shaded coffee and the stability of rainforest margins in northern Latin America", in Tscharntke, T., Leuschner, C., Zeller, M., Guherdja, E. and Bidin, A. (eds) *Stability of Tropical Rainforest Margins*, Springer-Verlag, Berlin, 225–261

Perfecto, I., Vandermeer, J. and Wright, A. (2009) *Nature's Matrix: Linking Conservation, Agriculture, Conservation and Food Sovereignty*, Earthscan, London

Persson, L. (1999) "Trophic cascades: abiding heterogeneity and the trophic level concept at the end of the road", *Oikos*, vol. 85, 385–397

Petit, L. J., Petit, D. R., Christian, D. G. and Powell, H. D. (1999) "Bird communities of natural and modified habitats in Panama", *Ecography*, vol. 22, no. 3, 292–304

Pfaff, A. S. (2001) "From deforestation to reforestation in New England, United States", in Palo, M. and Vanhanen, H. (eds) *World Forests from Deforestation to Transition?*, Klewer Academic Publisher, Dordrecht, Netherlands, 67–82

Phalan, B., Balmford, A., Green, R. E. and Scharlemann, J. P. (2011a) "Minimising the harm to biodiversity of producing more food globally", *Food Policy*, vol. 36, supp. 1, S62–S71

Phalan, B., Onial, M., Balmford, A. and Green, R. (2011b) "Reconciling food production and biodiversity conservation: land sharing and land sparing compared", *Science*, vol. 333, no. 6047, 1289–1291

Philibert, J. (2005) "One and a half century of diffusion: Fick, Einstein, before and beyond", *Diffusion Fundamentals*, vol. 2, no. 1, 1–10

Philpott, S. M. (2005a) "Changes in arboreal ant populations following pruning of coffee shade-trees in Chiapas, Mexico", *Agroforestry Systems*, vol. 64, no. 3, 219–224

Philpott, S. M. (2005b) "Trait-mediated effects of parasitic phorid flies (*Diptera: Phoridae*) on ant (*Hymenoptera: Formicidae*) competition and resource access in coffee agro-ecosystems", *Environmental Entomology*, vol. 34, no. 5, 1089–1094

Philpott, S. M. (2010) "A canopy dominant ant affects twig-nesting ant assembly in coffee agroecosystems", *Oikos*, vol. 119, no. 12, 1954–1960

Philpott, S. M. and Dietsch, T. (2003) "Coffee and conservation: a global context and the value of farmer involvement", *Conservation Biology*, vol. 17, no. 6, 1844–1846

Philpott, S. M. and Foster, P. F. (2005) "Nest-site limitation in coffee agroecosystems: artificial nests maintain diversity of arboreal ants", *Ecological Applications*, vol. 15, no. 4, 1478–1485

Philpott, S. M. and Armbrecht, I. (2006) "Biodiversity in tropical agroforests and the ecological role of ants and ant diversity in predatory function", *Ecological Entomology*, vol. 31, no. 4, 369–377

Philpott, S. M. and Bichier, P. (2012) "Effects of shade tree removal on birds in coffee agroecosystems in Chiapas, Mexico", *Agriculture, Ecosystems and Environment*, vol. 149, 171–180

Philpott, S. M., Greenberg, R., Bichier, P. and Perfecto, I. (2004) "Impacts of major predators on tropical agroforest arthropods: comparisons within and across taxa", *Oecologia*, vol. 140, no. 1, 140–149

Philpott, S. M., Perfecto, I. and Vandermeer, J. (2006a) "Effects of management system and season on arboreal ant diversity and abundance in coffee agroecosystems", *Biodiversity and Conservation*, vol. 15, no. 1, 139–155

Philpott, S. M., Uno, S. and Maldonado, J. (2006b) "The importance of ants and high-shade management to coffee pollination and fruit weight in Chiapas, Mexico", *Biodiversity and Conservation*, vol. 15, no. 1, 487–501

Philpott, S. M., Bichier, P., Rice, R. A. and Greenberg, R. (2007) "Field-testing ecological and economic benefits of coffee certification programs", *Conservation Biology*, vol. 21, no. 4, 975–985

Philpott, S. M., Arendt, W. J., Armbrecht, I., Bichier, P., Dietsch, T. V., Gordon, C., . . . and Zolotoff, J. (2008a) "Biodiversity loss in Latin American coffee landscapes: review of the evidence on ants, birds, and trees", *Conservation Biology*, vol. 22, no. 5, 1093–1105

Philpott, S. M., Bichier, P., Rice, R. A. and Greenberg, R. (2008b) "Biodiversity conservation, yield, and alternative products in coffee agroecosystems in Sumatra, Indonesia", *Biodiversity and Conservation*, vol. 17, no. 8, 1805–1820

Philpott, S. M., Lin, B. B., Jha, S. and Brines, S. J. (2008c) "A multi-scale assessment of hurricane impacts on agricultural landscapes based on land use and topographic features", *Agriculture, Ecosystems and Environment*, vol. 128, no. 1, 12–20

Philpott, S. M., Soong, O., Lowenstein, J. H., Pulido, A. L., Lopez, D. T., Flynn, D. F. and DeClerck, F. (2009) "Functional richness and ecosystem services: bird predation on arthropods in tropical agroecosystems", *Ecological Applications*, vol. 19, no. 7, 1858–1867

Philpott, S. M., Pardee, G. L. and Gonthier, D. J. (2012) "Cryptic biodiversity effects: importance of functional redundancy revealed through addition of food web complexity", *Ecology*, vol. 93, no. 5, 992–1001

Pimentel, D., Berardi, G. and Fast, S. (1983) "Energy efficiency of farming systems: organic and conventional agriculture", *Agriculture, Ecosystems and Environment*, vol. 9, no. 4, 359–372

Pimentel, D., Hepperly, P., Hanson, J., Douds, D. and Seidel, R. (2005) "Environmental, energetic, and economic comparisons of organic and conventional farming systems", *BioScience*, vol. 55, no. 7, 573–582

Pimm, S. L. (1982) *Food Webs*, Springer-Verlag, Netherlands

Pimm, S. L. and Lawton, J. H. (1978) "On feeding on more than one trophic level", *Nature*, vol. 275, no. 5680, 542–544

Pimm, S. L. and Rice, J. C. (1987) "The dynamics of multispecies, multi-life-stage models of aquatic food webs", *Theoretical Population Biology*, vol. 32, no. 3, 303–325

Pineda, E. and Halffter, G. (2004) "Species diversity and habitat fragmentation: frogs in a tropical montane landscape in Mexico", *Biological Conservation*, vol. 117, no. 5, 499–508

Pineda, E., Moreno, C., Escobar, F. and Halffter, G. (2005) "Frog, bat, and dung beetle diversity in the cloud forest and coffee agroecosystems of Veracruz, Mexico", *Conservation Biology*, vol. 19, no. 2, 400–410

Polis, G. A. and Holt, R. D. (1992) "Intraguild predation: the dynamics of complex trophic interactions", *Trends in Ecology and Evolution*, vol. 7, no. 5, 151–154

Polis, G. A. and Strong, D. R. (1996) "Food web complexity and community dynamics", *The American Naturalist*, vol. 147, no. 5, 813–846

Polis, G. A., Myers, C. A. and Holt, R. D. (1989) "The ecology and evolution of intraguild predation: potential competitors that eat each other", *Annual Review of Ecology and Systematics*, vol. 20, 297–330

Porter, S. D., Van Eimeren, B. and Gilbert, L. E. (1988) "Invasion of red imported fire ants (*Hymenoptera: Formicidae*): microgeography of competitive replacement", *Annals of the Entomological Society of America*, vol. 81, no. 6, 913–918

Post, D. M. (2002) "Using stable isotopes to estimate trophic position: models, methods, and assumptions", *Ecology*, vol. 83, no. 3, 703–718

Pretty, J. (2008) "Agricultural sustainability: concepts, principles and evidence", *Philosophical Transactions of the Royal Society B: Biological Sciences*, vol. 363, no. 1491, 447–465

Pretty, J. N., Noble, A. D., Bossio, D., Dixon, J., Hine, R. E., Penning de Vries, F. W. T. and Morison, J. I. L. (2006) "Resource-conserving agriculture increases yields in developing countries", *Environmental Science and Technology*, vol. 40, no. 4, 1114–1119

Pretty, J., Toulmin, C. and Williams, S. (2011) "Sustainable intensification in African agriculture", *International Journal of Agricultural Sustainability*, vol. 9, no. 1, 5–24

Pulliam, H. R. (1988) "Sources, sinks, and population regulation", *The American Naturalist*, vol. 132, no. 5, 652–661

Railsback, S. F. and Johnson, M. D. (2011) "Pattern-oriented modeling of bird foraging and pest control in coffee farms", *Ecological Modelling*, vol. 222, no. 18, 3305–3319

Raman, T. R. S. (2006) "Effects of habitat structure and adjacent habitats on birds in tropical rainforest fragments and shaded plantations in the Western Ghats, India", *Biodiversity and Conservation*, vol. 15, no. 4, 1577–1607

Ranganathan, J., Daniels, R. R., Chandran, M. S., Ehrlich, P. R., and Daily, G. C. (2008) "Sustaining biodiversity in ancient tropical countryside", *Proceedings of the National Academy of Sciences*, vol. 105, no. 46, 17852–17854

Rappole, J. H., King, D. I. and Rivera, J. H. V. (2003) "Coffee and conservation", *Conservation Biology*, vol. 17, no. 1, 334–336

Raup, D. M. (1994) "The role of extinction in evolution", *Proceedings of the National Academy of Sciences*, vol. 91, no. 15, 6758–6763

Raynolds, L. T., Murray, D. and Taylor, P. L. (2004) "Fair trade coffee: building producer capacity via global networks", *Journal of International Development*, vol. 16, no. 8, 1109–1121

Raynolds, L. T., Murray, D. and Heller, A. (2007) "Regulating sustainability in the coffee sector: a comparative analysis of third-party environmental and social certification initiatives", *Agriculture and Human Values*, vol. 24, no. 2, 147–163

Relyea, R. A. (2005) "The lethal impact of Roundup on aquatic and terrestrial amphibians", *Ecological Applications*, vol. 15, no. 4, 1118–1124

Renard, M. C. (2003) "Fair trade: quality, market and conventions", *Journal of Rural Studies*, vol. 19, no. 1, 87–96

Rice, R. A. (1999) "A place unbecoming: the coffee farm of northern Latin America", *Geographical Review*, vol. 89, no. 4, 554–579

Rice, R. A. and Ward, J. (1996) *Coffee, Conservation, and Commerce in the Western Hemisphere*, The Smithsonian Migratory Bird Center and the Natural Resources Defense Council, Washington, DC

Richter, A., Klein, A. M., Tscharntke, T. and Tylianakis, J. M. (2007) "Abandonment of coffee agroforests increases insect abundance and diversity", *Agroforestry Systems*, vol. 69, no. 3, 175–182

Ricketts, T. H. (2004) "Tropical forest fragments enhance pollinator activity in nearby coffee crops", *Conservation Biology*, vol. 18, no. 5, 1262–1271

Ricketts, T. H., Daily, G. C., Ehrlich, P. R. and Michener, C. D. (2004) "Economic value of tropical forest to coffee production", *Proceedings of the National Academy of Sciences of the United States of America*, vol. 101, no. 34, 12579–12582

Ricketts, T. H., Regetz, J., Steffan-Dewenter, I., Cunningham, S. A., Kremen, C., Bogdanski, A., . . . and Viana, B. F. (2008) "Landscape effects on crop pollination services: are there general patterns?", *Ecology Letters*, vol. 11, no. 5, 499–515

Riechert, S. E. and Lockley, T. (1984) "Spiders as biological control agents", *Annual Review of Entomology*, vol. 29, no. 1, 299–320

Riechert, S. E. and Hedrick, A. V. (1990) "Levels of predation and genetically based anti-predator behavior in the spider, *Agelenopsis aperta*", *Animal Behaviour*, vol. 40, no. 4, 679–687

Rietkerk, M. and van de Koppel, J. (2008) "Regular pattern formation in real ecosystems", *Trends in Ecology and Evolution*, vol. 23, no. 3, 169–175

Rietkerk, M., Boerlijst, M., van Langevelde, F., HilleRisLambers, R., de Koppel, J., Kumar, L., Prins, H. and de Roos, A. (2002) "Self-organization of vegetation in arid ecosystems", *The American Naturalist*, vol. 160, no. 4, 524–530

Rival, L. (2002) *Trekking Through History: The Huaorani of the Amazonian Ecuador*, Columbia University Press, New York

Robbins, C. S., Sauer, J. R., Greenberg, R. S. and Droege, S. (1989) "Population declines in North American birds that migrate to the Neotropics", *Proceedings of the National Academy of Sciences*, vol. 86, no. 19, 7658–7662

Robbins, C. S., Dowell, B. A., Dawson, D. K., Colón, J. A., Estrada, R., Sutton, A., Sutton, R. and Weyer, D. (1992) "Comparison of neotropical migrant landbird populations wintering in tropical forest, isolated forest fragments, and agricultural habitats", in Hagan, J. M. III and Johnston, D. W. (eds) *Ecology and Conservation of Neotropical Migrant Landbirds*, Smithsonian Institution Press, Washington, DC, 207–220

Roberts, D. L., Cooper, R. L. and Petit, L. J. (2000) "Flock characteristics of ant-following birds in premontane moist forest and coffee agroecosystems", *Ecological Applications*, vol. 10, no. 5, 414–1425

Robinson, R. A. and Sutherland, W. J. (2002) "Post-war changes in arable farming and biodiversity in Great Britain", *Journal of Applied Ecology*, vol. 39, no. 1, 157–176

Rocha, M. F., Passamani, M. and Louzada, J. (2011) "A small mammal community in a forest fragment, vegetation corridor and coffee matrix system in the Brazilian Atlantic Forest", *PLOS ONE*, vol. 6, no. 8, e23312

Roditakis, E., Couzin, I., Balrow, K. and Franks, N. (2000) "Improving secondary pick-up of insect fungal pathogen conidia by manipulating host behavior", *Annals of Applied Biology*, vol. 137, no. 3, 329–335

Rodríguez, J. M. (2004) *Informe Final: Bosque, Pago de Servicios Ambientales e Industria Foresta [Final Report: Forest, Payment for Environmental Services and Forest Industries]*, Comisión Nacional de Rectores and Defensoria de los Habitantes, San José, Costa Rica

Roebeling, P. and Ruben, R. (2001) "Technological progress versus economic policy as tools to control deforestation: the Atlantic Zone of Costa Rica", in Angelsen, A. and Kaimowitz, D. (eds) *Agricultural Technologies and Tropical Deforestation*, CABI Publishing, Oxon, UK, 135–152

Rohani, P. and Miramontes, O. (1995) "Host–parasitoid metapopulations: the consequences of parasitoid aggregation on spatial dynamics and searching efficiency", *Proceedings of the Royal Society of London. Series B: Biological Sciences*, vol. 260, no. 1359, 335–342

Rohani, P., Godfray, H. C. J. and Hassell, M. P. (1994) "Aggregation and the dynamics of host–parasitoid systems: a discrete-generation model with within-generation redistribution", *The American Naturalist*, vol. 144, no. 3, 491–509

Rohani, P., Lewis, T. J., Grünbaum, D. and Ruxton, G. D. (1997) "Spatial self-organisation in ecology: pretty patterns or robust reality?", *Trends in Ecology and Evolution*, vol. 12, no. 2, 70–74

Rojas, L., Godoy, C., Hanson, P., Kleinn, C. and Hilje, L. (2001) "Hopper (*Homoptera*: *Auchenorrhyncha*) diversity in shaded coffee systems of Turrialba, Costa Rica", *Agroforestry Systems*, vol. 53, no. 2, 171–177

Rooney, T. P., Wiegmann, S. M., Rogers, D. A. and Waller, D. M. (2004) "Biotic impoverishment and homogenization in unfragmented forest understory communities", *Conservation Biology*, vol. 18, no. 3, 787–798

Rosenheim, J. A., Kaya, H. K., Ehler, L. E., Marois, J. J. and Jaffee, B. A. (1995) "Intraguild predation among biological-control agents: theory and evidence", *Biological Control*, vol. 5, no. 3, 303–335

Rosset, P. M. and Martínez-Torres, M. E. (2012) "Rural social movements and agroecology: context, theory, and process", *Ecology and Society*, vol. 17, no. 3, art. 17. Available online at www.ecologyandsociety.org/vol17/iss3/art17/ (accessed August 6, 2014)

Roubik, D. D. W. (ed.) (1995) *Pollination of Cultivated Plants in the Tropics*, FAO, Rome

Roubik, D. D. W. (2002) "Tropical agriculture: the value of bees to the coffee harvest", *Nature*, vol. 417, no. 6890, 708–708

Rundlöf, M., Bengtsson, J. and Smith, H. G. (2008) "Local and landscape effects of organic farming on butterfly species richness and abundance", *Journal of Applied Ecology*, vol. 45, no. 3, 813–820

Russell, E. (2001) *War and Nature: Fighting Humans and Insects with Chemicals from World War I to Silent Spring*, Cambridge University Press, Cambridge

Salaman, P., Donegan, T. and Caro, D. (2009) "Check list to the birds of Colombia", *Conservación Colombiana*, vol. 8, 1–89

Saldaña-Vázquez, R. A., Sosa, V. J., Hernández-Montero, J. R. and López-Barrera, F. (2010) "Abundance responses of frugivorous bats (*Stenodermatinae*) to coffee cultivation and selective logging practices in mountainous central Veracruz, Mexico", *Biodiversity and Conservation*, vol. 19, no. 7, 2111–2124

Saldaña-Vázquez, R. A., Castro-Luna, A. A., Sandoval-Ruiz, C. A., Hernández-Montero, J. R. and Stoner, K. E. (2013) "Population composition and ectoparasite prevalence on bats (*Sturnira ludovici*; *Phyllostomidae*) in forest fragments and coffee plantations of central Veracruz, Mexico", *Biotropica*, vol. 45, no. 3, 351–356

Salinas-Zavala, C. A., Douglas, A. V. and Diaz, H. F. (2002) "Interannual variability of NDVI in northwest Mexico: associated climatic mechanisms and ecological implications", *Remote Sensing of Environment*, vol. 82, nos 2–3, 417–430

Santos Barrera, G., Pacheco, J., Mendoza-Quijano, F., Bolaños, F., Cháves, G., Daily, G. C., Ehrlich, P. R. and Ceballos, G. (2008) "Diversity, natural history and conservation of amphibians and reptiles from the San Vito region southwestern Costa Rica", *Revista de Biología Tropical*, vol. 56, no. 2, 755–778

Sanz, J. J. (2001) "Experimentally increased insectivorous bird density results in a reduction of caterpillar density and leaf damage to Pyrenean oak", *Ecological Research*, vol. 16, no. 3, 387–394

Scheffer, M. (2009) *Critical Transitions in Nature and Society*, Princeton University Press, Princeton, NJ

Scheffer, M., Rinaldi, S., Gragnani, A., Mur, L. R. and Van Nes, E. H. (1997) "On the dominance of filamentous cyanobacteria in shallow, turbid lakes", *Ecology*, vol. 78, no. 1, 272–282

Scheffer, M., Carpenter, S., Foley, J. A., Folke, C. and Walker, B. (2001) "Catastrophic shifts in ecosystems", *Nature*, vol. 413, no. 6856, 591–596

Scheffer, M., Bascompte, J., Brock, W. A., Brovkin, V., Carpenter, S. R., Dakos, V., . . . and Sugihara, G. (2009) "Early-warning signals for critical transitions", *Nature*, vol. 461, no. 7260, 53–59

Scheffer, M., Carpenter, S., Lenton, T. M., Bascompte, J., Brock, W., Dakos, V., van de Koppel, J., van de Leemput, I., Levin, S., van Nes, E. H., Pascual, M. and Vandermeer, J. (2012) "Anticipating critical transitions", *Science*, vol. 338, no. 6105, 344–348

Schmitz, O. J. and Sokol-Hessner, L. (2002) "Linearity in the aggregate effects of multiple predators in a food web", *Ecology Letters*, vol. 5, no. 2, 168–172

Schmitz, O. J., Hambäck, P. A. and Beckerman, A. P. (2000) "Trophic cascades in terrestrial systems: a review of the effects of carnivore removals on plants", *The American Naturalist*, vol. 155, no. 2, 141–153

Schnell, S. M. (2007) "Food with a farmer's face: community-supported agriculture in the United States", *Geographical Review*, vol. 97, no. 4, 550–564

Schoener, T. W. (1989) "Food webs from the small to the large: the Robert H. MacArthur Award Lecture", *Ecology*, vol. 70, no. 6, 1559–1589

Schoener, T. W. and Spiller, D. A. (1987) "Effect of lizards on spider populations: manipulative reconstruction of a natural experiment", *Science*, vol. 236, no. 4804, 949–952

Schoener, T. W., Spiller, D. A. and Losos, J. B. (2002) "Predation on a common *Anolis* lizard: can the food-web effects of a devastating predator be reversed?", *Ecological Monographs*, vol. 72, no. 3, 383–407

Schroeder, R. (1951) "Resultados obtenidos de una investigación del micro-clima en un cafetal", *Cenicafé*, vol. 2, 13–43

Schulz, A. and Wagner, T. (2002) "Influence of forest type and tree species on canopy ants (*Hymenoptera*: *Formicidae*) in the Budongo Forest Reserve, Uganda", *Oecologia*, vol. 133, no. 2, 224–232

Sekercioglu, C. H. (2006) "Increasing awareness of avian ecological function", *Trends in Ecology and Evolution*, vol. 21, no. 8, 464–471

Sen, A. K. (1962) "An aspect of Indian agriculture", *Economic Weekly*, vol. 14, nos 4–6, 243–246

Seufert, V., Ramankutty, N. and Foley, J. A. (2012) "Comparing the yields of organic and conventional agriculture", *Nature*, vol. 485, no. 7397, 229–232

Siebert, S. F. (2002) "From shade- to sun-grown perennial crops in Sulawesi, Indonesia: implications for biodiversity conservation and soil fertility", *Biodiversity and Conservation*, vol. 11, 1889–1902

Simberloff, D. (1986) "Design of nature reserves", in Usher, M. B. (ed) *Wildlife Conservation Evaluation*, Springer-Verlag, Netherlands, 315–337

Simberloff, D. and Abele, L. G. (1982) "Refuge design and island biogeographic theory: effects of fragmentation", *The American Naturalist*, vol. 120, no. 1, 41–50

Skelly, D. K., Werner, E. E. and Cortwright, S. A. (1999) "Long-term distributional dynamics of a Michigan amphibian assemblage", *Ecology*, vol. 80, no. 7, 2326–2337

Skinner, G. J. and Allen, G. W. (1996) *Naturalists' Handbook 24: Ants*, Slough, Richmond, UK

Slingo, J. M., Challinor, A. J., Hoskins, B. J. and Wheeler, T. R. (2005) "Introduction: food crops in a changing climate", *Philosophical Transactions of the Royal Society B: Biological Sciences*, vol. 360, no. 1463, 1983–1989

Solis-Montero, L., Flores-Palacios, A. and Cruz-Angón, A. (2005) "Shade-coffee plantations as refuges for tropical wild orchids in central Veracruz, Mexico", *Conservation Biology*, vol. 19, no. 3, 908–916

Solomon, S., Qin, D., Manning, M., Chen, Z., Marquis, M., Averyt, K. B., Tignor, M. and Miller, H. L. (eds) (2007) *Contribution of Working Group I to the Fourth Assessment Report of the Intergovernmental Panel on Climate Change*, Cambridge University Press, Cambridge

Somarriba, E., Harvey, C. A., Samper, M., Anthony, F., González, J., Staver, C. and Rice, R. A. (2004) "Biodiversity conservation in neotropical coffee (*Coffea arabica*) plantations", in Schroth, G., da Fonseca, G. A. B., Harvey, C. A., Gascon, C., Vasconcelos, H. L. and Izac, A. M. N. (eds) *Agroforestry and Biodiversity Conservation in Tropical Landscapes*, Island Press, Washington, DC, 198–226

Sosas, V. J., Hernández-Salazar, E., Hernández-Conrique, D. and Castro-Luna, A. (2008) 'Murcielagos', in Manson, R., Hernandez-Ortiz, V., Gallina, S. and Mehltreter, K. (eds) *Agroecosistemas Cafetaleros de Veracruz: Biodiversidad, Manejo y Conservacion*, Instituto Nacional de Ecología, Mexico, 181–191

Soto-Pinto, L., Perfecto, I, Castillo-Hernandez, J. and Caballero-Nieto, J. (2000) "Shade effect on coffee production at the northern Tzeltal zone of the state of Chiapas, Mexico", *Agriculture, Ecosystems and Environment*, vol. 80, no. 1, 61–69

Soto-Pinto, L., Romero-Alvarado, Y., Caballero-Nieto, J. and Segura Warnholtz, G. (2001) "Woody plant diversity and structure of shade-grown-coffee plantations in Northern Chiapas, Mexico", *Revista de Biología Tropical*, vol. 49, no. 2, 977–987

Sourzat, P. (2009) "The truffle and its cultivation in France", *Acta Botanica Yunnanica*, Supplement, vol. 10, no. 6, 72–80

Spiller, D. A. and Schoener, T. W. (1990) "A terrestrial field experiment showing the impact of eliminating top predators on foliage damage", *Nature*, vol. 347, no. 6292, 469–472

Spiller, D. A. and Schoener, T. W. (1994) "Effects of top and intermediate predators in a terrestrial food web", *Ecology*, vol. 75, no. 1, 182–196

Stanhill, G. (1990) "The comparative productivity of organic agriculture", *Agriculture, Ecosystems and Environment*, vol. 30, no. 1, 1–26

Starkel, L. (1972) "The role of catastrophic rainfall in the shaping of the relief of the lower Himalaya (Darjeeling Hills)", *Geographia Polonica*, vol. 21, 103–147

Staver, A. C., Archibald, S. and Levin, S. A. (2011) "The global extent and determinants of savanna and forest as alternative biome states", *Science*, vol. 334, no. 6053, 230–232

Staver, C., Guharay, F., Monterroso, D. and Muschler, R. G. (2001) "Designing pest-suppressive multistrata perennial crop systems: shade-grown coffee in Central America", *Agroforestry Systems*, vol. 53, no. 2, 151–170

Steffan-Dewenter, I., Münzenberg, U., Bürger, C., Thies, C. and Tscharntke, T. (2002) "Scale-dependent effects of landscape context on three pollinator guilds", *Ecology*, vol. 83, no. 5, 1421–1432

Steffan-Dewenter, I., Potts, S. G. and Packer, L. (2005) "Pollinator diversity and crop pollination services are at risk", *Trends in Ecology and Evolution*, vol. 20, no. 12, 651–652

Stigter, C. J., Mohammed, A. E., Nasr Al-Amin, N. K., Onyewotu, L. O. Z., Oteng'i, S. B. B. and Kainkwa, R. M. R. (2002) "Agroforestry solutions to some African wind problems", *Journal of Wind Engineering and Industrial Aerodynamics*, vol. 90, no. 10, 1101–1114

Stork, N. E. and Brendell, M. J. D. (1990) "Variation in the insect fauna of Sulawesi trees with season, altitude and forest type", in Knight, W. J. and Holloway, J. D. (eds) *Insects and the Rain Forests of Southeast Asia*, (Wallacea) Royal Entomological Society, London, 173–190

Stouffer, D. B., Camacho, J., Jiang, W. and Amaral, L. A. N. (2007) "Evidence for the existence of a robust pattern of prey selection in food webs", *Proceedings of the Royal Society B: Biological Sciences*, vol. 274, no. 1621, 1931–1940

Strong, D. R. (1992) "Are trophic cascades all wet? Differentiation and donor-control in speciose ecosystems", *Ecology*, vol. 73, no. 3, 747–754

Study of Critical Environmental Problems (SCEP) (1970) *Man's Impact on the Global Environment*, MIT Press, Cambridge, MA

Styrsky, J. D. and Eubanks, M. D. (2007) "Ecological consequences of interactions between ants and honeydew-producing insects", *Proceedings of the Royal Society B: Biological Sciences*, vol. 274, no. 1607, 151–164

Swift, M. J. and Anderson, J. M. (1994) "Biodiversity and ecosystem function in agroecosystems", in Shulze, E. D. and Mooney, H. A. (eds) *Biodiversity and Ecosystem Function*, Springer-Verlag, New York, 15–42

Tanabe, K. and Namba, T. (2005) "Omnivory creates chaos in simple food web models", *Ecology*, vol. 86, no. 12, 3411–3414

Teare, I. D., Kanemasu, E. T., Powers, W. L. and Jacobs, H. S. (1973) "Water-use efficiency and its relation to crop canopy area, stomatal regulation and root distribution", *Agronomy Journal*, vol. 65, no. 2, 207–211

Tejeda-Cruz, C. and Sutherland, W. J. (2004) "Bird responses to shade coffee production", *Animal Conservation*, vol. 7, no. 2, 169–179

Tejeda-Cruz, C. and Gordon, C. E. (2008) "Aves", in Manson, R., Hernandez-Ortiz, V., Gallina, S. and Mehltreter, K. (eds) *Agroecosistemas Cafetaleros de Veracruz Biodiversidad y Conservacion*, Instituto Nacional de Ecología, Mexico, 149–160

Teodoro, A., Muñoz, V. A., Tscharntke, T., Klein, A. M. and Tyianakis, J. M. (2010) "Early succession arthropod community changes on experimental passion fruit plant patches along a land-use gradient in Ecuador", *Agriculture, Ecosystems and Environment*, vol. 140, no. 1, 14–19

Thom, R. (1972) *Stabilité Structurelle et Morphogéneses: Essai d'une Théorie Générale de Modèles*, W. A. Benjamin, Inc., Readings, New York

Thomlinson, J. R., Serrano, M. I., Lopez, T. M., Aide, T. M. and Zimmerman, J. K. (1996) "Land-use dynamics in a post-agricultural Puerto Rican landscape (1936–1988)", *Biotropica*, vol. 28, no. 4a, 525–536

Tillberg, C. V., Holway, D. A., LeBrun, E. G. and Suarez, A. V. (2007) "Trophic ecology of invasive Argentine ants in their native and introduced ranges", *Proceedings of the National Academy of Sciences*, vol. 104, no. 52, 20856–20861

Tilman, D. and Kareiva, P. (1997) *Spatial Ecology*, Princeton University Press, Princeton, NJ

Torres, J. A. (1984) "Diversity and distribution of ant communities in Puerto Rico", *Biotropica*, vol. 16, no. 4, 296–303

Tscharntke, T., Klein, A. M., Kruess, A., Steffan-Dewenter, I. and Thies, C. (2005) "Landscape perspectives on agricultural intensification and biodiversity–ecosystem service management", *Ecology Letters*, vol. 8, no. 8, 857–874

Tscharntke, T., Leuschner, C., Zeller, M., Guherdja, E. and Bidin, A. (eds) (2007) *Stability of Tropical Rainforest Margins: Linking Ecological, Economic and Social Constraints of Land Use and Conservation*, Springer-Verlag, Berlin

Tscharntke, T., Clough, Y., Bhagwat, S. A., Buchori, D., Faust, H., Hertel, D., . . . and Wanger, T. C. (2011) "Multifunctional shade-tree management in tropical agroforestry landscapes: a review", *Journal of Applied Ecology*, vol. 48, no. 3, 619–629

Tscharntke, T., Clough, Y., Wanger, T. C., Jackson, L., Motzke, I., Perfecto, I., Vandermeer, J. and Whitbread, A. (2012a) "Global food security, biodiversity conservation and the future of agricultural intensification", *Biological Conservation*, vol. 151, no. 1, 53–59

Tscharntke, T., Tylianakis, J. M., Rand, T. A., Didham, R. K., Fahrig, L., Batary, P., . . . and Westphal, C. (2012b) "Landscape moderation of biodiversity patterns and processes: eight hypotheses", *Biological Reviews*, vol. 87, no. 3, 661–685

Turcotte, D. L. and Malamud, B. D. (2004) "Landslides, forest fires, and earthquakes: examples of self-organized critical behavior", *Physica A: Statistical Mechanics and its Applications*, vol. 340, no. 4, 580–589

Turing, A. M. (1952) "The chemical basis of morphogenesis", *Philosophical Transactions of the Royal Society of London B: Biological Sciences*, vol. 237, no. 641, 37–72

Tylianakis, J. M., Klein, A. M. and Tscharntke, T. (2005) "Spatiotemporal variation in the diversity of *Hymenoptera* across a tropical habitat gradient", *Ecology*, vol. 86, no. 12, 3296–3302

UNCTAD-UNEP (2008) *Organic Agriculture and Food Security in Africa*, UNEP-UNCTAD Capacity-Building Taskforce on Trade, Environment and Development, United Nations, New York and Geneva

Uno, S. (2007) "Effects of management intensification on Coccids and parasitic Hymenopterans in coffee agroecosystems in Mexico", PhD thesis, University of Michigan, Ann Arbor, MI

Urrutia-Escobar, M. X. and Armbrecht, I. (2013) "Effect of two agroecological management strategies on ant (*Hymenoptera: Formicidae*) diversity on coffee plantations in southwestern Colombia", *Environmental Entomology*, vol. 42, no. 2, 194–203

Valenzuela-Gonzalez, J., Quiroz-Robledo, L. and Martínez-Tlapa, D. L. (2008) 'Hormigas (Insecta: *Hymenoptera: Formicidae*)', in Manson, R., Hernandez-Ortiz, V., Gallina, S. and Mehltreter, K. (eds) *Agroecosistemas Cafetaleros de Veracruz Biodiversidad y Conservacion*, Instituto Nacional de Ecología, Mexico, 107–121

Van Bael, S. A. and Brawn, J. D. (2005) "The direct and indirect effects of insectivory by birds in two contrasting Neotropical forests", *Oecologia*, vol. 145, no. 4, 658–668

Van Bael, S. A., Brawn, J. D. and Robinson, S. K. (2003) "Birds defend trees from herbivores in a Neotropical forest canopy", *Proceedings of the National Academy of Sciences*, vol. 100, no. 14, 8304–8307

Van Bael, S. A. V., Philpott, S. M., Greenberg, R., Bichier, P., Barber, N. A., Mooney, K. A. and Gruner, D. S. (2008) "Birds as predators in tropical agroforestry systems", *Ecology*, vol. 89, no. 4, 928–934

van de Koppel, J., Rietkerk, M. and Weissing, F. J. (1997) "Catastrophic vegetation shifts and soil degradation in terrestrial grazing systems", *Trends in Ecology and Evolution*, vol. 12, no. 9, 352–356

van der Ploeg, J. D. (2009) *The New Peasantries: Struggles for Autonomy and Sustainability in an Era of Empire and Globalization*, Routledge, London

van Noordwijk, M., Lusiana, B., Villamor, G., Purnomo, H. and Dewi, S. (2011) "Feedback loops added to four conceptual models linking land change with driving forces and actors", *Ecology and Society*, vol. 16, no. 1, res. 1. Available online at www.ecologyandsociety.org/vol16/iss1/resp1/ (accessed June 24, 2014)

Vandermeer, J. (1989) *The Ecology of Intercropping*, Cambridge University Press, Cambridge

Vandermeer, J. (1990) "Notes on agroecosystem complexity: chaotic price and production trajectories deducible from simple one-dimensional maps", *Journal of Biological Agriculture and Horticulture*, vol. 6, no. 4, 293–304

Vandermeer, J. (1993) "Loose coupling of predator–prey cycles: entrainment, chaos, and intermittency in the classic MacArthur consumer–resource equations", *The American Naturalist*, vol. 141, no. 5, 687–716

Vandermeer, J. (1996) "Seasonal isochronic forcing of Lotka–Volterra equations", *Progress of Theoretical Physics*, vol. 96, no. 1, 13–28

Vandermeer, J. (1997) "Syndromes of production: an emergent property of simple agroecosystem dynamics", *Journal of Environmental Management*, vol. 51, no. 1, 59–72

Vandermeer, J. (2004) "Coupled oscillations in food webs: balancing competition and mutualism in simple ecological models", *The American Naturalist*, vol. 163, no. 6, 857–867

Vandermeer, J. (2006a) "Omnivory and the stability of food webs", *Journal of Theoretical Biology*, vol. 238, no. 3, 497–504

Vandermeer, J. (2006b), "Oscillating populations and biodiversity maintenance", *BioScience*, vol. 56, no. 12, 967–975

Vandermeer, J. (2008) "The niche construction paradigm in ecological time", *Ecological Modeling*, vol. 214, nos 2–4, 385–390

Vandermeer, J. (2010) "The inevitability of surprise in agroecosystems", *Ecological Complexity*, vol. 8, 377–382

Vandermeer, J. (2013) "Forcing by rare species and intransitive loops creates distinct bouts of extinction events conditioned by spatial pattern in competition communities", *Theoretical Ecology*, vol. 6, no. 4, 395–404

Vandermeer, J. and Perfecto, I. (1997) "The agroecosystem: a need for the conservation biologist's lens", *Conservation Biology*, vol. 11, no. 3, 591–592

Vandermeer, J. and Perfecto, I. (2005) *Breakfast of Biodiversity: The Political Ecology of Rain Forest Destruction*, Food First Books, Oakland, CA

Vandermeer, J. and Pascual, M. (2006) "Competitive coexistence through intermediate polyphagy", *Ecological Complexity*, vol. 3, no. 1, 37–43

Vandermeer, J. and Perfecto, I. (2006) "A keystone mutualism drives pattern in a power function", *Science*, vol. 311, no. 5763, 1000–1002

Vandermeer, J. and Perfecto, I. (2007) "The agricultural matrix and a future paradigm for conservation", *Conservation Biology*, vol. 21, no. 1, 274–277

Vandermeer, J. and Perfecto, I. (2012) "Syndromes of production in agriculture: prospects for social-ecological regime change", *Ecology and Society*, vol. 17, no. 4, art. 39. Available online at www.ecologyandsociety.org/vol17/iss4/art39/ (accessed June 22, 2014)

Vandermeer, J. and Yitbarek, S. (2012) "Self-organized spatial pattern determines biodiversity in spatial competition", *Journal of Theoretical Biology*, vol. 300, 48–56

Vandermeer, J. and Goldberg, D. E. (2013) *Population Ecology: First Principles*, second edition, Princeton University Press, Princeton, NJ

Vandermeer, J., van Noordwijk, M., Anderson, J., Ong, C. and Perfecto, I. (1998) "Global change and multi-species agroecosystems: concepts and issues", *Agriculture, Ecosystems and Environment*, vol. 67, no. 1, 1–22

Vandermeer, J., Stone, L. and Blasius, B. (2001) "Categories of chaos and fractal basin boundaries in forced predator–prey models", *Chaos, Solitons & Fractals*, vol. 12, no. 2, 265–276

Vandermeer, J., Evans, M. A., Foster, P., Höök, T., Reiskind, M. and Wund, M. (2002a) "Increased competition may promote species coexistence", *Proceedings of the National Academy of Sciences*, vol. 99, no. 13, 8731–8736

Vandermeer, J., Perfecto, I., Ibarra Nuñez, G., Philpott, S. M. and Garcia Ballinas, A. (2002b) "Ants (*Azteca* sp.) as potential biological control agents in shade coffee production in Chiapas, Mexico", *Agroforestry Systems*, vol. 56, no. 3, 271–276

Vandermeer, J., Perfecto, I. and Philpott, S. M. (2008) "Clusters of ant colonies and robust criticality in a tropical agroecosystem", *Nature*, vol. 451, no. 7177, 457–459

Vandermeer, J., Perfecto, I., and Liere, H. (2009) "Evidence for hyperparasitism of coffee rust (*Hemileia vastatrix*) by the entomogenous fungus, *Lecanicillium lecanii*, through a complex ecological web", *Plant Pathology*, vol. 58, no. 4, 636–641

Vandermeer, J., Perfecto, I. and Philpott, S. M. (2010a) "Ecological complexity and pest control in organic coffee production: uncovering an autonomous ecosystem service", *BioScience*, vol. 60, no. 7, 527–537

Vandermeer, J., Perfecto, I. and Schellhorn, N. (2010b) "Propagating sinks, ephemeral sources and percolating mosaics: conservation in landscapes", *Landscape Ecology*, vol. 25, no. 4, 509–518

Vantaux, A., Roux, O., Magro, A. and Orivel, J. (2012) "Evolutionary perspectives on myrmecophily in ladybirds", *Psyche: A Journal of Entomology*. Available online at http://dx.doi.org/10.1155/2012/591570 (accessed June 24, 2014)

Varón Devia, E. H. (2002) "Distribución espacio-temporal de hormigas con potencial como depredadoras de *Hypothenemus hampei* e *Hypsipyla grandella*, en sistemas agroforestales de café, en Costa Rica", MSc thesis, CATIE, Turrialba

Veddeler, D., Olschewski, R., Tscharntke, T. and Klein, A. M. (2008) "The contribution of non-managed social bees to coffee production: new economic insights based on farm-scale yield data", *Agroforestry Systems*, vol. 73, no. 2, 109–114

Vega, F. E., Infante, F., Castillo, A. and Jaramillo, J. (2009) "The coffee berry borer, *Hypothenemus hampei* (Ferrari) (*Coleoptera: Curculionidae*): a short review, with recent findings and future research directions", *Terrestrial Arthropod Reviews*, vol. 2, no. 2, 129–147

Vélez, M., Bustillo, A. E. and Posada, F. J. (2000) "Predación sobre *Hypothenemus hampei* (Ferrari) de las hormigas *Solenopsis* spp, *Pheidole* spp., y *Dorymyrmex* spp. durante el secado del café", in Vélez, M., Bustillo, A. E. and Posada, F. J. (eds) *Resúmenes XXVII Congreso Sociedad Colombiana de Entomología*, Medellín, Colombia, 17

Vélez, M., Bustillo, A. E. and Posada, F. J. (2001) "Hormigas de la zona central cafetería y perspectivas de su uso en el control de *Hypothenemus hampei* (Ferrari) (*Coleoptera: Scolytidae*)", in Vélez, M., Bustillo, A. E. and Posada F. J. (eds) *Resúmenes XXVIII Congreso Sociedad Colombiana de Entomología*, Pereira, Colombia, 51

Vélez, M., Bustillo, A. E. and Posada, F. J. (2003) "Depredación de *Hypothenemus hampei* por *Solenopsis geminata* y *Gnamptogenys* sp. (*Hymenoptera: Formicidae*)", in Sociedad Colombiana de Entomología (ed.) *Libro de Resúmenes XXX Congreso Sociedad Colombiana de Entomología*, Universidad Autónoma, Cali, Colombia, 75

Vitousek, P. M., Mooney, H. A., Lubchenco, J. and Melillo, J. M. (1997) "Human domination of Earth's ecosystems", *Science*, vol. 277, no. 5325, 494–499

Vogt, W. (1948) *Road to Survival*, Slodne Associates, New York

Völkl, W. and Kraus, W. (1996) "Foraging behaviour and resource utilization of the aphid parasitoid *Pauesia unilachni*: adaptation to host distribution and mortality risks", *Entomologia Experimentalis et Applicata*, vol. 79, no. 1, 101–109

Wake, D. B. and Vredenburg, V. T. (2008) "Are we in the midst of the sixth mass extinction? A view from the world of amphibians", *Proceedings of the National Academy of Sciences*, vol. 105 (supplement 1), 11466–11473

Walker, B. H. (1989) "Diversity and stability in biodiversity conservation", in Weston, D. and Pearl, M. (eds) *Conservation Biology for the Twenty-First Century*, Oxford University Press, Oxford, 121–130

Waller, J. M. (1982) "Coffee rust: epidemiology and control", *Crop Protection*, vol. 1, no. 4, 385–404

Waller, J. M., Bigger, M. and Hillocks, R. J. (eds) (2007) *Coffee Pests, Diseases and Their Management*, CABI Publishing, Oxon, UK

Wallerstein, I. (2001) "Braudel and Interscience: a preacher to empty pews?", *Review (Fernand Braudel Center)*, vol. 24, no.. 1, 3–12

Ward, H. M. (1882) "Researches on the life-history of *Hemileia vastatrix*, the fungus of the 'coffee-leaf disease'", *Journal of the Linnean Society of London, Botany*, vol. 19, no. 121, 299–335

Wardle, D. A. (1999) "Is 'sampling effect' a problem for experiments investigating biodiversity–ecosystem function relationships?", *Oikos*, vol. 87, no. 2, 403–407

Way, M. J. and Khoo, K. C. (1992) "Role of ants in pest-management", *Annual Review of Entomology*, vol. 37, 479–503

Weaver, P. L. and Birdsey, A. (1986) "Tree succession and management opportunities in coffee shade stands", *Turrialba*, vol. 36, no. 1, 47–58

Weis, T. (2004) "Restructuring and redundancy: the impacts and illogic of neoliberal agricultural reforms in Jamaica", *Journal of Agrarian Change*, vol. 4, no. 4, 461–491

Weiss, H. and Bradley, R. S. (2001) "What drives societal collapse?", *Science*, vol. 291, no. 5506, 609–610

Weisser, W. W. and Völkl, W. (1997) "Dispersal in the aphid parasitoid, *Lysiphlebus cardui* (Marshall)(*Hym., Aphidiidae*)", *Journal of Applied Entomology*, vol. 121, no. 1–5, 23–28

Werner, E. E., and Peacor, S. D. (2003) "A review of trait-mediated indirect interactions in ecological communities", *Ecology*, vol. 84, no. 5, 1083–1100

Werner, E. E., Yurewicz, K. L., Skelly, D. K. and Relyea, R. A. (2007) "Turnover in an amphibian metacommunity: the role of local and regional factors", *Oikos*, vol. 116, no. 10, 1713–1725

Whelan, C. J., Wenny, D. G. and Marquis, R. J. (2008) "Ecosystem services provided by birds", *Annals of the New York Academy of Sciences*, vol. 1134, no. 1, 25–60

Whetton, P. H., Fowler, A. M., Haylock, M. R. and Pittock, A. B. (1993) "Implications of climate change due to the enhanced greenhouse effect on floods and droughts in Australia", *Climatic Change*, vol. 25, nos 3–4, 289–317

Whitmore, T. M. and Turner, B. L. (1992) "Landscapes of cultivation in Mesoamerica on the eve of the conquest", *Annals of the Association of American Geographers*, vol. 82, no. 3, 402–425

William-Linera, G., and López-Gómez, A. M. (2008), "Estructura y diversidad de la vegetación leñosa", in Manson, R. H., Hernández-Ortiz, V., Gallina, S. and Mehltreter, K. (eds) *Agroecosistemas Cafetaleros de Veracruz: Biodiversidad, Manejo y Conservación*, Instituto Nacional de Ecología, Mexico, 55–68

William-Linera, G., López-Gómez, A. M. and Castro, M. A. (2005) "Complementariedad y patrones de bosque de niebla del centro de Veracruz (México)", in Halffter, G., Soberón, J., Koleff, P. and Melic, A. (eds) *Sobre Diversidad Biológica: El Significado de las Diversidades Alfa, Beta y Gamma (Monografías Tercer Milenio)*, Vol. 4, S. E. A., Zaragoza, Spain, 153–164

Williams, N. S., Morgan, J. W., McDonnell, M. J. and McCarthy, M. A. (2005) "Plant traits and local extinctions in natural grasslands along an urban–rural gradient", *Journal of Ecology*, vol. 93, no. 6, 1203–1213

Williams-Guillén, K. and Perfecto, I. (2010) "Effects of agricultural intensification on the assemblage of leaf-nosed bats (*Phyllostomidae*) in a coffee landscape in Chiapas Mexico", *Biotropica*, vol. 42, no. 5, 605–613

Williams-Guillén, K. and Perfecto, I. (2011) "Ensemble composition and activity levels of insectivorous bats in response to management intensification in coffee agroforestry systems", *PLOS ONE*, vol. 6, no. 1, e16502

Williams-Guillén, K., Perfecto, I. and Vandermeer, J. (2008) "Bats limit insects in a neotropical agroforestry system", *Science*, vol. 320, no. 5872, 70

Willig, M. R., Kaufman, D. M. and Stevens, R. D. (2003) "Latitudinal gradients of biodiversity: pattern, process, scale, and synthesis", *Annual Review of Ecology and Systematics*, vol. 34, 273–309

Willis, K. J., Gillson, L. and Brncic, T. (2004) "How 'virgin' is virgin rainforest?", *Science*, vol. 304, 402–403

Wilsey, B. J., Martin, L. M. and Polley, H. W. (2005) "Predicting plant extinction based on species-area curves in prairie fragments with high beta richness", *Conservation Biology,* vol. 19, no. 6, 1835–1841

Wilson, E. O. (1987a) "The arboreal ant fauna of Peruvian Amazon forests: a first assessment", *Biotropica*, vol. 19, no. 3, 245–251

Wilson, E. O. (1987b) "The little things that run the world", *Conservation Biology*, vol. 1, no. 4, 344–346

Wimberly, M. C. (2006) "Species dynamics in disturbed landscapes: when does a shifting habitat mosaic enhance connectivity?", *Landscape Ecology*, vol. 21, no. 1, 35–46

Winemiller, K. O. (1990) "Spatial and temporal variation in tropical fish trophic networks", *Ecological Monographs*, vol. 60, no. 3, 331–367

Wise, D. H. (1995) *Spiders in Ecological Webs*, Cambridge University Press, Cambridge

Wise, T. A. (2009) *Agricultural Dumping under NAFTA: Estimating the Costs of US Agricultural Policies to Mexican Producers*, Global Development and Environmental Institute, Tufts University, Medford, MA

Witter, M. (2004) "Yes we sell no bananas", *Harvard International Review*, vol. 26, no. 2, 84–85

Wollenweber, B., Porter, J. R. and Schellberg, J. (2003) "Lack of interaction between extreme high temperature events at vegetative and reproductive growth stages in wheat", *Journal of Agronomy and Crop Sciences*, vol. 189, no. 3, 142–150

Wormer, T. M. (1964) "The growth of the coffee berry", *Annals of Botany*, vol. 28, no. 1, 47–55

Wright, A. and Wolford, W. (2003) *To Inherit the Earth: The Landless Movement and the Struggle for a New Brazil*, Food First Books, Oakland, CA

Wu, J. and Levin, S. A. (1994) "A spatial patch dynamic modeling approach to pattern and process in an annual grassland", *Ecological Monographs*, vol. 64, no. 4, 447–464

Wu, J. and Levin, S. A. (1997) "A patch-based spatial modeling approach: conceptual framework and simulation scheme", *Ecological Modelling*, vol. 101, no. 2, 325–346

Wunderle, J. M. and Latta, S. C. (1996) "Avian abundance in sun and shade coffee plantations and remnant pine forest in the Cordillera Central, Dominican Republic", *Ornitologia Neotropical*, vol. 7, 19–34

Wunderle, J. M. and Latta, S. C. (1998) "Avian resource use in Dominican shade coffee plantations", *Wilson Bulletins*, vol. 110, no. 2, 271–281

Yachi, S. and Loreau, M. (1999) "Biodiversity and ecosystem productivity in a fluctuating environment: the insurance hypothesis", *Proceedings of the National Academy of Sciences*, vol. 96, no. 4, 1463–1468

Yitbarek, S., Vandermeer, J. H. and Allen, D. (2011) "The combined effects of exogenous and endogenous variability on the spatial distribution of ant communities in a forested ecosystem (*Hymenoptera: Formicidae*)", *Environmental Entomology*, vol. 40, no. 5, 1067–1073

Yodzis, P. (2000) "Diffuse effects in food webs", *Ecology*, vol. 81, no. 1, 261–266

Young, G. R. (1982) "Recent work on biological control in Papua New Guinea and some suggestions for the future", *International Journal of Pest Management*, vol. 28, no. 2, 107–114

Zimmerman, J. K., Aide, T. M. and Lugo, A. E. (2007) "Implications of land use history for natural forest regeneration and restoration strategies in Puerto Rico", in Hobbs, R. J. (ed.) *Old Fields: Dynamics and Restoration of Abandoned Farmland*, Island Press, Washington, DC, 51–74

INDEX